Energy, Climate and the Environment

The aim of this series is to provide texts which lay out the technical, environmental and political issues relating to proposed policies for responding to climate change. The focus is not primarily on the science of climate change, or on the technological detail, although there will be accounts of this, to aid assessment of the viability of various options. However, the main focus is the policy conflicts over which strategy to pursue. The series adopts a critical approach and attempts to identify flaws in emerging policies, propositions and assertions. In particular, it seeks to illuminate counter-intuitive assessments, conclusions and new perspectives. The intention is not simply to map the debates, but to explore their structure, their underlying assumptions and their limitations. The books in this series are incisive and authoritative sources of critical analysis and commentary, clearly indicating the divergent views that have emerged whilst also identifying the shortcomings of such views. The series does not simply provide an overview, but also offers policy prescriptions.

More information about this series at
http://www.palgrave.com/gp/series/14966

Michalis Mathioulakis
Editor

Aspects of the Energy Union

Application and Effects of European Energy Policies in SE Europe and Eastern Mediterranean

palgrave
macmillan

Editor
Michalis Mathioulakis
Greek Energy Forum
Athens, Greece

Energy, Climate and the Environment
ISBN 978-3-030-55980-9 ISBN 978-3-030-55981-6 (eBook)
https://doi.org/10.1007/978-3-030-55981-6

Cover credit: Carloscastilla/Alamy Stock Photo

This Palgrave Macmillan imprint is published by the registered company Springer Nature Switzerland AG
The registered company address is: Gewerbestrasse 11, 6330 Cham, Switzerland

Foreword

This book conveys the main themes of the "Energy Union Future Leaders Academy", an international educational programme created and annually hosted by the Greek Energy Forum since 2017 with the objective to disseminate knowledge on the pillars of the Energy Union. This initiative has already become an institution for the South-eastern European region and aspires to evolve into a beacon of insight and inspiration for talented young individuals that have the willingness and potential to lead the European energy sector into its future. It has now become clearer than ever that "regionality" is an integral component of contemporary energy policy. Only by working together, the European Member States and their citizens may achieve an efficient, sustainable and overall better common energy future. This has been the core of the vision of the Energy Union since its inception. Extending this vision even further, the European Commission is now moving forward with the introduction of the European Green Deal, a roadmap for achieving a carbon-neutral Europe by 2050. This objective entails a requirement to enhance economic competitiveness, decouple growth from unsustainable physical resource exploitation and ensure that no community or citizen is left behind.

Europe needs to turn the climate and environmental challenges ahead into opportunities that should make the Energy Transition fair and inclusive for all. Once again, the pillars for the success of the European Green Deal will be cooperation across borders, strong human capital and competent leadership. This is why initiatives like the "Energy Union Future Leaders Academy" deserve support and have received the official endorsement of the European Commission and of its former Vice President in charge of the Energy Union, Maros Sefcovic; because they expose talented young individuals interested in energy to the benefits of the Energy Union and through this to the very essence of regional cooperation and efficient policymaking. In this context, this publication provides a series of insightful presentations from academics, energy professionals and experts, on a broad range of critical issues that presently concern the European energy sector. I truly hope that this book will act as a source of knowledge, insight and inspiration for the readers who aspire to become promoters of the Energy Union and of the Energy Transition towards 2050.

Alexandros Lagakos
Co-Founder/Vice Chairman
Greek Energy Forum
Athens, Greece

Preface

This publication represents a multi-dimensional, comprehensive presentation of some of the most critical issues regarding the EU's Energy Union policy. The Energy Union is the driving force behind all the EU's objectives regarding energy and the environment. Applied European energy policies face several challenges ranging from the Geopolitics of Energy and Energy Regulation, to Climate Change, advancing Renewable and Gas technologies, and Consumer Empowerment structures. This book aims to provide an insightful look into some of these vital issues regarding the European energy sector with a special focus to the effects the Energy Union policy has in two sensitive regional systems, South-Eastern Europe, and the Eastern Mediterranean. The context of this publication spawns from a training programme by the Greek Energy Forum on the European Commission's Energy Union strategy. The Greek Energy Forum is an international energy think tank consisting of Greek energy professionals aspiring to establish an international platform of ideas and dialogue among energy professionals to enable change in the energy setting of SE Europe and the Eastern Mediterranean. The "Energy

Union Future Leaders Academy" training programme introduces university students from South-Eastern Europe to the significance of the European Energy Union and highlights the indispensable element of "regionality" in contemporary European energy policies. Following up on the basic concept of the Academy, this publication aims to address a series of issues essential for the future of energy systems in Europe. Part of the included analysis focuses on the effects of applied policies on the regional sub-systems of SE Europe and the Eastern Mediterranean.

The book aspires to offer the opportunity for readers to get a quality, hands-on approach on the Energy Union by the professionals and academics that interact with it on an everyday basis. Its purpose is to offer its readers the ability to receive a pragmatic presentation on the current situation and the challenges European energy policies face while applied across the wide spectrum of activities related to energy. Such challenges range from the wider geopolitical environment to the regulatory and technical framework that determines the success of EU's policies. Energy, being by definition a multi-disciplinary field, poses a challenge for readers of specific disciplinary background that need to grasp an overall understanding of the various aspects of this challenging sector. The aforementioned background that led to this publication determines the writing style of its chapters that are focused in providing a combination of practical and theoretical approach, resulting in a book that can be useful both to a more experienced audience as well as the ones less familiar with energy sector developments. Deriving from the initial concept of the "Energy Union Future Leaders Academy" programme, this publication received valuable scientific support from the leadership, executives and members of the Greek Energy Forum from its London, Athens and Brussels branches. The experience and expertise shared among the members of the Greek Energy Forum can only be matched by their enthusiasm and willingness to share and disseminate knowledge regarding the European and global energy sector. I would, therefore, like to acknowledge their support and extent my gratitude for their contribution in structuring and realising this publication. Last but not least, I would like to extend my warmest thanks and gratitude to the authors of this publication who were burdened with the difficult task of addressing admittedly challenging issues in a professionally and scientifically demanding, yet easily comprehendible manner.

Athens, Greece Michalis Mathioulakis

Acknowledgements This publication is an initiative of the Greek Energy Forum, an international energy think tank consisting primarily of Greek energy professionals. It is inspired by the Forum's "Energy Union Future Leaders Academy", a training programme that introduces university students and young professionals from around Europe to the significance of the European Energy Union policy.

About This Book

This publication consists of four parts, that include a total of fourteen chapters. The book is structured to deliver a broad educational output regarding the implementation of the Energy Union strategy while maintaining a pragmatic approach in analysing the various aspects of the policies in question. Furthermore, this publication helps create a connection between the EU's fundamental objectives regarding the structure of a common market and the challenges deriving from the European energy security framework on a national and supranational level. In addition, the book offers assistance in understanding the links between the von der Leyen Commission's priorities regarding Climate Change and its 'geopolitical' character, in comparison to the Juncker's 'political' Commission, and it's agenda for jobs, growth, fairness and democratic change.

Part I of the book contains three chapters addressing the security and geopolitical aspects of the Energy Union and analysing the relationship between energy security and national security. In Chapter 2, Christos Syriopoulos, from the Greek Ministry for Environment & Energy, examines the progress made towards the realisation of EU's energy security targets since the introduction of the Energy Union policy in 2015.

His analysis focuses on the political and economic aspects of the process. The chapter presents the overall situation of energy security in Europe while addressing its elements in close relation to the other four dimensions of the Energy Union, namely the internal market, energy efficiency, climate action & decarbonisation and research & innovation. In Chapter 3, Dr. Thrassy N. Marketos, from the University of Peloponnese, examines the security and political framework of the Eastern Mediterranean subsystem and its connection to the Energy Union's goals to diversify its supply routes. The chapter analyses the effects of the Libya war in the region, combined with a presentation of the interests of several international actors including Russia, Turkey, Egypt, Cyprus and the US and an estimation on the potential of gas flows to the EU from the Eastern Mediterranean. Furthermore, the analysis turns on China's role in the Eastern Mediterranean region and examines the analogy between the Pan-Asian Energy Club and the Eastern Mediterranean Gas Forum. Chapter 4 examines the fundamental elements of the relationship between energy and national security. It builds a connection between political realism, securitisation theory and energy security and identifies the framework and conditions under which energy infrastructure acquires strategic value for the stakeholders it involves. The chapter presents the wide context of energy security threats and examines the EU's vulnerabilities related to them. Additionally, it identifies the tools of the European regulatory and policy framework that, while designed to address internal issues, incorporate an 'externality' effect, thus influencing EU's relations with third countries in the energy sector.

Part II contains four chapters that explore and analyse the regional, regulatory and investment aspects of the Energy Union with a particular focus in South-Eastern Europe and the Balkans. In Chapter 5, Dr. Sotirios Manolkidis, from the Greek Regulatory Authority for Energy, addresses the issue of energy growth and sustainability for the Western Balkan Six Initiative countries within the Energy Community. It examines the significant energy challenges the region faces ever since the conflicts following the break of the former Yugoslavia in the 1990s regarding damaged or depleted energy infrastructure, energy poverty and unreliable energy supply. The analysis addresses the need for institutional reform within the Energy Community and argues for the further strengthening of its existing dispute settlement mechanism alongside policy changes and the introduction of new regulatory frameworks. In Chapter 6, Professor Panayiotis Glavinis, from the Aristotle University of Thessaloniki, together with Anna Konstantinidou, Afroditi Semkou, Elias Andreadis and Elias Kolovos, examine the extent of integration of the regional energy markets between the Member States of the EU and the countries of the wider Europe. The analysis further explores whether the non-EU energy markets are in a position to conform to the *acquis communautaire* and whether they are able to ensure the security of the energy supply to the EU in affordable prices. The chapter identifies the steps that are necessary to integrate these crucial regional energy markets into the EU common market regarding harmonisation of the legal framework, essential infrastructure and networks, and promotion and the mutual protection regimes for foreign direct investments. It also examines the role of the Energy Community, the TEN-E strategy and the Energy Charter Treaty as useful tools for the EU to foster the integration of the energy markets in the wider Europe. In Chapter 7, Dr. Maria Bozoudi, from the American College of Thessaloniki examines the investment framework regarding integrating the Balkan countries into the European energy market and infrastructure network and identifies the investment risks stemming from delays in institutional reforms. The analysis explores Western Balkan governments' commitment to follow the EU's energy and climate footsteps through targets for 20,230 mirroring those of the EU and identifies the links between 'soft measures' and interest for investment in the energy sector of the Western Balkans. Furthermore, the chapter examines the role of the Energy Community and its effectiveness in promoting progress in the transition efforts in the Western Balkans towards environmental sustainability in the energy sector. In Chapter 8, Dr. Gokce Mete Mete, from the International Energy Charter, and Janessa Goh Pei-Ru, deliberate on the role of the Energy Charter process in contributing towards the energy transition and energy security in South-East Europe in line with the EU's Energy Union vision. The analysis presents the elements that support the potential of the Energy Charter Treaty to aid the convergence among South-East European countries towards accelerating the energy transition and strengthening energy security. The chapter explores how the broader Energy Charter process and its investment protection provisions and provisions on energy efficiency and soft law instruments, can promote renewable energy investments, play a part in improving energy security, and promote conditions for overall investment and secure transit in the area.

Part III includes five chapters addressing the transitional aspects of the Energy Union towards green energy and a circular economy. It examines developments and the role of wind power generation, as well as the role of biofuels, renewable gases and LNG on the road to the energy transition. In Chapter 9, Alexandros Lagakos, from Blue Grid Gas & Power, provides an overview of the elements steering the adaptation, implementation and prospective expansion of the use of Liquefied Natural Gas as a marine fuel. The analysis explores LNG's ability to reduce greenhouse gas emissions in shipping. It presents the drivers for a transition to LNG in shipping regarding environmental performance, international environmental regulation and corporate social responsibility, and examines its overall potential to become the medium for the utilisation of renewable gases in the shipping industry. In Chapter 10, Constantine Levoyannis from the Brussels-branch of the Greek Energy Forum addresses the issue of renewable gases and their role and contribution in achieving climate neutrality by 2050 in the EU. The analysis examines the elements and merits from policies supporting a hybrid energy system that links electricity and gas, and the use of their competitive advantages on an environmental, energy efficiency and economic standpoint. Furthermore, the chapter examines the future of existing gas infrastructure through repurposing and redefining its role to decrease the cost of the energy transition as well as the role of bioenergy and hydrogen in the future of gas networks. In Chapter 11, Dr. Spyros Kiartzis, from Hellenic Petroleum, together with Katerina Tsita, and Evangelia Paschalidou, provide an insight into the future transformation of the energy markets and discuss technological, regulatory and financial challenges correlated to energy security. The analysis examines the various ways low-carbon transition shapes the profile of infrastructure investments, as well as the role of biofuels in the transformation of the transportation sector. Furthermore, the chapter addresses the technical, business and societal factors that determine the pace of energy transition, the necessity for commercialisation and scale-up of new technologies and the role of an integrated approach to policy, planning, management and development of energy systems in supporting a transition to decarbonising the energy supply-chain. In Chapter 12, Ivan Pineda from WindEurope argues that the best approach for Europe to become carbon neutral by 2050, thereby contributing to the 1.5°C goal of the Climate Paris Agreement, is through the renewables-based electrification of most sectors in the economy. The analysis also examines the challenges the deployment of renewable energy faces, regarding increasing the flexibility of the power system, and facilitating, permitting and plugging the financing gap an electricity market dominated by renewables poses for investment signals. The chapter relates renewables' future success with scaling up electrification, analyses the role of energy efficiency, electrification incentives, and CO2 pricing, and identifies the elements in bridging the gap between scale and implementation of renewable energy. In Chapter 13, Marios Papalexandrou from Aeolus Limited examines how the EU managed to transform the offshore wind industry from a few demonstration projects two decades ago to the mainstream generation technology of today. The analysis discusses offshore wind technology developments and trends from past to recent and future projects and presents best practices incorporated from countries leading offshore wind expansion in Europe. Finally, it highlights Europe's offshore wind potential and discusses future challenges allowing Europe to stay ahead of the curve in research and implementation of future offshore wind developments.

Part IV, the final part of this book consisting of two chapters, addresses specific technical aspects of the Energy Union regarding energy analytics, the transition from data acquisition to data-driven business models, and the role of equipment maintenance as a key factor for energy efficiency. In Chapter 14, Dr. Dimitrios I. Doukas, from NET2GRID, examines the ways that digital transformation in the energy sector, smart-metering and access to quality energy data put the end-user at the epicentre of the new European energy system. The analysis presents how smart meters are becoming the necessary tools for the end-users to increase their impact and leverage in the future grid by participating in demand response schemes and facilitating consumers' ability to switch energy providers. The chapter presents energy analytics alternatives for various parts of the smart grid with an emphasis on the residential-level sector and defines energy disaggregation as a concept and as the foundation layer on which most residential-level energy analytics services are built. Finally, it presents and categorises data-driven business models that allow energy companies to develop products and services that build stronger customer engagement. In Chapter 15, the final chapter of this publication, Dr. Vassilios Kappatos from the Greek Centre for Research and Technology, presents the benefits of energy efficiency and maintenance in industry and society, summarises the European energy efficiency and maintenance strategies, and presents the benefits of adequately maintaining equipment and systems to enhance energy efficiency. The analysis presents the ways energy efficiency can be achieved through the utilisation of efficient devices, the implementation of strategies that reduce energy consumption and the proper implementation of maintenance procedures. The chapter summarises the European energy efficiency and maintenance strategies and presents the benefits of maintenance in industry and society.

Overall, the fifteen chapters of this book are designed to offer an insightful look into the challenges that applied regulatory and political policies face in the European energy sector. With a broad spectrum of issues, spanning from the geopolitics of energy and energy regulation to implementing climate change, renewable energy production, new gas technologies and consumer empowerment structures, this book presents a pragmatic look into the Energy Union's efficiency in providing secure, sustainable, competitive and affordable energy for all Europeans.

Contents

Notes on Contributors

Dr. Maria Bozoudi With an academic background in politics and international affairs, Dr. Bozoudi's research work centres on economic diplomacy, global development and Southeast European transitions. She has worked extensively for regional trade, transport and investment facilitation initiatives under SECI and the Stability Pact for SEE. Among others, she has served as Executive Secretary of the Business Advisory Council for Southeast Europe and Eurasia, and Adjunct Professor at the American College of Thessaloniki. She is Senior Advisor at SEV Hellenic Federation of Industries.

Dr. Dimitrios I. Doukas has a solid academic background in electrical engineering and multiple years of international working experience. From 2011 to 2013, he was with ABB Corporate Research, Västerås, Sweden. Then, as an R&D engineer and manager for the Aristotle University of Thessaloniki, he was actively involved in numerous national and international projects. Since July 2017 he is with NET2GRID, currently serving as the company's CIO and the country manager in

Greece. Dr. Doukas is the author and co-author of more than 40 peer-reviewed international journal and conference papers whereas he is also a co-inventor of two patents.

Professor Dr. Panayotis Glavinis is a Professor of International Economic Law at the Aristotle University of Thessaloniki. He holds a Bachelor's Degree from the Faculty of Law of the same University, an LL.M. on International Law and a Ph.D. on "International Construction Contracts" (EJA, Paris 1993) from Paris 2 University, where he was also granted the *habilitation à diriger des recherches* for "The Settlement of Disputes arising out of Contracts between International Organizations and Private Parties" (LGDJ, Paris 1992). His recent book, *Mechanisms of Rebalancing of Power in the International Economic System* (Papazissis, Athens, 2019) received the prize of the Academy of Athens.

Dr. Vassilios Kappatos (Ph.D., M.Eng., C.Eng., FWeldl, MIEEE, MISCM) is the Head of Constructural and Infrastructure Research in Maritime and Air Transport Laboratory at the Hellenic Institute of Transport (HIT), Center for Research and Technology Hellas (CERTH), Greece. His research areas are Non-Destructive Evaluation, Structural Health Monitoring, Condition Monitoring, Structural Integrity, Pattern Recognition and Signal Processing. His research has been supported by the European Commission and other organisations. He has over 110 publications in International Journals, Conference Proceedings, Book Chapters and Articles. He is a reviewer of several international journals and conference proceedings.

Dr. Spyros Kiartzis is the New Technologies and Alternative Energy Sources Manager for the Hellenic Petroleum Group. He is working to implement a comprehensive plan that will increase the use of renewable energy and decrease Group's carbon footprint, promoting new technologies in the energy and transport sector. He is the Vice Chairman and CEO of Energiaki Pylou-Methonis SA, a wind park owner and member of the Board of Directors of ELPEDISON SA and Hellenic Petroleum Renewable Energy Sources SA.

Alexandros Lagakos is a co-founder of the Greek Energy Forum, having served as Chairman (2013–2017) and Vice-Chairman (2017–present). Today, he is the Chief Operating Officer of Blue Grid, a downstream LNG supplier in the Eastern Mediterranean. In the past, he held the positions of Freight Trader and Market Strategist (Thenamaris Ships Management), Senior Gas Analyst and Business Developer (Shell Energy Europe), Risk Analyst (Gazprom) and Energy Economics Consultant (Mott MacDonald). Alex is a member of the UN Gas Experts. He holds an M.Eng. in Electrical Engineering (Aristotle University of Thessaloniki) and an M.Sc. in Sustainable Energy Futures (Imperial College London).

Constantine Levoyannis is the Head of the Brussels branch of the Greek Energy Forum and is also EU Public Affairs & Communications Director for GRTgaz. A dual national of the UK and Greece, he holds a bachelor's degree in Law and a Master's degree in International Law & International Relations from the Brussels School of International Studies.

Professor Dr. Sotirios Manolkidis is a lawyer, visiting Professor at the International Hellenic University (2019–2021), Vice—President of the Hellenic Regulatory Authority for Energy (RAE), since 2015, ACER & CEER former Board Member (2016–2018). Born in Thessaloniki, Greece, he has graduated (LL.B.) from the Law School of the University of Thessaloniki and was awarded the title of Doctor of Laws of the European University Institute in Florence, Italy. Former Lecturer in Law at Cardiff Law School of the University of Wales (1995–1998).

Dr. Thrassy N. Marketos is a Jurist, was specialised in Public International Law in the University of Aix—Marseille III (France) and was nominated doctor in International Relations "Central Asia-China Energy Security interaction in Western Eurasia" by the Panteion University of Athens, Greece. He has worked for the Hellenic Ministry of Foreign Affairs and is a Eurasia Energy Security Geopolitics Visiting Professor in the University of Peloponnese, Greece. He published his doctoral thesis and the monograph *China's Energy Geopolitics: The Shanghai Cooperation Organization and Central Asia*, (Routledge, Taylor & Francis Books Ltd, 2009), as well as several papers referring to Eurasia Energy Geopolitics.

Michalis Mathioulakis is an Energy Strategy Analyst and the Academic Director of the Greek Energy Forum. He is a research associate at the Institute of International, European and Defence Analyses of the University of Macedonia in Thessaloniki, where he is responsible for the analysis of the energy sector in MENA and Eastern Mediterranean. He has substantial working experience in the financial sector as a broker and analyst, as well as the private education sector, teaching Finance and Risk Analysis in Executive M.B.A. programmes. He is currently providing consulting services regarding energy strategy and energy regulation for Law firms and Consulting firms associated with the energy sector. He holds a B.A. in Economics, a B.A. in Political Science and an M.Sc. in Strategic Studies from the University of Macedonia in Thessaloniki, as well as an M.Sc. in Energy Strategy, Law & Economics from the University of Piraeus. His research work at the Department of International and European Studies of the University of Macedonia in Greece focuses on Energy Strategy and the correlation between energy and geopolitics.

Dr. Gokce Mete Mete holds a Ph.D. from the Centre for Energy Petroleum, Mineral Law and Policy (CEPMLP) at the University of Dundee. Dr. Mete is the author of 18 published and upcoming publications. In her new book, *The Future of Gas in the EU: Subsidise or Decarbonise?* (Palgrave Macmillan, 2019), she carries out an experiential assessment of the impact of energy transitions on the future of natural gas in the EU energy mix and provides policy recommendations, including on the future of sector coupling, and explores alternatives pathways to promote incentives for technologies that can enable decarbonisation of the gas grid at scale. She is currently the Head of the Knowledge Centre at the International Energy Charter in Brussels, an intergovernmental organisation with over 90 signatories across the globe, whose work is dedicated to facilitating international cooperation in the energy sector.

Marios Papalexandrou was born in Thessaloniki, Greece. He holds a B.Sc. in Mechanical Engineering by the University of Western Macedonia, Greece and an M.Sc. in Sustainable Energy Engineering by KTH, Sweden. He has contributed to scientific publications for photovoltaics, biofuels and offshore wind. He is working for offshore wind since 2007 for more than EUR5bn of projects. Marios is the director of Aeolus wind consultancy and creator of the game Age of Renewables, promoting renewables to young generations. He is also a guest lecturer in Offshore Wind at Kingston University, London and at IST University, Lisbon since 2018.

Evangelia Paschalidou studied Law at the Aristotle University of Thessaloniki. She holds an LL.M. in Economy and Business Law, with a special interest in environmental law issues. Currently, she is pursuing her M.Sc. in Environmental Policy and Management at the International Institute for Industrial Environmental Economics (IIIEE) of Lund University in Sweden.

Janessa Goh Pei-Ru was a litigation lawyer in Singapore since 2013 and obtained her LL.M. in international law at Erasmus University Rotterdam in 2019. Her research focuses on international law issues concerning energy, climate change and the environment.

Iván Pineda is the Director of Public Affairs for Windeurope. He was also the Head of Policy Analysis and the lead technical expert in wind power integration to the grid. WindEurope is one of the most influential trade associations in Brussels. Before that, he worked for six years as a technical project manager for Procter & Gamble in London and worked as a finance manager at Carrefour. He studied in Mexico and Canada obtaining his B.Sc. in Industrial Engineering. He was awarded the prestigious Chevening Scholarship, and he obtained his M.Sc. in Sustainable Energy Technology at Imperial College London in 2010.

Christos Syriopoulos was born in Athens in 1978 and has studied Political Science (Bachelor) and International & European Studies (M.Sc.). In 2006 graduated from the Greek National School for Public Administration and started working for the Ministries for Development and Environment & Energy (Head of Unit for EU Energy Affairs). From 2015 until early 2020, he was representing Greece to the EU on energy issues (as the energy attaché in the Permanent Representation to the EU).

Katerina Tsita received her Diploma in Electrical & Computer Engineering from the Aristotle University of Thessaloniki and her Ph.D. from the University of Western Macedonia, Greece. Currently, she is teaching at the Department of Economics at the Aristotle University of Thessaloniki.

List of Figures

1

Introduction

Michalis Mathioulakis

The end of 2019 marked the approval of the new European Commission, headed by Ursula von der Leyen, that took office in December of the same year. Announcing the new Commission's agenda, president von der Leyen revealed, among others, two ambitious and distinct goals among its top priorities. She declared the new Commission's commitment to make Europe the first climate-neutral continent by 2050 through the implementation of the 'European Green Deal' roadmap, and pledged to lead a 'geopolitical Commission' where reinforcing the EU's role as a relevant international actor, would be a crucial priority.[1] Placing the environment and geopolitics among its the top priorities, the new

[1] "The von der Leyen Commission's priorities for 2019–2024", European Parliament Briefing, January 2020, https://www.europarl.europa.eu/RegData/etudes/BRIE/2020/646148/EPRS_BRI(2020)646148_EN.pdf.

M. Mathioulakis (✉)
University of Macedonia, Thessaloniki, Greece

Greek Energy Forum, Athens, Greece

M. Mathioulakis (ed.), *Aspects of the Energy Union*,
Energy, Climate and the Environment,
https://doi.org/10.1007/978-3-030-55981-6_1

Commission seems to be signalling a drastic change from the former Commission where the then President, Jean-Claude Juncker, claimed to lead a 'political Commission' with an agenda for jobs, growth, fairness and democratic change.[2] However different at first glance, these two agendas are linked through a very interesting component that is no other than their energy policy. One of Junker's Commission top priorities was the formation of 'A Resilient Energy Union with a Forward-Looking Climate Change Policy'.[3] The deployment of the 'European Green Deal' is therefore firmly based on the foundations laid by the 'Energy Union' strategy launched in February 2015 that, to a great extent, are still in effect and determine the extended aspects of EU's energy policy and regulatory framework. The objective of the Energy Union is to create the conditions under which European citizens can benefit from secure, sustainable, competitive and affordable energy.[4] However easily phrased, such an objective constitutes an extremely challenging task that involves a structural transformation of the European energy system, including the production, procurement, transportation, distribution, storage, pricing, trading and consumption of energy. In addition, it involves structural changes regarding the energy mix, i.e. the combination of energy sources used in all human activity, the flow of energy that enters and exits the EU, and the interconnection of the energy policy with EU's Trade, Competition and External Relations policies. The essence of the Energy Union can be linked directly to the core objective of the European Union itself originating back to the Treaty of Rome in 1957 and the provisions of Article 2 regarding the establishment of a common market.[5] The Energy Union provides the conditions to create a functioning common market with solidarity between the Member States in their

[2]"A New Start for Europe: My Agenda for Jobs, Growth, Fairness and Democratic Change", Jean-Claude Juncker Opening Statement in the European Parliament Plenary Session, July 15, 2014, https://ec.europa.eu/commission/sites/beta-political/files/juncker-political-guidelines-speech_en.pdf.

[3]"A New Start for Europe: My Agenda for Jobs, Growth, Fairness and Democratic Change".

[4]"Energy Union", European Commission, accessed December 18, 2019, https://ec.europa.eu/energy/topics/energy-strategy/energy-union_en.

[5]"Treaty establishing the European Economic Community", European Union—EUR-Lex, accessed December 18, 2019, https://eur-lex.europa.eu/legal-content/EN/TXT/?uri=LEGISSUM%3Axy0023.

efforts to secure energy for the citizens. This can be achieved via an integrated European energy system, where energy flows freely within and between the Member States. Such a system needs to be functioning under competition and efficient use of resources, under an effective regulatory framework, and with energy markets that support a sustainable, low-carbon and climate-friendly economy. It needs to place the European citizens at its core, taking ownership of the energy transition, and benefit from new technologies to reduce their bills and participate actively in a decentralised energy market. Such an ambitious plan requires colossal structural changes that affect almost every aspect of everyday life within the EU.

To achieve its goals, the Commission outlined five fundamental dimensions of the Energy Union strategy; five critical elements that hold the tools to lead the EU towards the necessary transformation of its energy system. These tools are designed to enhance energy security, sustainability and competitiveness within the Union; they are rigidly interconnected and intended to reinforce each other. The five dimensions of the Energy Union relate to (a) energy security, solidarity and trust between the Member States, (b) a fully integrated internal energy market, (c) energy efficiency and moderation of demand, (d) decarbonising the European economy and (e) research, innovation and competitiveness in the European energy market.[6] The implementation of the Energy Union is a challenging task that faces a broad spectrum of issues spanning from regulatory and security concerns to the differences and particularities between various regions of the EU, and from challenges regarding the transition to a new energy mix to the technical aspects and difficulties facing such a transformation. EU's energy regulatory framework focuses primarily on the structure and effective functioning of free-market and competition conditions in the European energy sector. Based on this framework, the Commission strives to transform EU's energy systems towards an energy mix in which clean, renewable and sustainable energy holds an ever-growing share in order to reach its Paris Agreement targets to limit CO2 emissions and limit global warming effects. Wind and solar power lead this effort in the EU, while alternative technologies for

[6]"Energy Union".

power generation and energy storage supplement the overall struggle for a cleaner environment. However critical, implementing a growing renewable energy share into the European electricity sector presents several challenges regarding the stability and resilience of European energy systems. Furthermore, the role of natural gas and renewable gases in assisting the transition towards lower CO2 emissions and the integration of multiple alternative technologies are essential for the secure transformation of the European energy sector. The effort for energy security is also linked directly to the successful implementation of policies that encourage energy savings and the efficient use of the available energy resources. Energy efficiency, therefore, holds a vital role in shaping the future of energy consumption and is, in turn, directly affected by several technical aspects of the Energy Union policy. The overall success of the Energy Union is also linked to the positive effects it can generate for third countries within the EU neighbourhood area as well as its interaction with other international regulatory frameworks. EU's participation in international agreements like the Energy Charter Treaty, and international organisation like the Energy Community, bring the European Union and its neighbouring countries from the former Soviet Union, the Western Balkans and, on a broader framework, the South-Eastern Europe and Eastern Mediterranean regions, closer together to create an integrated pan-European energy market.

The two later sub-regional systems, South-Eastern Europe and the Eastern Mediterranean present particular interest and hold a distinctive role in EU's efforts for energy security. Exposed to specific security, geographical and structural conditions, the two regional systems entail the potential both for opportunities—regarding the enhancement of European energy security—and threats, regarding their exposure to political pressures from third countries. In its new strategic agenda for 2019–2024 released in June 2019, the European Council provided the new framework of its strategic objectives focusing on five main priorities that include protecting citizens and freedoms, developing a strong and vibrant economic base, building a climate-neutral, green, fair and social Europe, and promoting European interests and values on the global stage. Analysing a wide range of perceived threats and challenges, the Council's Strategic Agenda identifies, among others Energy Security, promoting political reform, the rule of law, economic convergence and good neighbourly relations in the Western Balkans, strengthening the European Neighbourhood Policy (ENP) to the east and to the south of EU, the extension of Trans-European Networks and the Energy Community with physical and digital connections, and supporting the sovereignty, independence and territorial integrity of states, the inviolability of borders and the peaceful settlement of disputes.[7]

Out of the wide span of security challenges and objectives set in the EU's Global Strategy documents, the ones related to the regional subsystems of SE Europe and Eastern Mediterranean seem to carry common characteristics regarding the action and tools needed in order to address them. These include the strengthening of the internal resilience and external cooperation between states in the region that keep and promote European principles and the rule of European and International law against threats by state and non-state actors. Regarding the regional system of SE Europe, a special focus relates to the subsystem of the Western Balkans, and its energy relations with other EU member states in the region, namely Greece, Bulgaria and Romania. The Western Balkans in particular, face long-enduring security and energy challenges. Conflicts following the break-up of the former Yugoslavia in the 1990s left most of its energy infrastructure damaged, while the ongoing political turmoil discouraged investments in new energy infrastructure. The region suffers from the lack of coherent policies to address energy security, diversification of energy sources and energy poverty. The wider Balkan region, including its EU Member States, is considered by the EU as essential to the overall integration of European infrastructure networks. All the Western Balkans countries have committed to increasing their share of renewable energy by 2030 to reach between 25 and 40% of their energy mix, as part of their obligations under the Energy Community Treaty. Parallel to the situation in SE Europe, the second subsystem under consideration, the Eastern Mediterranean, is

[7]"A new strategic agenda 2019–2024", European Council press release, June 20, 2019, https://www.consilium.europa.eu/en/press/press-releases/2019/06/20/a-new-strategic-agenda-2019-2024/.

inherently overloaded with a political dynamic characterised by competition, shifting power differentials and states striving for maximising their influence and power. The existing state rivalries have been intensified by the discovery of natural gas which appears to augment long-term security dilemmas bound to also involve third powers with interests in the region and its resources. The Eastern Mediterranean natural gas reserves are concentrated in the Levantine Basin, and include four major offshore gas fields; the Israeli Tamar and Leviathan fields, the Cypriot Aphrodite field and the Egyptian Zohr field.

Overall, on a regional level, EU's energy objectives for SE Europe and the Eastern Mediterranean relate primarily to issues of energy security and the internal energy market. However, the EU's overall policies on climate change also bear a direct effect on its regional interests in the two subsystems. In particular, the EU's commitment to reduce CO2 emissions, lead to policies aiming to rapidly reduce and eliminate coal-fuelled power production. Taking into consideration that coal-fuelled power production has been one of the major locally acquired sources of power production, the decision for its elimination bears the question of its replacement. Although the Commission's objective is to replace coal-fuelled production with Renewable Energy Resources (RES) production, limitations in electricity storage technology, combined with an unstable environment for new investments, pose a challenge for the rapid expansion of RES in the two subsystems and raise the question on the role of natural gas as a transition fuel.

Regarding EU's energy security, SE Europe and the Eastern Mediterranean, hold a crucial role primarily regarding EU's efforts to decrease its dependence from Russian gas through the development of the Southern Gas Corridor for the supply of gas from the Caspian and the Middle East, the expansion of liquefied natural gas (LNG) infrastructure to ensure the liquidity and diversity of the regional energy markets, and the linking of European markets with the Eastern Mediterranean through electricity and gas interconnections.[8] EU's Energy Union policy offers a comprehensive tool for regional cooperation in the Balkans and the

[8]"A Framework Strategy for a Resilient Energy Union with a Forward-Looking Climate Change Policy", European Commission communication, February 25, 2015, https://eur-lex.europa.eu/legal-content/EN/TXT/?uri=COM:2015:80:FIN.

Eastern Mediterranean where new energy networks and interconnections strengthen economic activity between states and improve economic output, thus offering the conditions for stronger internal resilience and external cooperation between states in the region. All the above elements, affect and shape the European energy sector in a continuous play of cooperation, collision and interconnection, thus raising the significance of a comprehensive and efficient framework to combine them. The Energy Union offers such a framework and has been the driving force for change in the European energy sector. It is the link between EU's fundamental drive towards a free-market and competition economy, its commitment to secure the availability, affordability and efficient use of energy resources and its resolve to protect the environment and secure a better future for all European citizens.

The chapters that follow examine these aspects of the Energy Union policy and the ways its implementation affects everyday life for European citizens. The book begins with a presentation of the overall framework where energy policies are applied, with three chapters addressing the security and geopolitical aspects of the Energy Union and analysing the relationship between energy security and national security. It continues with a regional focus where four chapters explore and analyse the regional, regulatory and investment aspects of the Energy Union with a particular focus in South-Eastern Europe and the Balkans. The transitional aspects of the Energy Union towards green energy and a circular economy are addressed in the following five chapters, examining current developments and the role of wind power generation, as well as the role of biofuels, renewable gases and LNG on the road to the energy transition. The book concludes with two chapters addressing specific technical aspects of the Energy Union regarding energy analytics, the transition from data acquisition to data-driven business models, and the role of equipment maintenance as a key factor for energy efficiency.

The broad spectrum of influence energy holds in all aspects of economic activity, renders the Energy Union as one of the most crucial policies in the European regulatory and political framework. The success in achieving its objectives will prove vital for the overall success of the European integration project and the European way of life.

1.1 About the Book

This publication consists of four parts, that include a total of fourteen chapters. The book is structured to deliver a broad educational output regarding the implementation of the Energy Union strategy while maintaining a pragmatic approach in analysing the various aspects of the policies in question. Furthermore, this publication helps create a connection between the EU's fundamental objectives regarding the structure of a common market and the challenges deriving from the European energy security framework on a national and supranational level. In addition, the book offers assistance in understanding the links between the von der Leyen Commission's priorities regarding Climate Change and its 'geopolitical' character, in comparison to the Juncker's 'political' Commission, and it's agenda for jobs, growth, fairness and democratic change.

Part I of the book contains three chapters addressing the security and geopolitical aspects of the Energy Union and analysing the relationship between energy security and national security. In Chapter 2, Christos Syriopoulos, from the Greek Ministry for Environment & Energy, examines the progress made towards the realisation of EU's energy security targets since the introduction of the Energy Union policy in 2015. His analysis focuses on the political and economic aspects of the process. The chapter presents the overall situation of energy security in Europe while addressing its elements in close relation to the other four dimensions of the Energy Union, namely the internal market, energy efficiency, climate action & decarbonisation and research & innovation. In Chapter 3, Dr. Thrassy N. Marketos, from the University of Peloponnese, examines the security and political framework of the Eastern Mediterranean subsystem and its connection to the Energy Union's goals to diversify its supply routes. The chapter analyses the effects of the Libya war in the region, combined with a presentation of the interests of several international actors including Russia, Turkey, Egypt, Cyprus and the US and an estimation on the potential of gas flows to the EU from the Eastern Mediterranean. Furthermore, the analysis turns on China's role in the Eastern Mediterranean region and examines the analogy between the Pan-Asian Energy Club and the Eastern Mediterranean

Gas Forum. Chapter 4 examines the fundamental elements of the relationship between energy and national security. It builds a connection between political realism, securitisation theory and energy security and identifies the framework and conditions under which energy infrastructure acquires strategic value for the stakeholders it involves. The chapter presents the wide context of energy security threats and examines the EU's vulnerabilities related to them. Additionally, it identifies the tools of the European regulatory and policy framework that, while designed to address internal issues, incorporate an 'externality' effect, thus influencing EU's relations with third countries in the energy sector.

Part II contains four chapters that explore and analyse the regional, regulatory and investment aspects of the Energy Union with a particular focus in South-Eastern Europe and the Balkans. In Chapter 5, Dr. Sotirios Manolkidis, from the Greek Regulatory Authority for Energy, addresses the issue of energy growth and sustainability for the Western Balkan Six Initiative countries within the Energy Community. It examines the significant energy challenges the region faces ever since the conflicts following the break of the former Yugoslavia in the 1990s regarding damaged or depleted energy infrastructure, energy poverty and unreliable energy supply. The analysis addresses the need for institutional reform within the Energy Community and argues for the further strengthening of its existing dispute settlement mechanism alongside policy changes and the introduction of new regulatory frameworks. In Chapter 6, Professor Panayiotis Glavinis, from the Aristotle University of Thessaloniki, together with Anna Konstantinidou, Afroditi Semkou, Elias Andreadis and Elias Kolovos, examine the extent of integration of the regional energy markets between the Member States of the EU and the countries of the wider Europe. The analysis further explores whether the non-EU energy markets are in a position to conform to the *acquis communautaire* and whether they are able to ensure the security of the energy supply to the EU in affordable prices. The chapter identifies the steps that are necessary to integrate these crucial regional energy markets into the EU common market regarding harmonisation of the legal framework, essential infrastructure and networks, and promotion and the mutual protection regimes for foreign direct investments. It also examines the role of the Energy Community, the TEN-E strategy and the Energy Charter Treaty as useful tools for the EU to foster the integration of the energy markets in the wider Europe. In Chapter 7, Dr. Maria Bozoudi, from the American College of Thessaloniki examines the investment framework regarding integrating the Balkan countries into the European energy market and infrastructure network and identifies the investment risks stemming from delays in institutional reforms. The analysis explores Western Balkan governments' commitment to follow the EU's energy and climate footsteps through targets for 20,230 mirroring those of the EU and identifies the links between 'soft measures' and interest for investment in the energy sector of the Western Balkans. Furthermore, the chapter examines the role of the Energy Community and its effectiveness in promoting progress in the transition efforts in the Western Balkans towards environmental sustainability in the energy sector. In Chapter 8, Dr. Gökçe Mete, from the International Energy Charter, and Janessa Goh Pei-Ru, deliberate on the role of the Energy Charter process in contributing towards the energy transition and energy security in South-East Europe in line with the EU's Energy Union vision. The analysis presents the elements that support the potential of the Energy Charter Treaty to aid the convergence among South-East European countries towards accelerating the energy transition and strengthening energy security. The chapter explores how the broader Energy Charter process and its investment protection provisions and provisions on energy efficiency and soft law instruments, can promote renewable energy investments, play a part in improving energy security, and promote conditions for overall investment and secure transit in the area.

Part III includes five chapters addressing the transitional aspects of the Energy Union towards green energy and a circular economy. It examines developments and the role of wind power generation, as well as the role of biofuels, renewable gases and LNG on the road to the energy transition. In Chapter 9, Alexandros Lagakos, from Blue Grid Gas & Power, provides an overview of the elements steering the adaptation, implementation and prospective expansion of the use of Liquefied Natural Gas as a marine fuel. The analysis explores LNG's ability to reduce greenhouse gas emissions in shipping. It presents the drivers for a transition to LNG in shipping regarding environmental performance, international environmental regulation and corporate social responsibility, and examines its overall potential to become the medium for the utilisation of renewable gases in the shipping industry. In Chapter 10, Constantine Levoyannis from the Brussels-branch of the Greek Energy Forum addresses the issue of renewable gases and their role and contribution in achieving climate neutrality by 2050 in the EU. The analysis examines the elements and merits from policies supporting a hybrid energy system that links electricity and gas, and the use of their competitive advantages on an environmental, energy efficiency and economic standpoint. Furthermore, the chapter examines the future of existing gas infrastructure through repurposing and redefining its role to decrease the cost of the energy transition as well as the role of bioenergy and hydrogen in the future of gas networks. In Chapter 11, Dr. Spyros Kiartzis, from Hellenic Petroleum, together with Katerina Tsita, and Evangelia Paschalidou, provide an insight into the future transformation of the energy markets and discuss technological, regulatory and financial challenges correlated to energy security. The analysis examines the various ways low-carbon transition shapes the profile of infrastructure investments, as well as the role of biofuels in the transformation of the transportation sector. Furthermore, the chapter addresses the technical, business and societal factors that determine the pace of energy transition, the necessity for commercialisation and scale-up of new technologies and the role of an integrated approach to policy, planning, management and development of energy systems in supporting a transition to decarbonising the energy supply-chain. In Chapter 12, Ivan Pineda from WindEurope argues that the best approach for Europe to become carbon neutral by 2050, thereby contributing to the 1.5°C goal of the Climate Paris Agreement, is through the renewables-based electrification of most sectors in the economy. The analysis also examines the challenges the deployment of renewable energy faces, regarding increasing the flexibility of the power system, and facilitating, permitting and plugging the financing gap an electricity market dominated by renewables poses for investment signals. The chapter relates renewables' future success with scaling up electrification, analyses the role of energy efficiency, electrification incentives, and CO2 pricing, and identifies the elements in bridging the gap between scale and implementation of renewable energy. In Chapter 13, Marios Papalexandrou from Aeolus Limited examines how the EU managed to transform the offshore wind industry from a few demonstration projects two decades ago to the mainstream generation technology of today. The analysis discusses offshore wind technology developments and trends from past to recent and future projects and presents best practices incorporated from countries leading offshore wind expansion in Europe. Finally, it highlights Europe's offshore wind potential and discusses future challenges allowing Europe to stay ahead of the curve in research and implementation of future offshore wind developments.

Part IV, the final part of this book consisting of two chapters, addresses specific technical aspects of the Energy Union regarding energy analytics, the transition from data acquisition to data-driven business models, and the role of equipment maintenance as a key factor for energy efficiency. In Chapter 14, Dr. Dimitrios I. Doukas, from NET2GRID, examines the ways that digital transformation in the energy sector, smart-metering and access to quality energy data put the end-user at the epicentre of the new European energy system. The analysis presents how smart meters are becoming the necessary tools for the end-users to increase their impact and leverage in the future grid by participating in demand response schemes and facilitating consumers' ability to switch energy providers. The chapter presents energy analytics alternatives for various parts of the smart grid with an emphasis on the residential-level sector and defines energy disaggregation as a concept and as the foundation layer on which most residential-level energy analytics services are built. Finally, it presents and categorises data-driven business models that allow energy companies to develop products and services that build stronger customer engagement. In Chapter 15, the final chapter of this publication, Dr. Vassilios Kappatos from the Greek Centre for Research and Technology, presents the benefits of energy efficiency and maintenance in industry and society, summarises the European energy efficiency and maintenance strategies, and presents the benefits of adequately maintaining equipment and systems to enhance energy efficiency. The analysis presents the ways energy efficiency can be achieved through the utilisation of efficient devices, the implementation of strategies that reduce energy consumption and the proper implementation of maintenance procedures. The chapter summarises the European energy efficiency and maintenance strategies and presents the benefits of maintenance in industry and society.

Overall, the fifteen chapters of this book are designed to offer an insightful look into the challenges that applied regulatory and political policies face in the European energy sector. With a broad spectrum of issues, spanning from the geopolitics of energy and energy regulation to implementing climate change, renewable energy production, new gas technologies and consumer empowerment structures, this book presents a pragmatic look into the Energy Union's efficiency in providing secure, sustainable, competitive and affordable energy for all Europeans.

Part I

Security & Geopolitical Aspects of the Energy Union

2

Has the Energy Union Strategy Delivered Concrete Solutions to Europe's Energy Security Question?

Christos Syriopoulos

2.1 Introduction

From the very beginning of the European Union's history in the 50s, energy has been an integral part of the European integration process. From the Treaties regulating coal and nuclear energy in 1951 and 1957, respectively, to measures mitigating the effects of difficulties in the supply of crude oil and petroleum products in the 60s and 70s[1] and to market liberalisation packages in the 90s and 00s, participants in the European political process have been trying to address challenges and promote their

The views and opinions expressed in this paper are those of the author and do not necessarily reflect the official policy or position of the Hellenic Republic.

[1] Council Directive 73/238/EEC on measures to mitigate the effects of difficulties in the supply of crude oil and petroleum products, 1973, OJ (L 228), pp. 1–2.

C. Syriopoulos (✉)
Ministry for Environment and Energy, Hellenic Republic, Athens, Greece
e-mail: syriopoulosch@prv.ypeka.gr

M. Mathioulakis (ed.), *Aspects of the Energy Union*,
Energy, Climate and the Environment,
https://doi.org/10.1007/978-3-030-55981-6_2

political and economic aspirations. In this gradual process, the role of the EU in the energy sector evolves striving to fulfil, at times, contradicting targets, i.e. energy security, sustainability, competitiveness, affordability. A watchful analyst, when trying to understand EU energy policies and legislation should always bear in mind a complex political, economic and institutional set of parameters. EU Treaties are setting the scene of the process (EU institutions, allocation of competences between MS and Institutions, legislative procedures, etc.), but the actual dynamics that create the policy output vary from intra-state or intra-European issues, like the lack of sufficient indigenous energy resources, or diverging economic interests across market participants and other stakeholders, to global questions like climate change, regional conflicts and world security and defence considerations. This chapter attempts to take stoke of the progress made towards the realisation of EU's energy security targets since 2015, the year that the European Commission presented its Energy Union Strategy. It is not the intention of the writer to underestimate legal and institutional issues. Still, mainly for methodological reasons, the analysis that follows will concentrate on the political and economic aspects. Furthermore, the intention is to present the general situation of energy security, and as a consequence, the text may lack details. Lastly, since the Energy Union Strategy is made up of five closely related and mutually reinforcing dimensions, energy security will be discussed in close relation to the other four dimensions (internal market, energy efficiency, climate action/decarbonisation, research & innovation.[2]

2.2 The EU Energy Security Situation Until 2015

After the oil crises of the 1970s, Europe for more than 30 years, had not faced any significant enduring disruption of its energy supply. This resulted in low prioritisation of energy security in the national and European political agenda and fewer adopted energy security measures in

[2] European Commission Communication on "A Framework Strategy for a Resilient Energy Union with a Forward-Looking Climate Change Policy", COM (2015), 80 final (February 25, 2015), [Hereinafter Energy Union Communication].

comparison to the period after 2006, the year of the first gas supply crisis. Furthermore, the discussion concentrated mainly on the impact of energy dependency on the EU economic stance in the global context and not on the potentially political nature of the issue. For instance Commissions Green Paper "Towards a European strategy for the security of energy supply" in 2000,[3] that presents the main aspects of a long-term energy strategy was initiated by the dramatic rise in oil prices (crude oil price has tripled between March 1999 and November 2000), which could undermine the recovery of the European economy. In historical terms, one should note that two out of the three Treaties that established the European Communities in the 50s were related to energy (ECSC, Euratom), in an attempt to ensure systematic supplies of coal and nuclear energy in the Community. The Treaty establishing the European Economic Community did not establish a common energy policy, nor did the Maastricht and Amsterdam Treaties. In this respect, until the Treaty of Lisbon, energy issues had been dealt with through Treaties legal bases concerning the internal market, environmental policy, taxation, etc.

The Lisbon Treaty included a specific article on energy (Art. 194 of the Treaty on the Functioning of the European Union), providing for co-decision and qualified majority voting, which sets out four main aims of the EU's energy policy: to ensure the functioning of the energy market, to ensure the security of supply in the Union, to promote energy efficiency and energy saving and develop new and renewable forms of energy and to promote the interconnection of energy networks. Nevertheless, during the 00s, the EU had already tried to address energy security concerns. In this respect, Council Directive 2004/67/EC[4] established for the first time a legal framework at EU level to safeguard the security of natural gas supply. This Directive, among others, established the Gas Coordination Group, as a useful exchange information mechanism. The European Council in December 2006[5] recognised that the Union had to work to ensure the long-term security of energy supply,

[3]European Commission Green Paper "Towards a European Strategy for the Security of Energy Supply", COM (2000), 769 final (November 29, 2000).

[4]Council Directive 2004/67/EC concerning measures to safeguard security of natural gas supply, 2004, OJ (L 127).

[5]European Council Conclusions 14/15 December 2006, 16879/06 (December 14/15, 2006).

defined some areas for further work (para 34) and endorsed (para 36) the Network of Energy Security Correspondents (NESCO), which would be an essential tool for collecting and processing existing geopolitical and energy-related information and would also provide an early warning tool to ensure the security of energy supply. Moreover, the third energy package contained essential means to enhance the EU's security of energy supply. For instance Commission's rights to assess whether the control of an EU Gas Transmission Operator(s) by a person or persons from a third country or third countries will put at risk the security of energy supply to the EU.[6] These instruments were complemented by the EU-Russia energy dialogue (and an Early Warning Mechanism), EU energy relations with Norway, the EU-Ukraine Memorandum of Understanding (MoU) on Energy, as well as other energy dialogues with producing and transit countries. Further to gas, energy security measures were taken regarding electricity and oil. In electricity, Directive 2005/89/EC set out the necessary actions that the Member States should take in order to ensure the security of electricity supply in general.[7] In oil, the Directives 2006/67/EC[8] and 2009/119/EC[9] laid down rules aimed at ensuring a high level of security of oil supply in the Union. The EU Member States would have to maintain minimum stocks of crude oil and petroleum products (emergency and specific stocks), and, in addition, they should prepare contingency plans in case of a shortage. Furthermore, coordination between EU and International Energy Agency (IEA) was envisaged, as well as ways to secure coherence between EU and IEA stock-holding practices. Lastly, the EU in 2008 launched a policy to protect its critical infrastructure with the adoption of a directive establishing a procedure

---

[6]European Parliament and Council Directive 2009/73/EC concerning common rules for the internal market in natural gas and repealing Directive 2003/55/EC, 2009, OJ (L 211), Article 11, [Hereinafter Gas Directive].

[7]European Parliament and Council Directive 2005/89/EC concerning measures to safeguard security of electricity supply and infrastructure investment, 2006, OJ (L 33).

[8]Council Directive 2006/67/EC imposing an obligation on Member States to maintain minimum stocks of crude oil and/or petroleum products, 2006, OJ (L 217).

[9]Council Directive 2009/119/EC imposing an obligation on Member States to maintain minimum stocks of crude oil and/or petroleum products and repealing Directive 2004/67/EC, 2009, OJ (L 265).

for the identification and designation of European Critical Infrastructures (ECIs) and a common approach to the assessment of the need to improve the protection of such infrastructures.

The understanding of the notion of energy security changed dramatically for the EU after the temporary disruptions of gas supplies that strongly hit some of the eastern EU Member States (MS) in 2006 and 2009. In January 2006 gas supplies to the EU were restored after one day of disruptions and available storage capacities easily made up the shortfall. But the situation in January 2009 was unique in the history of EU gas supplies. An unresolved commercial dispute between Naftogaz and Gazprom had, as a result, Gazprom to stop supplying gas for Ukrainian consumption on the 1st of January 2009, while gas to be transited through Ukraine for European consumption continued to be delivered to Ukraine. From the 1st to the 6th of January there were growing reductions of gas volumes towards the EU, and from the 7th to the 20th of January, all supplies from Russia through Ukraine to the EU were cut. In its analysis of the 2009 crisis, the European Commission[10] concluded that "a repeat of the January 2009 gas supply crisis, from a similar or different cause, cannot be ruled out, (…), the EU energy security strategy must be strengthened, (…) the crisis highlighted weaknesses in each stage of the gas supply security approach at EU level". The central idea for the next steps in energy security was the formalisation of ad hoc measures used to address the disruptions of gas supplies in January 2009. The most critical selected policy tools to rapidly adverse this negative situation was the Gas Regulation and the new TEN-E Regulation. The new TEN-E Regulation[11] sets the objective of the development and interoperability of trans-European energy networks and envisages the establishment of a list of priority projects for the EU, called Projects of Common Interest (PCI). These projects have a "priority status" at the national level to

[10]European Commission staff working document—Accompanying document to the Proposal for a Regulation of the European Parliament and of the Council concerning measures to safeguard security of gas supply and repealing Directive 2004/67/EC—The January 2009 gas supply disruption to the EU: an assessment, SEC(2009) 977 final, (July 16, 2009).

[11]European Parliament and Council Regulation 347/2013 on guidelines for trans-European energy infrastructure and repealing Decision No. 1364/2006/EC and amending Regulations (EC) No. 713/2009, (EC) No. 714/2009 and (EC) No. 715/2009, 2013, OJ (L 115). [Hereinafter TEN-E Regulation].

ensure rapid administrative treatment and may be granted with financing from the Connecting Europe Facility (CEF).[12] The Gas Regulation No. 994/2010 included measures to safeguard the security of gas supply in terms of preparation and mitigation. Member States were required to draw up plans including preventive and emergency measures and had to meet several obligations regarding infrastructure capacity and gas supply.[13] Despite the above measures, the European energy security stress tests conducted in the summer of 2014 showed that a possible severe disruption of gas supplies from the east would still have a major impact throughout the EU, particularly in Eastern Europe.[14] The stress tests were based on "non-cooperative" and "cooperative scenarios". In the cooperative scenario, the effects of the disruption were significantly dampened in those Member States and Energy Community Contracting Parties most affected. This vividly revealed the need for solidarity and cooperation among EU MS and Energy Community contracting parties. Furthermore, it showed the strategic importance of diversification of routes and sources, since Greece was almost unaffected in the non-cooperative scenario (due to the diversification that Revithousa LNG Terminal provides), but in the cooperative scenario would also experience some non-negligible shortfall. Furthermore, the overall energy security situation as presented by the European Commission in mid-2014 was,

---

[12]Regulation 347/2013 was the outcome of discussions that started after the 2009 crisis. A communication from the Commission entitled "Energy Infrastructure Priorities for 2020 and Beyond—A Blueprint for an Integrated European Energy Network" COM(2010), 677 final (November 17, 2010), called for a new energy infrastructure policy to optimise network development at European level for the period up to 2020 and beyond. There were also Energy Council Conclusions [Council Conclusions on "Energy 2020: A strategy for competitive, sustainable and secure energy", 6207/1/11 (February 28, 2011)], a European Parliament resolution on energy infrastructure priorities for 2020 and beyond [European Parliament resolution on energy infrastructure priorities for 2020 and beyond, P7_TA(2011)0318, (July 5, 2011)] and European Council Conclusions [European Council Conclusions 4 February 2011, 2/11, (February 4, 2011)] underlining the need to modernise and expand Europe's energy infrastructure and to interconnect networks across borders. In addition, European Council insisted that no Member State should remain isolated from the European gas and electricity networks after 2015 or see its energy security jeopardised by lack of the appropriate connections.

[13]European Parliament and Council Regulation 994/2010 concerning measures to safeguard security of gas supply and repealing Council Directive 2004/67/EC, 2010, OJ (L 295).

[14]European Commission Communication, on "The Short-Term Resilience of the European Gas System Preparedness for a Possible Disruption of Supplies from the East During the Fall and Winter of 2014/2015", COM (2014) 654 final (October 16, 2014).

at the least, disappointing.[15] The EU imported 53% of the energy it consumed. There was an immense energy import dependency to crude oil (almost 90%) and natural gas (66%, with two-thirds of these imports coming from countries outside of the European Economic Area). The situation was even worse for less integrated and connected regions of the EU, such as the Baltic states and Eastern Europe. The most pressing energy security of supply issue was the strong dependence from a single external supplier. Six Member States (the Baltic States, Finland, Slovakia and Bulgaria) depended from Russia as the single external supplier for their entire gas imports, and two (the Czech Republic and Austria) had very concentrated imported gas supplies.[16] Last, for electricity, three Member States (Estonia, Latvia and Lithuania) were dependent on one external operator for the operation and balancing of their electricity network. In addition to the above, the situation in the Crimean crisis that started in 2014 has acted as a wake-up call for Europe to pursue even more actively its strategy of increased energy independence from Russia.[17]

2.3 Initial Negotiation Positions of the Main Actors in 2015

In the first months of 2015, the strategic agenda of the energy sector for the next five years had already been formulated. The October 2014

[15]European Commission Communication, on "European Energy Security Strategy", COM (2014) 330 final (May 28, 2014).

[16]European Commission, "staff Working Document, In-depth Study of European Energy Security Accompanying the document Communication from the Commission to the Council and the European Parliament: European Energy Security Strategy," SWD(2014) 330 final/2 (May 28, 2014).

[17]For the implications of the Crimean Crisis for the EU energy security and an analysis with possible alternative gas routes and sources bypassing Russia see Pasquale De Micco, "The EU's Energy Security Made Urgent by the Crimean Crisis", In depth analysis, European Parliament (April 2014).

European Council[18] agreed upon the 2030 energy and climate targets (binding EU target of an at least 40% domestic reduction in greenhouse gas emissions by 2030 compared to 1990, an EU target of at least 27% for the share of renewable energy consumed in the EU in 2030 and an indicative target at the EU level of at least 27% for improving energy efficiency in 2030 compared to projections of future energy consumption based on the current criteria), the future governance of the sector (there would be EU and not national targets) and addressed the issues of completing the internal market and achieving energy security. On energy security, besides the usual language in favour of critical gas interconnections, better EU coordination in external energy relations, strengthening regional cooperation (regarding Energy Community), the European Council (EUCO) "encouraged the Member States and involved companies to provide relevant information to the Commission and seek its support throughout negotiations, including on the ex-ante assessment of the intergovernmental agreements' compatibility with EU legislation and its energy security priorities". In February 2015 the European Commission presented the Energy Union Strategy,[19] and on March 2015 the European Council,[20] mandated the Commission to implement the new Strategy. On Energy Security the European Council called for "reinforcing the legislative framework for the security of supply for electricity and gas; energy security can also be strengthened by robust grids, increased energy efficiency and having recourse to indigenous resources as well as safe and sustainable low carbon technologies", "ensuring full compliance with EU law of all agreements related to the buying of gas from external suppliers, notably by reinforcing transparency of such agreements and compatibility with EU energy security provisions" and "using all external policy instruments to establish strategic energy partnerships with increasingly important producing and transit countries". Donald Tusk, the President of the European Council, had already (as Poland's Prime minister) acknowledged in April 2014 that the Energy Union is necessary to secure EU's gas supply because

[18]European Council Conclusions 23/24 October 2014, 169/14 (October 23–24, 2014).
[19]Energy Union Communication, supra note 3.
[20]European Council Conclusions 19/20 March 2015, 11/15.

its current dependence on Russian energy makes Europe weak.[21] The President of European Commission Jean-Claude Juncker in his political guidelines included "A resilient Energy Union with a forward-looking climate change policy" as one of his ten priorities.[22] He referred to "Geopolitical events that reminded us that Europe relies too heavily on fuel and gas imports", he called for a "need to diversify EU's energy sources, and reduce the high energy dependency of several Member States" and warned that "if the price for energy from the East becomes too expensive, either in commercial or in political terms, Europe should be able to switch very swiftly to other supply channels". The European Parliament debated in plenary session and welcomed the Strategy on the 25th of February 2015.[23] The Members of the European Parliament broadly supported the strategy's key priorities, such as security of energy supply, completion of the internal energy market, continuous decarbonisation and focus on research and innovation in energy fields.

The promotion of interconnections was a priority for all groups. In the same time, the Greens noticed a lack of climate ambition, particularly on energy efficiency. Last—and most importantly for the scope of this chapter—there were divergent views on the role of the Commission in intergovernmental energy agreements (IGAs), with GUE/NGL and NI opposing its compulsory involvement. It should also be noted that Commission's Vice President Mr Maroš Šefčovič, in his closing remarks, defended the Commission's proposals for ex-ante compatibility checks of IGAs, and more transparency in long-term gas contracts and diversification of sources and routes, for reasons of EU energy security. The first discussions among the EU MS on the Energy Union Strategy took place during the preparations for the March 2015 European Council.

---

21 Donald Tusk, "A United Europe can end Russia's Energy Stranglehold" *Financial Times*, April 21, 2014.

22 Jean-Claude Juncker, "A New Start for Europe: My Agenda for Jobs, Growth, Fairness and Democratic Change, Political Guidelines for the next European Commission", Strasbourg, July 15, 2014.

23 The European Parliaments debate of the 25th February 2015 is available in EP's webpage: https://www.europarl.europa.eu/doceo/document/CRE-8-2015-02-25-ITM-013_EN.html, accessed February 7, 2020.

MS broadly supported the notion of an Energy Union Strategy with five closely interrelated and mutually reinforcing dimensions. Nevertheless, there were significantly divergent views on several issues. The most important of them were (i) the climate ambition, i.e. whether the EU should set higher, than those of October 2014 EUCO, RES and energy efficiency 2030 targets, (ii) how to administer the Governance of the Energy Union, i.e. who would be in the driver's seat Commission or MS and (iii) energy security and the future role of Commission. On the third issue, there were already expressed strong reservations by several MS on proposals concerning ex-ante compatibility checks of IGAs and more transparency in long-term gas contracts. In the same time, a block of MS was formulated that strongly supported these proposals. Last but not least, other critical issues were raised by some MS such as the lack of adequate interconnections, especially in isolated regions of the Union and the future role of nuclear energy in the energy mix. In parallel, academia[24] and stakeholders, such as companies[25] and NGOs, were also dealing with the issue of the Energy Union strategy putting forward numerous proposals and many times diverging opinions. Climate NGOs such as WWF and Friends of the Earth posed severe reservations on the strategy's aspects dealing with fossil fuels and energy security.[26]

[24]For example see Sami Andoura and Jean-Arnold Vinois, "From the European Energy Community to the Energy Union, A Policy Proposal for the Short and Long Term", Notre Europe Jacques Delors Institute, January 2015, also Simone Tagliapietra and George Zachmann, "Rethinking the Security of the European Union's Gas Supply", Bruegel, January 2016.

[25]See EURELECTRIC, Eurogas, EPIA and EWEA Joint Statement on "The Energy Union, a fresh push for the Internal Energy Market", February 23, 2015, https://eurogas.org/website/wp-content/uploads/2018/03/Joint_Statement_on_the_Energy_Union.pdf. accessed February 7, 2020.

[26]For example, see Friends of the Earth comment "Europe's Climate and Energy Plans Frustrate", February 25, 2015, https://www.foeeurope.org/Europe-climate-energy-plans-frustrate-250215, accessed February 7, 2020, where it is stated that "It Stresses the Need to Shift the Economy Away from Fossil Fuels, (...). But this Vision is Blurred by Attempts to Placate the Fossil Industry by Diversifying Gas Supplies". Also, WWF briefing paper, "The Energy Union, Ensuring a Vision without Blind Spots", February 2015, where it is proposed that "Different gas Supplies Must not Mean More Gas Supplies".

2.4 Proposed Changes Related to Energy Security, Negotiations and Final Output of the Legislative Procedure

2.4.1 The 2016 Energy Security Package

In February 2016 the European Commission presented an energy security package consisting of two legislative proposals (Security of Gas Supply Regulation, Decision on Intergovernmental Agreements in energy) and two strategies (Liquefied natural gas (LNG) and gas storage strategy, Heating and Cooling strategy). The package in total was aiming at (i) addressing the risk of gas supply disruptions by introducing a solidarity principle among the Member States, (ii) enhancing transparency and ensuring full compliance with EU law on intergovernmental agreements and company gas supply contracts, (iii) reducing the physical and regulatory barriers for liquefied natural gas (LNG) access to EU markets and (iv) decreasing fossil fuels consumption and increasing the use of renewables in heating and cooling.

2.4.1.1 Security of Gas Supply Regulation

The existing Gas Regulation No 994/2010 included measures to safeguard the security of gas supply in terms of preparation and mitigation. The proposed new Regulation introduced[27] a solidarity principle among MS as a last resort, a regional approach when designing security of supply measures and reinforced cooperation with EU neighbours (Energy Community Contracting Parties), as well as additional transparency measures for certain security of supply relevant contracts which will have to be automatically notified by the natural gas companies to the Commission and the MS upon their conclusion or modification. MS and gas companies would need to take up new responsibilities. MS would have to prepare Risk Assessments, Preventive Action Plans and

[27] European Commission, Proposal for a Regulation concerning measures to safeguard the security of gas supply and repealing Regulation 994/2010, COM (2016) 52 final—2016/030 (COD), (February 16, 2016).

Emergency Plans at a regional level and agree on the technical, legal and financial arrangements to apply the solidarity principle and include them in the Emergency Plans. Natural gas companies would have to notify the Member States and the Commission about some security of supply relevant contracts upon their conclusion or modification. The Commission would have a greater overall coordination role and a more robust set of tools to assess energy security risks. Among the issues that created the most significant controversy during the negotiations was the application of the solidarity principle across borders (i.e. how to compensate country economic actors for gas volumes redirected to economic actors in nearby MS or Energy Community contracting parties) and the notification of gas contracts to Commission. For the second issue, a compromise was reached when objecting MS, the Commission and the European Parliament (whose Reporter, Chair of ITRE and former Polish Prime Minister Jerzy Buzek vividly backed the initial proposal and proposed to raise its ambition further)[28] decided that natural gas undertakings shall notify the competent, concerned authority (and not the Commission) on the details of gas supply contracts with a cross-border dimension and a duration of more than one year. It was also decided that, under duly justified circumstances, the Commission may request the natural gas undertaking to provide the contract.[29]

2.4.1.2 Decision on Intergovernmental Agreements (IGAs) with Third Countries in Energy

The notification of IGAs to European Commission was envisaged already by 2012 (Decision No 994/2012/EU).[30] However, Decision No 994/2012/EU proved ineffective in terms of ensuring compliance of intergovernmental agreements with Union law. In its new

[28]European Parliament Report No. A8-0310/2016 (October 20, 2016).

[29]European Parliament and Council Regulation 2017/1938 concerning measures to safeguard the security of gas supply and repealing Regulation 994/2010, 2017, OJ (L 280).

[30]European Parliament and Council Decision 994/2012/EU establishing an information exchange mechanism with regard to intergovernmental agreements between Member States and third countries in the field of energy, 2012, OJ (L 299).

legislative proposal, the European Commission explained that[31] "intergovernmental agreements often contain no appropriate termination or adaptation clauses which would allow the Member States to bring the intergovernmental agreement in compliance with Union law within a reasonable period of time. Furthermore, the positions of the signatories have already been fixed, which creates political pressure not to change any aspect of the agreement". In addition, Commission's analysis of all notified IGAs showed that around one-third of the ex-post notified agreements related to energy infrastructure or energy supply contained provisions that were not compliant with EU law and none of them had been successfully renegotiated. The proposed Decision mainly introduced mandatory ex-ante compatibility check of intergovernmental agreements by the Commission. MS would have to notify their draft IGAs to the Commission before concluding them, and they would not sign IGAs until the Commission has issued its opinion. MS would also have to inform Commission when starting new negotiations for an IGA and they can ask Commission to participate in the negotiations. The Decision also included ex-post assessment by the Commission to non-legally binding instruments. Once again, discussions within the Council were difficult, as a significant number of MS would not accept to lose their sovereign right to conclude IGAs at will. On the other side of the table, the Commission firmly supported its proposal, backed by those MS traditionally suspicious of Russia, who welcomed the Commission's participation in this process. The European Parliament (with a reporter from Poland once again) out bided the initial Commission's proposal leading to a more ambitious final compromise. The final Decision[32] creates an ex-ante notification mechanism, which includes oil IGAs (further to gas, as in the initial proposal). At the same time, electricity IGAs will be subject to ex-post compliance checks. Furthermore,

[31] European Commission, Proposal for a Decision on establishing an information exchange mechanism with regard to intergovernmental agreements and non-binding instruments between Member States and third countries in the field of energy and repealing Decision No. 994/2012/EU, COM (2016) 53 final—2016/031 (COD) (February 16, 2016).

[32] European Parliament and Council Decision 2017/684 on establishing an information exchange mechanism with regard to intergovernmental agreements and non-binding instruments between Member States and third countries in the field of energy, and repealing Decision No. 994/2012/EU, 2017, OJ (L 99).

MS shall not sign, ratify or agree to any agreement before the Commission informs about its doubts and shall take an utmost account of the Commissions' opinion before signing, ratifying or agreeing to an IGA. Last, regarding the assistance from the Commission, it can be either invited by a MS to assist in the negotiation of an IGA as an observer or, if it considers it to be necessary, the Commission can request to participate. It should be noted that IGAs is not a very common policy tool in the energy sector in the last decade. Still, on the other hand, this Decision can halt future IGAs and also creates a legal precedent for other policy areas. Finally, there is only one known case of the implementation of these new provisions; the EastMed pipeline IGA signed in January 2020, where the Commission was asked to participate in the negotiations conducted during 2018. It seems that all relevant provisions were successful and implemented on time.

2.4.1.3 Liquefied Natural Gas (LNG) and Gas Storage Strategy

The 2006 and 2009 gas crises had shown that EU MS that have access to LNG import terminals and gas storage facilities were more resilient to possible supply interruptions than those dependent on a single gas supplier. Furthermore, the global LNG market is growing with the entrance of new suppliers and the construction of new LNG terminals in consumer countries. According to IEA[33] between 2018 and 2023, around 140 bcm of liquefaction capacity will be added, increasing global capacity by almost 30%. More than half of that expansion (over 80 bcm) takes place in the United States, followed by Australia (30 bcm) and Russia (15 bcm). Based on the proposed strategy and in order to improve the access of all Member States to LNG and storage as an alternative source of gas, the EU would need to[34] build the necessary infrastructure to complete the internal market, allowing all MS to access international LNG markets, complete, in regulatory terms, the internal

[33]IEA, Gas market Report 2018.

[34]European Commission Communication on "EU Strategy for Liquefied Natural Gas and Gas Storage", COM (2016) 49 final (February 2, 2016).

gas market, use storage facilities more efficiently and work closely with international partners, to promote free, liquid and transparent global LNG markets. In addition to LNG, storage of gas has two main functions for the energy market. Firstly, it balances fluctuations in daily and seasonal demand and secondly it provides the security of supply of gas in case of potential disruptions or particularly high (seasonal) demand. In this respect, with the LNG and storage strategy, the Commission proposed to use in a more coordinated manner, several policy tools and initiatives. In real terms the strategy intended to make use of the TEN-E Regulation (Regulation 347/2013) and regional cooperation mechanisms (Central East South Europe Gas Connectivity group, Baltic Energy Market Interconnection Plan, South-West Europe high-level group) to further promote missing interconnection links, LNG terminals and underground storage facilities, as well as to resolve all the regulatory issues which fragment the market or pose barriers for users to access LNG terminals and storage facilities. In addition to that, the European Commission would take up the role of conducting international dialogues with LNG current and potential suppliers (especially with the USA, Australia, Qatar) and major LNG importers such as Japan.

2.4.1.4 Heating and Cooling Strategy

The fourth and last element of the package was the heating and cooling strategy.[35] The main aspirations of the strategy were to decrease energy leakage from buildings, maximise efficiency and boost the share of renewables.

2.4.2 Clean Energy for All Europeans Package

On the 30th of November 2016, the European Commission, following EU's commitment to reduce greenhouse gas emissions by 40% by 2030, adopted its new legislative proposals on energy efficiency, renewables

[35] European Commission Communication on "An EU Strategy on Heating and Cooling", COM (2016) 51 final (February 02, 2016).

and electricity market design.[36] The legislative proposals had three main objectives: to prioritise energy efficiency, to achieve a leading role for the EU in the global promotion of renewable energy and to provide a fair deal for consumers. The clean energy package included two proposals on energy efficiency, three on the organisation of the electricity market, one on RES and one on Energy Union Governance. The package also contained several Commission Regulations on eco-design of products that would require Council approval. Finally, the package included various Communications and Reports that accompanied the legislative proposals such as assessing the state of the electricity market, energy efficiency best practices, reporting on energy prices and costs in Europe and a Communication on accelerating innovation in the clean energy technologies. Although energy security was not one of the three goals of the package, the various proposals had significant implications for the energy security of the EU. The market design included four proposals, and only one was directly relevant to energy security [Proposal for a Directive on common rules for the internal market in electricity (recast), Proposal for a Regulation on the electricity market (recast), Proposal for a Regulation establishing a European Union Agency for the Cooperation of Energy Regulators (recast) and Proposal for a Regulation on risk preparedness in the electricity sector]. The Commission decided that a new Regulation on risk preparedness in the electricity sector was necessary[37] since Directive 2005/89, failed to address emergency-related aspects, "i.e. how to make sure that Member States are aware and duly prepared to all kind of security of supply risks, that they clarify roles and responsibilities in case of emergency and that they take into consideration the potential cross-border impact when adopting safeguard measures)".[38] At the same time, essential aspects of the electricity

[36]European Commission Communication, "Clean Energy for all Europeans", COM (2016) 860 final (November 30, 2016).

[37]European Commission Proposal for a Regulation on risk-preparedness in the electricity sector and repealing Directive 2005/89/EC, COM (2016) 862 final—2016/0377 (COD) (November 30, 2016).

[38]European Commission Staff working document on "Evaluation Report Covering the Evaluation of the EU's Regulatory Framework for Electricity Market Design and Consumer Protection in the Fields of Electricity and Gas Evaluation of the EU Rules on Measures to Safeguard Security of Electricity Supply and Infrastructure Investment (Directive 2005/89) Accompanying the

market design proposals had critical implications for energy security. For instance, among others, the Electricity Regulation was targeting at regulating the national Capacity Mechanisms with the view to address mainly market distortions and secondarily energy security considerations. Several MS and the European Commission seemed to prefer a more market-based approach, which envisaged "price signals" as the best incentive for new investments and not national Capacity Mechanisms.[39] In any case, and beyond any political preferences (market vs state intervention), the new market design which envisages a decentralised electricity system (i.e. numerous small renewable installations instead of big fossil-fuelled electricity production Units), poses new challenges for energy security.[40] In the Commissions interim report of the sector inquiry on Capacity Mechanisms it is mentioned that "While the shift towards more renewable energy production is an intended development, it poses a challenge to the security of supply if the result is the closure of, or lack of incentives to invest in, flexible power plants which are still needed to back up intermittent wind and solar renewable generation".[41]

Document Proposal for a Directive of the European Parliament and of the Council on Common Rules for the Internal Market in Electricity (recast) Proposal for a Regulation of the European Parliament and of the Council on the Electricity Market (recast) Proposal for a Regulation of the European Parliament and of the Council establishing a European Union Agency for the Cooperation of Energy Regulators (recast) Proposal for a Regulation of the European Parliament and of the Council on Risk Preparedness in the Electricity Sector", SWD(2016)/412 final—2016/0379 (COD) (February 30 2016), p. 7.

[39] For Capacity Mechanisms see European Commission "Final Report of the Sector Inquiry on Capacity Mechanisms", Brussels, 30.11.2016, COM (2016) 752 final (November 30, 2016).

[40] See Agency for the Cooperation of Energy Regulators Recommendation No. 05/2014 on the regulatory response to the future challenges emerging from developments in the internal energy market, "Energy Regulation: A Bridge to 2025 Conclusions Paper" (November 19 2014).

[41] European Commission, "Interim Report of the Sector Inquiry on Capacity Mechanisms", p. 8, https://ec.europa.eu/competition/sectors/energy/capacity_mechanism_report_en.pdf, accessed February 6, 2020.

2.5 External Dimension of the Energy Union and the Gas Directive Amendment

2.5.1 The External Dimension of the Energy Union

The Energy Union was envisaged to have a strong external dimension. The last of the fifteen actions of the Energy Union Strategy called for the use of all external policy instruments in order to speak with one voice on energy and climate, to strengthen EU energy cooperation with third countries and to make full use of the EU's external trade policy to promote access to energy resources and foreign markets for European energy technology and services.[42] Furthermore, there were specific references to Algeria and Turkey; Azerbaijan and Turkmenistan; the Middle East; Africa as important producing and transit countries, Norway, the EU's second-largest supplier of crude oil and natural gas, United States and Canada, and also to the reframing of energy relationship with Russia, at least when the conditions would be right. Particular attention would be paid to upgrading the Strategic Partnership on energy with Ukraine. Last, the Commission would propose to strengthen the Energy Community and the energy relationships with the European Neighbourhood Partnership (ENP) countries.[43] Besides the rather ambitious declarations for a robust external energy dimension, the actual policy outcome is far from the set goals. But we should not blame the EU entirely for this. The situation in Crimea and the general strained relationship with Russia didn't allow for the "reframing of the energy relationship". The illegal actions of Turkey in the Eastern Mediterranean made the EU-Turkey energy dialogue impossible. New initiatives in the Middle East in the general political context of the war in Syria and the US-Iran relations would be difficulty administered. On the other hand, in the global fight against climate change, the EU has done quite well, despite the significant shift of the USA policy on this issue.[44] The process of

[42] Energy Union Communication, supra note 3, p. 19.

[43] Energy Union Communication, supra note 3, pp. 6–7.

[44] For EU actions against climate change see: European Commission "Fourth Report on the State of the Energy Union", Brussels, COM (2019) 175 final (April 9, 2019), p. 22.

integrating the energy markets in the south and east borders of the EU further evolved through the Energy Community, the Eastern Partnership and regional cooperation mechanisms such as the Central and South-Eastern Europe energy connectivity (CESEC) established in July 2015. The Energy Community further assisted the contracting parties to adopt the key elements of the EU's energy and climate acquis, but the update of the Energy Community Treaty (including the adoption of solidarity clause among EU MS and EC contracting parties in the frame of the new Security of Gas Regulation) has not yet finalised. Through the Eastern Partnership, the EU, among others, started an effort to help the Eastern Partners to diversify gas routes and sources through the promotion of the use of liquefied natural gas (LNG).

In November 2018, members of the Eastern Partnership energy panel agreed to set up a regional sub-sectoral network on LNG markets, aiming to assess the economic and political opportunities for these markets (4 meetings were organised until late 2019).[45] CESEC proved to be a useful tool to help to resolve missing links in infrastructures and to address regulatory barriers in the South-eastern energy markets (CESEC High-Level Group has endorsed a list of priority projects and an annually revised action plan to address regulatory issues). At the same time, the EU promoted new gas interconnection projects in order to diversify its sources and routes. In this respect, the Southern Gas Corridor will start the first gas deliveries in 2020, while other projects are following, such as the EastMed pipeline (the IGA between Greece, Cyprus and Israel was signed in January 2020, CEF funds will be used for the front end engineering, and design—FEED and a final investment decision could be made during 2022, after a market test). In Ukraine, during the last five years, the EU had significant results. The European Commission facilitated several rounds of trilateral talks between Ukraine and the Russian Federation aimed at ensuring the uninterrupted transit of natural gas from Russia via Ukraine, which were successfully finalised in December

[45] See European Commission's web page: https://ec.europa.eu/energy/en/topics/international-cooperation/key-partner-countries-and-regions/neighbourhood-east/meetings-eastern-partnership, accessed February 8, 2020.

2019.[46] Also, the Commission and the Energy Community Secretariat, along with some EU MS and the United States offered tremendous help to Ukraine to modernise its energy market, adjust to the EU acquis and diversify its energy supplies. At the same time, however, the transit of Russian gas to Europe could still bypass Ukraine as Turk-Stream is already working and Nord Stream II will soon be constructed. Last, the relations with the United States were complicated because of diverging views between the US and EU on climate issues, but at the same time, energy imports from the US flourished in the last three years. After a President Juncker and President Trump Joint Statement on the 25th of July 2018, when it was agreed to strengthen EU–U.S. strategic cooperation concerning energy, US LNG exports to the EU have been increasing substantially (between July 2018 and May 2019 exports to the EU increased by 272%).[47]

2.5.2 The Gas Directive Amendment

In May 2019 the Council and the European Parliament agreed upon the amendment of the Gas Directive 2009/73/EC,[48] proposed by the Commission in November 2017.[49] The amended Directive aims at ensuring that the rules governing the EU's internal gas market (ownership unbundling, third-party access, non-discriminatory tariffs and transparency requirements) apply to gas transmission lines between a member state and a third country, up to the border of the member state's territory

46 Vice-President Maroš Šefčovič Statement on the positive outcome of trilateral gas talks, December 20, 2019, https://ec.europa.eu/commission/presscorner/detail/en/STATEMENT_19_6841, accessed February 2, 2020.

47 EU press release, "U.S. Liquefied Natural Gas Exports up by 272% as EU and U.S. Host High-Level Business-to-Business Energy Forum", May 2, 2019, https://ec.europa.eu/commission/presscorner/detail/en/IP_19_2313, accessed February 8, 2020.

48 Gas Directive, supra note 7.

49 European Commission, Proposal for a Directive amending Directive 2009/73/EC concerning common rules for the internal market in natural gas, COM (2017) 660 final 2017/0294 (COD) (November 8, 2017).

and territorial sea.[50] During negotiations in the Council and the European Parliament, MS had diverging views mainly due to the possible impact of the amended Directive to Nord Stream II pipeline. The outcome of the negotiations was a delicate compromise of two (plus one) diverging groups of member states, which was at the last moment reached after a common proposal by France and Germany.[51] Two years earlier, in June 2017, the Commission had adopted a request to the Council for a mandate to negotiate with the Russian Federation the key principles for the operation of the Nord Stream II gas pipeline project. The proposal reasoned that Nord Stream II should not operate in a legal void or according to a third country's energy laws and in this respect the Commission needed a mandate to negotiate a special legal framework, which would take into account fundamental principles stemming from international and EU energy law (transparency in pipeline operation, non-discriminatory tariff-setting, an appropriate level of non-discriminatory third-party access and a degree of separation between activities of supply and transmission).[52] After a request backed by a few MS, in September 2017 the Council's legal service delivered its assessment rejecting the Commission's proposal on the grounds that Directive 2009/73/EC could not apply to the Nord Stream II pipeline.[53] After the

---

[50] European Parliament and Council Directive 2019/692 amending Directive 2009/73/EC concerning common rules for the internal market in natural gas, 2019, OJ (L 117).

[51] During the Gas Directive amendment negotiations, France kept a general scrutiny reservation, without expressing any official view on this legislative file, until the last moment. In 07.02.2019, the day before the COREPER I meeting, where the decision on the proposal would be mabe, the French foreign ministry spokeswoman stated that "France Intends to Support the Adoption of Such a Directive" see Anca Gurzu, "France and Germany Split over Nord Stream 2 Pipeline", Politico, February 2, 2019, https://www.politico.eu/article/france-and-germany-split-over-nord-stream-2-pipeline-russian-gas-regulation/,accessed February 2, 2020.

In COREPER I the bilateral Franco-German compromise on the Gas Directive was supported by almost all member countries. See Joshua Posaner, Anca Gurzu, Paola Tamma, "Franco-German Alliance SURVIVES Nord Stream 2 scare", Politico, February 8, 2019, https://www.politico.eu/article/france-germany-alliance-survives-nord-stream-2-scare/, accessed February 2, 2020.

[52] European Commission press release, "Commission Seeks a Mandate from Member States to Negotiate with Russia an Agreement on Nord Stream 2", June 9, 2017. https://ec.europa.eu/commission/presscorner/detail/en/IP_17_1571, accessed February 9, 2020.

EU relevant documents are classified EU restricted.

[53] See Katja Yafimava, "The Council Legal Service's Assessment of the European Commission's Negotiating Mandate and what it Means for Nord Stream 2", The Oxford Institute of energy

Council's Legal Service negative assessment and taking under account the Commission's will to go ahead[54] backed by a significant number of MS,[55] the only available solution seemed to be the amendment of Directive 2009/73/EC. On the other hand, it has been argued that the EU has the exclusive competence to negotiate and conclude international agreements like the one for Nord Stream II. In addition, it was argued that if the Court of Justice of the EU was to be asked, it might have had a different view on this case from the Council's Legal Service.[56] Of course, from a policy perspective, it should be taken into consideration that there wasn't time for Commission to infringe Council's Legal Service negative assessment, as Nord Stream II was already being constructed. The discussions in the European Institutions for both proposals (the negotiation mandate and the amended Directive) and their outcome exposed the limits of the EU exercise to "speak with one voice". The European Commission and the European Parliament were strongly in favour of regulating the Nord Stream II project.[57] The European Commission, after the failure of its proposal to acquire a negotiation mandate, proposed to amend the

studies, October 2017, https://www.oxfordenergy.org/wpcms/wp-content/uploads/2017/10/The-Council-Legal-Services-assessment-of-the-European-Commissions-negotiating-mandate-and-what-it-means-for-Nord-Stream-2.pdf, accessed February 9, 2020.

[54]Euractive, "Juncker: No EU Unity for Russia Nord Stream 2 Talks", October 20, 2017, https://www.euractiv.com/section/energy/news/juncker-no-eu-unity-for-russia-nord-stream-2-talks/, accessed February 9, 2020.

[55]In March 2016 Eight EU governments signed a letter objecting to the Nord Stream-2 project. According to Reuters the letter was signed by the prime ministers of the Czech Republic, Estonia, Hungary, Latvia, Poland, Slovakia and Romania and the president of Lithuania. See Andrius Sytas "EU Leaders Sign Letter Objecting to Nord Stream-2 Gas Link", Reuters, March 16, 2016, https://uk.reuters.com/article/uk-eu-energy-nordstream-idUKKCN0WI1YV, accessed February 9, 2020.

[56]See Chryssa. Tramountana, "What is the Scope of the EU External Competence in the Field of Energy Today?" Notre Europe Jacques Delors Institute, December 2019, https://institutdelors.eu/wp-content/uploads/2019/12/AD_1219_Tramountana_JAV.pdf accessed March 4, 2020.

[57]See European Parliament resolution on EU strategy for liquefied natural gas and gas storage, P8 TA (2016)0406, October 25, 2016: "32. Expresses Concern at the Proposed Doubling of Capacity of the Nord Stream Pipeline, and the Counterproductive Effects this would have on Energy Security and Diversification of Supply Sources and the Principle of Solidarity Among Member States; Highlights the Geopolitical Implications of the Project and the Underlying Principles of a Fully Integrated, Secure, Competitive and Sustainable Energy Union, Stressing that as such it should not Benefit from EU Financial Support or from Derogations from EU law; Underlines that a Doubling of the Capacity of the Nord Stream Pipeline would give One Company a Dominant Position on the European Gas Market, which Should be Avoided;".

Gas Directive. The European Parliament, when adopted its negotiation position, made the Commission's initial proposal even stricter.[58] Still, a year later it accepted the Council's compromise in just one negotiation round (avoiding any further delays that might allow Nord Stream II to be finalised under no new EU regulation). MS were divided into two (plus one) diverging groups: (i) those who had national interests to support the pipeline (because their national companies were shareholders of the Nord Stream II AS, or their market would benefit[59]) and were against Commission's proposals, (ii) those who were against the project and supported the proposed legislation, i.e. countries traditionally suspicious of Russia or MS who could lose their status as transit countries and (iii) MS with existing, planned or under construction gas infrastructure projects at the EU borders, which might be negatively affected by the new Directive. Lastly, France was keeping a general scrutiny reservation, without expressing any official view on this legislative file, until the last moment. Beyond all these considerations, these discussions can be seen as another episode of a longstanding general politically and diplomatically strained relationship between the West and Russia.[60] It should be noted that after the entry into force of the amended Gas Directive, there are still open issues and space for new disputes. Firstly there are doubts about the full application of the Directive by Germany and its national regulatory authority (Bundesnetzagentur). Secondly, Nord Stream II is already seeking the Gas Directive annulment by the Court of

[58] European Parliament Report, A8-0143/2018, April 11, 2019.

[59] Georg Zachmann, Nord Stream 2 means gains for Germany but pain for Europe, Bruegel, June 23, 2017.

[60] There are numerous US public statements against Nord Stream II and in favour of Commission's proposals. For instance, see: Anca Gurzu, "Trump's Russian pipeline policy

Opposing the Nord Stream 2 project finds favour with Central Europe", Politico, June 19, 2017, https://www.politico.eu/article/trumps-russian-pipeline-policy/?utm_source=POLITICO.EU&utm_campaign=c4fbc5192c-EMAIL_CAMPAIGN_2017_06_23&utm_medium=email&utm_term=0_10959edeb5-c4fbc5192c-189583977, accessed February 2, 2020.

Justice of the EU[61] and also it submitted a case to the dispute settlement mechanism of the Energy Charter Treaty.[62]

2.6 Conclusions and Some Thoughts for the Future

There is no doubt that since the introduction of the Energy Union Strategy in 2015, the EU's energy security was significantly enhanced. With the new legislation in place, markets can function better, and enhanced regional cooperation can raise gas deliveries in a crisis. There is considerable improvement of diversification of supplies via LNG and new gas infrastructure. In addition, measures to safeguard the security of gas supply required by Regulation 2017/1938 have dramatically improved Europe's resilience to possible gas supply disruptions. At the same time, it seems that negotiations for the Energy Union, especially those for the security of supply, have exhausted the legal limits of the current EU Treaties, as well as, MS tolerance to transfer new powers from the national to the European level. A broader discussion, including all policy sectors (energy, migration, security, defence, etc.) in the highest decision-making level, may be necessary to further reallocate powers between national and European level in energy. The paradox situation where generally Eurosceptic MS call for more robust EU energy security policies while traditionally pro-European MS express strong reservations, cannot drive in the future the EU energy policies for much longer. In addition to all the above, it is clear that there are limits for the EU ambition to "speak with one voice" in the international energy stage. This is determined by multiple factors, such as diverging political goals for the MS in the international diplomatic stage, MS special national interests in their foreign relations with third countries and differences in the energy mix due to economic interests or different national indigenous

[61] Case T-526/19: Action brought on 25 July 2019—Nord Stream 2 v Parliament and Council OJ C 305 (September 9, 2019).

[62] Energy Charter Web Page. List of Investment dispute settlement case. https://www.energycharter.org/what-we-do/dispute-settlement/all-investment-dispute-settlement-cases/, accessed March 4, 2020.

resources. Most importantly, the EU has to provide the right answers promptly. If not, there is a strong possibility that international developments might exceed its ability to resolve overwhelming questions even in its own territory or right outside its borders [for instance, the US initiative Partnership for Transatlantic Energy Cooperation (P-TEC) could replace EU's regional initiatives in South and East Europe]. The European Commission proved once again that it has the leading role in the process of formulating EU's energy security policy. At the same time, however, MS can push the brakes anytime they sense that critical national interests are at stake. The European Parliament, especially when it shares common views with the Commission, can play a significant role in overpassing MS reservations. Finally, the negotiations on energy security issues prove that individuals, who embody the European Institutions, often have multiplier effects on the orientation of the outcome of the policy formulation process.

References

Agency for the Cooperation of Energy Regulators Recommendation No. 05/2014 on the regulatory response to the future challenges emerging from developments in the internal energy market, "Energy Regulation: A Bridge to 2025 Conclusions Paper" (November 19, 2014).

Andoura, Sami, Jean-Arnold, Vinois. "From the European Energy Community to the Energy Union, A Policy Proposal for the Short and Long Term", Notre Europe Jacques Delors Institute, January 2015.

Case T-526/19: Action brought on 25 July 2019—Nord Stream 2 v Parliament and Council, OJ C 305 (September 9, 2019).

Council Conclusions on "Energy 2020: A Strategy for Competitive, Sustainable and Secure Energy", 6207/1/11 (February 28, 2011).

Council Directive 2004/67/EC concerning measures to safeguard security of natural gas supply, 2004, OJ (L 127).

Council Directive 2006/67/EC imposing an obligation on member states to maintain minimum stocks of crude oil and/or petroleum products, 2006, OJ (L 217).

Council Directive 73/238/EEC on measures to mitigate the effects of difficulties in the supply of crude oil and petroleum products, 1973, OJ (L 228).

Council Directive 2009/119/EC imposing an obligation on member states to maintain minimum stocks of crude oil and/or petroleum products and repealing Directive 2004/67/EC, 2009, OJ (L 265).

De Micco, Pasquale. "The EU's Energy Security Made Urgent by the Crimean Crisis", In depth analysis, European Parliament (April 2014).

Donald, Tusk, "A United Europe Can End Russia's Energy Stranglehold" *Financial Times*, April 21, 2014.

Energy Charter Web Page. "List of Investment Dispute Settlement Case". https://www.energycharter.org/what-we-do/dispute-settlement/all-investment-dispute-settlement-cases/.

Euractive, "Juncker: No EU Unity for Russia Nord Stream 2 Talks", October 20, 2017, https://www.euractiv.com/section/energy/news/juncker-no-eu-unity-for-russia-nord-stream-2-talks/, accessed 9 February 2020.

EURELECTRIC, Eurogas, EPIA and EWEA. Joint Statement on "The Energy Union, a Fresh Push for the Internal Energy Market", February 23, 2015, https://eurogas.org/website/wp-content/uploads/2018/03/Joint_Statement_on_the_Energy_Union.pdf, accessed February 7, 2020.

European Commission "Final Report of the Sector Inquiry on Capacity Mechanisms", Brussels, 30.11.2016, COM (2016) 752 final (November 30, 2016).

European Commission "Fourth Report on the State of the Energy Union", Brussels, COM (2019) 175 final (April 9, 2019).

European Commission Communication on "Energy Infrastructure Priorities for 2020 and Beyond—A Blueprint for an Integrated European Energy Network", COM (2010), 677 final (November 17, 2010).

European Commission Communication on "European Energy Security Strategy", COM (2014) 330 final (May 28, 2014).

European Commission Communication, on "The Short-term Resilience of the European Gas System Preparedness for a Possible Disruption of Supplies from the East During the Fall and Winter of 2014/2015", COM (2014) 654 final (October 16, 2014).

European Commission Communication, on "A Framework Strategy for a Resilient Energy Union with a Forward-Looking Climate Change Policy", COM (2015), 80 final (February 25, 2015).

European Commission Communication on "An EU Strategy on Heating and Cooling", COM (2016) 51 final (February 02, 2016).

European Commission Communication on “EU Strategy for Liquefied Natural Gas and Gas Storage”, COM (2016) 49 final (February 2, 2016).
European Commission Communication, “Clean Energy for all Europeans”, COM (2016) 860 final (November 30, 2016).
European Commission Green Paper “Towards a European Strategy for the Security of Energy Supply”, COM (2000), 769 final (November 29, 2000).
European Commission Press Release, “Commission Seeks a Mandate from Member States to Negotiate with Russia an Agreement on Nord Stream 2”, June 9, 2017, https://ec.europa.eu/commission/presscorner/detail/en/IP_17_1571, accessed February 9, 2020.
European Commission press release, “U.S. Liquefied Natural Gas Exports up by 272% as EU and U.S. Host High-Level Business-to-Business Energy Forum”, May 2, 2019, https://ec.europa.eu/commission/presscorner/detail/en/IP_19_2313, accessed February 8, 2020.
European Commission Proposal for a Regulation on Risk-preparedness in the Electricity Sector and Repealing Directive 2005/89/EC, COM (2016) 862 final—2016/0377 (COD) (November 30, 2016).
European Commission staff working document—Accompanying Document to the Proposal for a Regulation of the European Parliament and of the council concerning measures to safeguard security of gas supply and repealing Directive 2004/67/EC—The January 2009 gas supply disruption to the EU: An assessment, SEC(2009) 977 final (July 16 2009).
European Commission Staff working document on “Evaluation Report Covering the Evaluation of the EU’s Regulatory Framework for Electricity Market Design and Consumer Protection in the Fields of Electricity and Gas Evaluation of the EU Rules on Measures to Safeguard Security of Electricity Supply and Infrastructure Investment (Directive 2005/89) Accompanying the Document Proposal for a Directive of the European Parliament and of the Council on common rules for the internal market in electricity (recast) Proposal for a Regulation of the European Parliament and of the Council on the electricity market (recast) Proposal for a Regulation of the European Parliament and of the Council establishing a European Union Agency for the Cooperation of Energy Regulators (recast) Proposal for a Regulation of the European Parliament and of the Council on risk preparedness in the electricity sector”, SWD(2016) 412 final—2016/0379 (COD) (February 30, 2016).
European Commission, “Interim Report of the Sector Inquiry on Capacity Mechanisms”, https://ec.europa.eu/competition/sectors/energy/capacity_mechanism_report_en.pdf, accessed February 6, 2020.

European Commission, Staff Working Document, In-depth Study of European Energy Security Accompanying the document Communication from the Commission to the Council and the European Parliament: European energy security strategy, SWD(2014) 330 final/2 (May 28, 2014).

European Commission, Proposal for a Decision on Establishing an Information Exchange Mechanism with Regard to Intergovernmental Agreements and Non-binding Instruments Between Member States and Third Countries in the Field of Energy and Repealing Decision No. 994/2012/EU, COM(2016) 53 final—2016/031 (COD), (February 16, 2016).

European Commission, Proposal for a Directive Amending Directive 2009/73/EC Concerning Common Rules for the Internal Market in Natural Gas, COM (2017) 660 FINAL 2017/0294 (COD) (November 8, 2017).

European Commission, Proposal for a Regulation concerning measures to safeguard the security of gas supply and repealing Regulation 994/2010, COM (2016) 52 final—2016/030 (COD) (February 16, 2016).

European Commission's web page: https://ec.europa.eu/energy/en/topics/international-cooperation/key-partner-countries-and-regions/neighbourhood-east/meetings-eastern-partnership, accessed February 8 2020.

European Council Conclusions 14/15 December 2006, 16879/06 (December 15, 2006).

European Council Conclusions 19/20 March 2015, 11/15 (March 20, 2015).

European Council Conclusions 23/24 October 2014, 169/14 (October 24, 2014).

European Council Conclusions 4 February 2011, 2/11 (February 4, 2011).

European Parliament and Council Directive 2009/73/EC concerning common rules for the internal market in natural gas and repealing Directive 2003/55/EC, 2009, OJ (L 211).

European Parliament and Council Decision 2017/684 on establishing an information exchange mechanism with regard to intergovernmental agreements and non-binding instruments between member states and third countries in the field of energy, and repealing Decision No. 994/2012/EU, 2017, OJ (L 99).

European Parliament and Council Decision 994/2012/EU establishing an information exchange mechanism with regard to intergovernmental agreements between member states and third countries in the field of energy, 2012, OJ (L 299).

European Parliament and Council Directive 2005/89/EC concerning measures to safeguard security of electricity supply and infrastructure investment, 2006, OJ (L 33).

European Parliament and Council Directive 2019/692 amending directive 2009/73/EC concerning common rules for the internal market in natural gas, 2019, OJ (L 117).
European Parliament and Council Regulation 2017/1938 concerning measures to safeguard the security of gas supply and repealing Regulation 994/2010, 2017, OJ (L 280).
European Parliament and Council Regulation 347/2013 on guidelines for trans-european energy infrastructure and repealing Decision No. 1364/2006/EC and amending regulations (EC) No. 713/2009, (EC) No. 714/2009 and (EC) No. 715/2009, 2013, OJ (L 115).
European Parliament and Council Regulation 994/2010 concerning measures to safeguard security of gas supply and repealing Council Directive 2004/67/EC, 2010, OJ (L 295).
European Parliament Report No. A8-0310/2016 on the proposal for a Regulation of the European Parliament and of the Council concerning measures to safeguard the security of gas supply and repealing Regulation (EU) No. 994/2010 (October 20, 2016).
European Parliament Report, A8-0143/2018 (April 11, 2019).
European Parliament resolution on energy infrastructure priorities for 2020 and beyond, P7_TA (2011)0318 (July 5, 2011).
European Parliament resolution on EU strategy for liquefied natural gas and gas storage, P8_TA (2016)0406 (October 25, 2016).
European Parliaments debate of the 25th February 2015, https://www.europarl.europa.eu/doceo/document/CRE-8-2015-02-25-ITM-013_EN.html, accessed February 7, 2020.
Friends of the Earth "Europe's Climate and Energy Plans Frustrate", February 25, 2015, https://www.foeeurope.org/Europe-climate-energy-plans-frustrate-250215, accessed February 7, 2020.
Gurzu, Anca, "Trump's Russian Pipeline Policy Opposing the Nord Stream 2 Project Finds Favor with Central Europe", Politico, June 19, 2017.
Gurzu. Anca. "France and Germany Split Over Nord Stream 2 Pipeline", Politico, February 2, 2019, https://www.politico.eu/article/france-and-germany-split-over-nord-stream-2-pipeline-russian-gas-regulation/, accessed February 2, 2020.
https://www.politico.eu/article/trumps-russian-pipeline-policy/?utm_source=POLITICO.EU&utm_campaign=c4fbc5192c-EMAIL_CAMPAIGN_2017_06_23&utm_medium=email&utm_term=0_10959edeb5-c4fbc5192c-189583977, accessed February 2, 2020.
IEA, Gas market Report 2018.

Juncker, Jean-Claude. "A New Start for Europe: My Agenda for Jobs, Growth, Fairness and Democratic Change, Political Guidelines for the Next European Commission", Strasbourg, July 15, 2014.
Joshua, Posaner, Anca, Gurzu, Paola, Tamma. "Franco-German Alliance Survives Nord Stream 2 Scare", Politico, February 8, 2019, https://www.politico.eu/article/france-germany-alliance-survives-nord-stream-2-scare/, accessed February 2, 2020.
Šefčovič, Maroš. "Statement on the Positive Outcome of Trilateral Gas Talks", December 20, 2019, https://ec.europa.eu/commission/presscorner/detail/en/STATEMENT_19_6841, accessed February 2, 2020.
Sytas, Andrius. "EU Leaders Sign Letter Objecting to Nord Stream-2 Gas Link", Reuters, March 16, 2016, https://uk.reuters.com/article/uk-eu-energy-nordstream-idUKKCN0WI1YV, accessed February 9, 2020.
Tagliapietra, Simone, George, Zachmann. "Rethinking the Security of the European Union's Gas Supply", Bruegel, January 2016.
Tramountana, Chryssa. "What is the Scope of the EU External Competence in the Field of Energy Today?" Notre Europe Jacques Delors Institute, December 2019, https://institutdelors.eu/wp-content/uploads/2019/12/AD_1219_Tramountana_JAV.pdf accessed March 4, 2020.
Vinois, Jean-Arnold, Thierry Bros, "Russian Gas Pipelines and the European Union: Moving form a Love-Hate Relationship with 'Adults in the Room'?" Notre Europe Jacques Delors Institute, Policy paper No. 247, December 2019.
WWF briefing paper, "The Energy Union, Ensuring a Vision Without Blind Spots", February 2015.
Yafimava, Katja. "The Council Legal Service's Assessment of the European Commission's Negotiating Mandate and what it Means for Nord Stream 2", The Oxford Institute of energy studies, October 2017, https://www.oxfordenergy.org/wpcms/wp-content/uploads/2017/10/The-Council-Legal-Services-assessment-of-the-European-Commissions-negotiating-mandate-and-what-it-means-for-Nord-Stream-2.pdf, accessed February 9, 2020.
Zachmann, Georg. Nord Stream 2 Means Gains for Germany but Pain for Europe, Bruegel, June 23, 2017.

3

Energy Geopolitics Crossovers in Central & Eastern Mediterranean at the Prize of the Energy Union Policy

Thrassy N. Marketos

3.1 Introduction

From 2008 onwards, the security and political order of the Eastern Mediterranean (EM) subsystem, as it is operating since the onset of the Cold War, collapsed due to worsening relations between the two most important regional allies of the US, Turkey and Israel, and continuing turmoil in the Arab world. That order was replaced by a proxy civil war in Syria and Libya, a geopolitical rivalry between Turkey, Israel, Cyprus and Greece, as well as between the West and international actors like Russia, Turkey and Iran, who enthusiastically aspire to review the regional status quo in Western Eurasia. In addition, EM is becoming an increasingly important area for the international security system because of its natural resources, as the interests of regional actors interact with major international actors such as the US, France, Germany, the European Union and the North Atlantic Treaty Organization (NATO). The

T. N. Marketos (✉)
University of Peloponnese, Tripoli, Greece

M. Mathioulakis (ed.), *Aspects of the Energy Union*,
Energy, Climate and the Environment,
https://doi.org/10.1007/978-3-030-55981-6_3

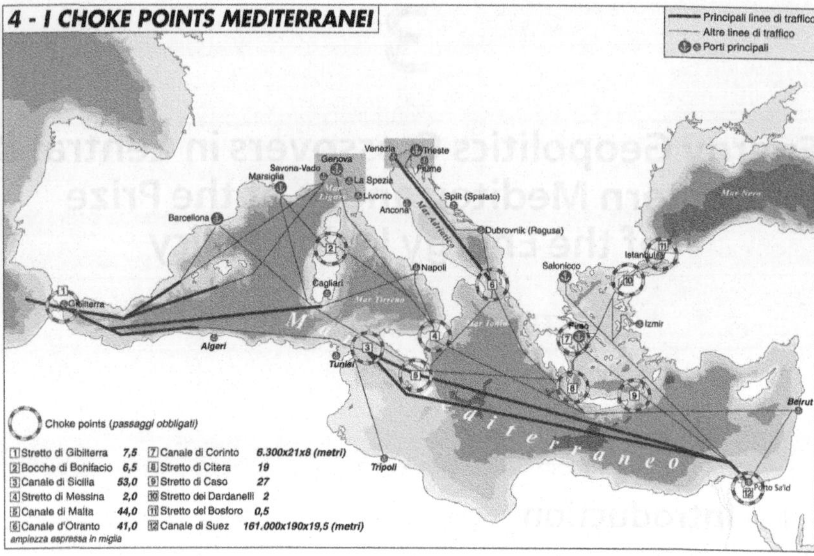

Fig. 3.1 Mediterranean Sea chock points (*Source* Laura Canali, LIMES)

discovery of offshore gas reserves in the EM since 2010 can provide the EU with a highly needed energy diversification from Russian and Middle East energy resources, as well as greater regional security. In 1987, the US Army College coined the acronym 'VUCA' to describe the context of Volatility, Uncertainty, Complexity and Ambiguity created with the end of the Cold War. Today, there is no more appropriate concept than that to describe the geopolitical situation of the Mediterranean region. The interests of EM's bordering states cross with unusual activism from the Gulf countries—primarily United Arab Emirates, Saudi Arabia and Qatar—as well as Russia. Libya and the Mediterranean (central and eastern), are the primary theatres in which new alliances are formed in a tangle that has been further complicated by the intertwining of various dynamics taking place in the area.[1]As seen here in Fig. 3.1, the Mediterranean sea includes eight distinct maritime chokepoints, where ship navigation and, therefore, trade activity, are congested.

[1]Leonardo Bellodi, Il Mare dei Destini Incrociati (The Sea of Crossed Destinies), (Rome: LIMES, www.limesonline.com, 2020), 1.

For the EU, the Energy Union strategy (25.02.2015), is strongly linked to the Eastern Mediterranean in its efforts to diversify its supply routes; a key element in decreasing EU Member States' dependence from a single supplier of natural gas, and ensuring a secure and affordable supply of energy to all Europeans. Linking the Energy Union strategy to the EM fully corresponds to its following five closely related and mutually reinforcing dimensions: (a) energy security through solidarity and cooperation between EU countries, (b) a fully integrated internal energy market enabling the free flow of energy through the EU through adequate infrastructure and without technical or regulatory barriers, (c) improved energy efficiency, which will reduce dependence on energy imports, lower emissions, and drive jobs and growth, (d) commitment to apply the Paris Agreement regarding the climate change and the decarbonising of the economy, while retaining its leadership in the area of renewable energy, and (e) supporting breakthroughs in low-carbon and clean energy technologies by prioritising research and innovation to drive the energy transition and improve competitiveness.

3.2 Eastern Mediterranean Region Positioning in the Energy Geopolitics Conundrum of EU Energy Union Strategy

3.2.1 Libya War Energy Perplexions in the Eastern Mediterranean

In Libya, similar to Syria, foreign powers have intervened in a proxy civil war ravaging the opposition between Tripolitania and Cyrenaica. On the one hand, Qatar supports the Tripoli-confined and internationally recognised GNA government. On the other, the Emirates (UAE), Saudi Arabia and Egypt support LNA Marshal Halifa Haftar army, based in Cyrenaica. These forces aim to contain the presence of Islamic extremism in the North Africa region. Natural gas is the factor that connects the Libyan Central Mediterranean theatre to the Eastern Mediterranean.

Turkey and Libya, agreeing to delimitate their respective EEZ (November 2019), traced a line dividing 'Mare Nostrum'. Even though the GNA presidential council in Tripoli hastened to declare that the Agreement is legitimate according to UNCLOS provisions, even Al Jazeera, the voice of Qatar, an allied country of al-Sarrag GNA government who shares certain closeness with Erdogan to the Muslim Brotherhood, expressed reservations on the question if the Turkey-Libya agreement follows the provisions of the International Law of the Sea. The US and the EU have condemned the Agreement, supporting the rights of the other coastal states: Greece, Cyprus, Israel and Egypt. Turkish's president Erdogan alignment with Libya's fledging GNA government and the signed EEZ deal lead to more profound cooperation within the East Med Gas Forum (EMGF) countries (Egypt, Israel, Greece, Cyprus, Italy, Jordan and the Palestinian Authority), seen here in Fig. 3.2. Moreover, the

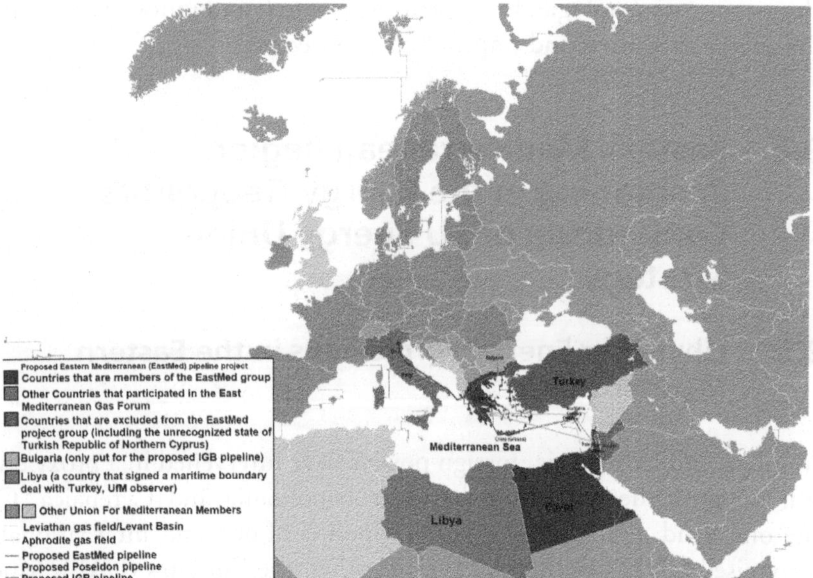

Fig. 3.2 Eastern Mediterranean Gas Forum countries (EMGF) (*Source* Adam Hegazy)

EMGF participants stepped up their efforts to increase and institutionalise their alignment not only on energy and economic issues but also on a military level. The aggressive Turkish policy in the Eastern Mediterranean and Libya brought some EU Member States like France and Italy, closer to the EMGF project. Paris has even officially asked to become a member. The above lead to the assumption that the Turkish President's Neo-Ottoman aspirations seem to start backfiring.

Turkish military support to Libya's GNA government, and the reported transport of Syrian Islamic extremist fighters to the country, lead Russia to become wary of Ankara's power projections, siding itself with its interests in Egypt, UAE, Saudi Arabia, Syria and Libya. Turkey itself, harshly contested the presence of US-based ExxonMobil in Cypriot EEZ but said nothing about the fact that ExxonMobil operates there in partnership with Qatar Petroleum. This should be assessed in relation to the $ 15 billion that Qatar provided to ease Turkey's financial crisis, whose currency sank five times against the US dollar.[2] EMGF parties could monetise on the situation, using Moscow and European interests to constrain Turkey and block its encroachment in the East Med and Libya. At the same time, EMGF parties could use the Libyan focus of European powers at present, to entice not only France but also Italy, UK and others to be involved in the offshore gas developments and security structure. All major energy companies are present in Egypt and Cyprus. Thus, by further integrating EMGF in the economic-geopolitical strategies of the EU (and NATO), Libya's future could be an asset for gathering support.[3]

3.2.2 Principal Actors

Such a perspective aligns with the afore-mentioned Energy Union strategy, aiming to diversify supply routes, decrease dependence from a single supplier of natural gas and ensure the secure and affordable supply

---

[2]Leonardo Bellodi, Ibid.

[3]Cyril Widdershoven and Rob van der Stel, Libya Conference's Shaky Results could Force Europe to Back EMGF, (Elshardt, The Netherlands: Verocy, verocy.com for www.financialmirror.com), 2020, 2.

of energy to all EU Member States. In particular, EU wants to create a Mediterranean gas hub in the South of Europe to help diversify its energy suppliers and routes (Fourth State of the Energy Union Report/April 2019), in accordance with the Regulation on the Governance of the Energy Union and Climate Action (EU 2018/1999/24.12.2018). The latter stipulates the importance of meeting EU's 2030 energy and climate targets and sets goals conforming to the United Nations Framework Convention on Climate Change and the Paris Agreement (12.12.2015, signed in 2016). Russia and Turkey seek a presence and a leading role in the Middle East and North Africa to escape the international isolation imposed on them. Russian President Putin, hit by severe international sanctions, needs to promote the idea of 'great Russia'. At the same time, Turkish President Erdogan, who lost the municipal elections in Istanbul in 2019, relies on the concept of the restoration of the Ottoman Empire, of which Libya, Syria, Egypt and most of Middle East were part of (Libya belonged to the Ottoman Empire from 1551 to 1912). Moscow and Ankara seek to ensure that they are prominent actors when all parties sit at the negotiating table of the Libyan crisis. Following the success in Syria, Moscow wants to make its weight felt in Libya and would see with great favour a permanent naval settlement on the strategic North African coasts. Turkey itself, complains that the Russian presence in Libya would cut it out of the Mediterranean, which it finds unacceptable. In this game of hydrocarbons exports, there is an omnipresent risk for the European Union to end-up partially replacing Russia's 'divide and rule' energy policy, with Turkey's arrogant and over self-esteem perception as the optimum and only worthy regional energy hub.[4] This is a realisation that needs to be promptly addressed by all EU Member States, including Italy, which, within its balancing objectives between the warring factions in the Libya proxy war, hasn't signed—through participating in the project—the EastMed pipeline accord (Athens, 02.01.2020), even if condemning the EEZ delimitating agreement between Turkey and the Libyan Tripoli-based, UN-backed, Government of National Accord (GNA)(27.11.2019). Natural gas discoveries in the Levant basin have,

[4]Thrassy N. Marketos, Turbulence in Europe's Energy Security Map (in Greek), (Athens: Foreign Affairs Hellenic Edition, www.foreignaffairs.com), 2015, 15.

indeed, given formidable and unexpected leverage to Egypt, a fact that certainly does not please Turkey, a Muslim Brotherhood supporter and sworn enemy of President's al-Sisi government in Cairo. It is noteworthy also to mention that Turkey and Egypt both aspire to acquire the informal role of leaders of Sunni Muslims in the world. With the disintegration of the Ottoman Empire, Turkey turned its gaze to the West, to the Aegean, while Egypt looked east, to the Arab world. During the Cold War, President Nasser of Egypt sided with the Soviet Union, while Turkey entered NATO. In 2012, tensions eased with the rise to power of Mohamed Morsi. The proximity of both countries to the Muslim Brotherhood cemented their relationship; the commercial exchange quickly grew from 300 million to 5 billion dollars. In 2013, when Morsi was dismissed by his successor Abdel Fattah el-Sisi, the latter convened a trilateral summit with Cyprus and Greece to promote the export of gas from the Cypriot coasts to Egypt, since prior to the discovery by Italian ENI of the immense Zohr deposit, Cairo was running out of gas. This move did not please the Turkish President.

With the maritime EEZ delimitation agreement with Libya, Turkey has the opportunity to thwart Egypt's plans, including the EastMed pipeline project. If Tripoli falls to Marshall Khalifa Haftar, Commander of the Libyan National Army (LNA), this Agreement would indeed be revoked. So, Ankara and Cairo will continue to support, with all means, opposing leaders in the Libyan drama. The Libyan Government of National Accord (GNA), in a desperate move to hold into power, signed in November 2019 an illegal agreement to trace common EEZ with Turkey, giving Ankara ownership of gas fields in the area, where it plans to conduct seismic research and drilling. It should be expected that the international community unanimously condemns and revokes this Agreement since between the two countries interlap major and medium-size Greek islands, including Crete. By refusing to these inhabited islands the globally recognised right to possess continental shelf according to the UNCLOS provisions, Turkey, a NATO member, violated Greece's and Cyprus' rights regarding EEZs, thus setting a framework of continuous energy-related frictions with two EU Member States. Despite the fact that UN Security Council in 2011 asked United Nation's member countries to abide by the arms embargo it declared, Ankara decided to help GNA government in Libya by sending hundreds of mercenaries of Islamic conviction from the Syrian battleground and numerous arms to fight against LNA forces as they advance on the capital. LNA leader Khalifa Haftar, in turn, blocked off Libya's oil production coming from wells under his control, depriving so the GNA government in Tripoli of much-needed cash. It is worth noting that this government is supported by a variety of militias, including Jihadist groups linked to al-Qaeda and the Muslim Brotherhood. Germany convened an intergovernmental summit in Berlin, also gathering representatives from all sides in this proxy war and several international observers. Nothing but void calling on both Libyan sides to de-escalate hostilities was heard. Also evident, was Turkey's inability to attract support by any North African or European country on its involvement. Turkey's ongoing conflict with all its neighbours in the EM, and frictions with NATO regarding the installation of Russian missile defence systems, is pushing France, Greece and Italy to side with LNA, a Kremlin's ally.[5] Ankara's policy provides help to Russian strategy applied in Europe, which consists of using energy as a tool for conducting foreign policy. By combining the route of Gazprom pipelines NordStream and TurkStream 1 and 2 (the latter also dubbed as BalkanStream), the Kremlin encloses the eastern part of Germany towards the Czech Republic, down to Slovakia, Hungary, Serbia and Bulgaria, offering itself a zone of influence related to the critical element of energy, in areas that according to the post-Soviet Kremlin reasoning are crucial for Russia's defence against the West.[6] Analyst Michalis Mathioulakis remarks that these two 'pipelines, the NordStream 2 and the second line of the TurkStream, have been subject to sanctions by the United States through the 2020 National Defence Authorization Act. Washington sees the two projects as what they are: a part of Kremlin's strategy to create national and regional dependence on Russian energy flows and leverage this dependence to exert political, economic, and military influence in Europe'.[7] Turkey is actively assisting Russia's cause

[5]Cyril Widdershoven and Rob van der Stel, Libya's Haftar Moving towards a Military Victory as Europe Pushes for Peace, (Elshadt: Verocy www.verocy.com for Al-Arabiya), 2020, 3.

[6]Michalis Mathioulakis, Russian Energy-Related Buffer Zones in Europe, (https://www.linkedin.com/pulse/russian-energy-related-buffer-zones-europe-michalis-mathioulakis/, 2020), 2.

[7]Ibid.

to avoid EM gas concurrence in the European market, by using the maritime Eastern Mediterranean area's delimitation deal with the head of the GNA government, as a pretext to block Greece, Israel and Cyprus plan to construct the EastMed Pipeline project, which targets the EU market. By striking the delineation deal with Sarraj (GNA), Erdogan is seeking first, to cancel any possible cooperation between Greece, Cyprus, Israel and Egypt in EM; second, to exert pressure on Athens and Nicosia; third, to totally restrain any in-depth offshore energy exploration in the region; fourth to assure getting back some of the multibillion-dollar Libyan debt to Turkish companies and even more worrying, fifth to grab a part of the oil and gas wealth that Libya holds, seeking more hydrocarbon exports to Turkey. Countering these Turkish goals, Greece has stroke with Egypt an Exclusive Economic Zone (EEZ) demarkation deal, deposited to the United Nations Organization according to the Internnational Law of the Sea (August 2020).

3.2.3 Ankara's Bet

But what is behind Turkey's active pursuit of hydrocarbon resources? Ankara, according to Dr. Thomas O' Donnell, 'fears' to become 'over-dependent' on energy supplies from Russia—via TurkStream—and Iran. To this end, Ankara has applied illicit aggression towards Cyprus, claiming offshore gas deposits located between Turkey and Cyprus, on the pretext that in doing so is following a contract that its state-run TRAO company has struck with Turkish Cypriot community on the island, to defend their rights on the island's energy wealth. Simply put, Turkey aspires to eliminate the possibility that Cyprus, Israel and Egypt market EM gas. One can argue with certainty that, to achieve peace in Libya, an international breakthrough agreement is needed, addressing the marketing of the country's vast oil reserves and offshore hydrocarbon potential. Further on, as Daniele Santoro argues, 'Libya is today the cornerstone of Anatolia's defence; the necessary outpost to remove any threat from the peninsula and consolidate the grip on the maritime space that separates Turkey from North Africa.[8] This requires (…) to control more or less directly the territories from which it is possible to project influence on the maritime spaces that surround the peninsula. Among which Libya is particularly essential to bring advanced plateaus closer to Anatolia, Crete and Cyprus. The control of the territorial ends of the 'blue homeland' (MaviVatan), presupposes the availability of a powerful navy, and, first of all, awareness of the strategic importance of the sea for defensive purposes'.[9]

3.2.4 Egypt-Turkey Relations

Egypt's potential role as a regional gas hub is vastly supported by Italy, arguing that it complements the geography of alliances that is consolidating in the region following the Arab riots that started in 2011. This geography sees Israel, Saudi Arabia, the United Arab Emirates and al-Sisi's Egypt affected to limit the rise of the Muslim Brotherhood, seen as a pawn in the regional ambitions of Turkey and Qatar. The latter would be penalised by the Egyptian development of gas from the Levant, which, thanks to its unique geographical position, could subtract market shares from the LNG of the Gulf Emirate, both in the Atlantic-Mediterranean basin and in the Indian-Pacific basin. It should be noted that in addition to the Israeli-Egyptian agreements (Delek and Noble Energy companies to supply the Egyptian Dolphinus Holding 64 bcm in a period of ten years), the UAE acquired 10% of Zohr deposit from the Italian company ENI. Following these announcements, Cyprus said it wants Aphrodite's deposit gas to be supplied to the Egyptian liquefaction terminals of Damietta and Idku. US' position during the Obama administration, was mainly inspired by the notion of economic peace, according to which the joint development of the region's resources would have favoured a cooperative approach between coastal countries. If past administration has been successful in using gas to bring Israel, Jordan and Egypt closer

---

[8] Daniele Santoro, Ankara (Ri)discovers the Sea, (Rome: Limes, www.limesonline.com, 2020), 3.

[9] Marco Giuli, Egypt at the Center of Gas Geopolitics in the Eastern Mediterranean, (Rome: Limes, www.limesonline.com, 2018), 1.

together, it has failed to achieve the same result concerning the dispute over Cyprus or relations between Israel and Turkey.

The Kremlin is particularly interested in strengthening energy cooperation with Egypt and has signed Agreement (Rosatom) for the construction of a 4.8 GW nuclear power plant by 2022, same as has done in various Middle East countries, from Jordan to Turkey. In 2017, Rosneft entered the development of the Zohr field by acquiring 30% from ENI. Novatek also obtained—within a consortium composed of ENI and Total—an exploitation license in Lebanese waters. Moscow uses these contracts to obtain multiple benefits: first to remain part of international consortia, especially with European companies, to increase the complexity of selected European governments towards US sanctions; second, to frustrate EU attempts to diversify gas supplies by participating in these projects; third, to encourage the international expansion of Rosneft and Novatek to compensate for their reduced domestic possibilities due to Gazprom's monopoly on the export of Russian gas via pipeline; fourth, to bring Arab countries leaders interested in energy cooperation, closer in a multi-sectoral foreign policy, especially at a time when the regional role of the US seems more transactional and less predictable. However, despite the increased presence, one cannot hypothesise a future Russian commitment to guaranteeing maritime security necessary for the energy development of the region, which must necessarily depend on coastal countries. Marco Giuli, from the Institute of European Studies, argues that in this context, Egypt remains in a position of strategic centrality. He adds that to consolidate it and reap political benefits, Cairo needs to reduce domestic demand to free up additional resources for exports, while offering guarantees of a more stable political environment together with Israel.[10] Cairo and Abu Dhabi are categorically opposite to Ankara's expansionist moves in Libya. The same applies to Erdogan's increasing pressure exercised directly on Cyprus, at least for Egypt. In the aftermath of stopping and threatening international oil giant companies, such as ENI and Total, violating Cyprus EEZ with military vessels, Turkey moves further on intending to station military drones on the occupied Turkish-Cypriots inhabited part of the island

[10]Marco Giuli, 2.

and installing a naval base and a gas pipeline connecting the so-called—recognised only by Ankara—'Turkish Republic of North Cyprus' and Turkey.[11]

3.2.5 US Policy Towards the Republic of Cyprus

Contrary to Turkish interests, the US Congress came together to help Exxon Mobil through legislation that could remake the geopolitics of the Middle East and Europe. The US stipulates its willing to provide backing for the development of natural gas resources offshore Israel and Cyprus. Also, promises support for the construction of pipelines and LNG terminals, including the creation of a US Eastern Mediterranean Energy Center in the region run by the US Department of Energy. The American Congress has also repealed the prohibition of weapons transfers to Cyprus dating back in 1987, promotes greater US military assistance to Greece and Cyprus, and instructs the US to maintain its newly situated predator drone fleet in the region. The whole policy refers to the 'Eastern Mediterranean Security and Energy Partnership Act'. The Cypriot government, encouraged by European countries (France, Italy) military vessels deployment to block Turkey's approach, aims first to organise an offshore deep-water pipeline strategy, supported by the European Union and financial backers such as the EIB. EM coastal countries, at least Cyprus, Greece and Israel, intend to implement a multibillion-euro pipeline approach that would target their supply to Southeast Europe. This energy-thirsty region's demand will only grow, but at the same time, Russia, several ex-Soviet Union countries, but also Iran and Iraq are watchful for the same markets. The EastMed pipeline would be directly competing with all Turkish-based and Russian-inspired pipeline projects, already a possible cause for conflict.

The EastMed pipeline project is, primarily, politically motivated, inspired to connect politically, economically and militarily Israel to the European Union for strategic reasoning, a concept also backed by Washington. Another option, as Cyril Widdershoven and Rob van der Stel,

[11]Cyril Widdershoven and Rob van der Stel, Cyprus Energy Future Depends on Ankara, (Elshardt: Verocy EastMed Insights www.verocy.com for www.financialmirror.com, 2019), 11.

argue, would be to explore and produce offshore gas a part of which will solve local necessities, and the rest of it would be transferred via pipeline to the existing Israeli-Egyptian gas infrastructure.[12] The latter option is more attractive both from a commercial and strategic point of view, promotes regional cooperation, and takes advantage of the global LNG market opportunities. Cyprus gas needs are growing, while current demand is covered by environmentally unfriendly sources of energy. The most applicable option for Aphrodite's island lies in substituting these sources of energy by using its potential reserves and supplementing it in the short term with additional LNG imports. The need for large-scale LNG options, as it's currently being implemented by LNG FSRU contracts, should be reconsidered, as small-scale LNG availability in the region (Egypt, Qatar and Algeria) could be sufficient and cost-attractive too. Finally, a third option lies in the proposal of Greek oil company Energean to supply Israeli gas from the Karish North field. A relatively small gas pipeline could be built between the Energean for the European Union field and Cyprus, needing less than $350 million investment.

3.2.6 Can Eastern Mediterranean Gas Reach the EU Market?

However, the abundance of gas in Europe and the challenging pricing terrain, combined with US supply and Russian reserves, lead us to assume, parallel to Harry Tzimitras, that the EU, for the time being, may not be, in realistic terms, a final market destination for EM hydrocarbons. Political problems in many EU member states, what many consider as the European crisis, lack of common aspect between the EU Commission and some member states regarding energy policy—such as the NordStream affair—and EU far-reaching consequence crisis (such as Brexit), all make Europe a challenging destination, facilitator or driver of developments. In the meantime, US is absent or little engaged in the EM, and also the fact that there is hardly one American foreign policy

[12]Cyril Widdershoven and Rob van der Stel, Ibid., 12.

for the moment—rather a variety of foreign policies exercised by agencies that lack structured coordinating. As a consequence, Washington is restrained in its leverage on Turkey, and cannot promote solutions to regional problems or cooperative answers for the region's energy reserves. This has greatly helped Turkey to project power in the broader region from the Aegean to the EM, including Cyprus, all the way to Syria and Libya. As Harry Tzimitras explains, 'Turkey seems to be renegotiating its place in the West—if not actively seeking alternatives. To this end, it is projecting political and military power, sending a message to the broader West that no planning in and for the region can realistically take place without Turkey'. Although 'on the Cyprus issue, the solution of which is central to any realistic prospects for inclusive exploitation of EM gas resources', prospects are slim for the moment, 'energy could moreover become the catalyst for cooperation in the broader region and also render Cyprus an excellent hub and a connector itself. To this effect, Cyprus has the distinct advantage of being a country in the region, but not *of* the region'.[13]

3.3 China's Role in the EU Eastern Mediterranean Region

3.3.1 Go East Versus Go West Policy

Chinese leaders and strategists have often criticised the US' strategy of encircling and containing a rising China.[14] As Christina Lin notes, 'Beijing views that its eastern flank is already surrounded by anti-Chinese alliances forged by the US through defence treaties with Japan, South Korea, Australia, Philippines and Thailand, in addition to defence cooperation with Taiwan, Singapore and Indonesia'. Figure 3.3 that follows shows the positions of US Naval bases and other military installations

[13] Harry Tzimitras, Eastern Mediterranean Prospects: in Need of a Pragmatic Geopolitical Assessment, (Rome: Instituto Affari Internationali, www.iai.it, 2020), 3.

[14] Ye Zicheng, Inside China's Grand Strategy: The Perspective from the People's Republic, ed. Steven I. Levine and Guili Liu, (Lexington, Kentucky: The University Press of Kentucky, 2011).

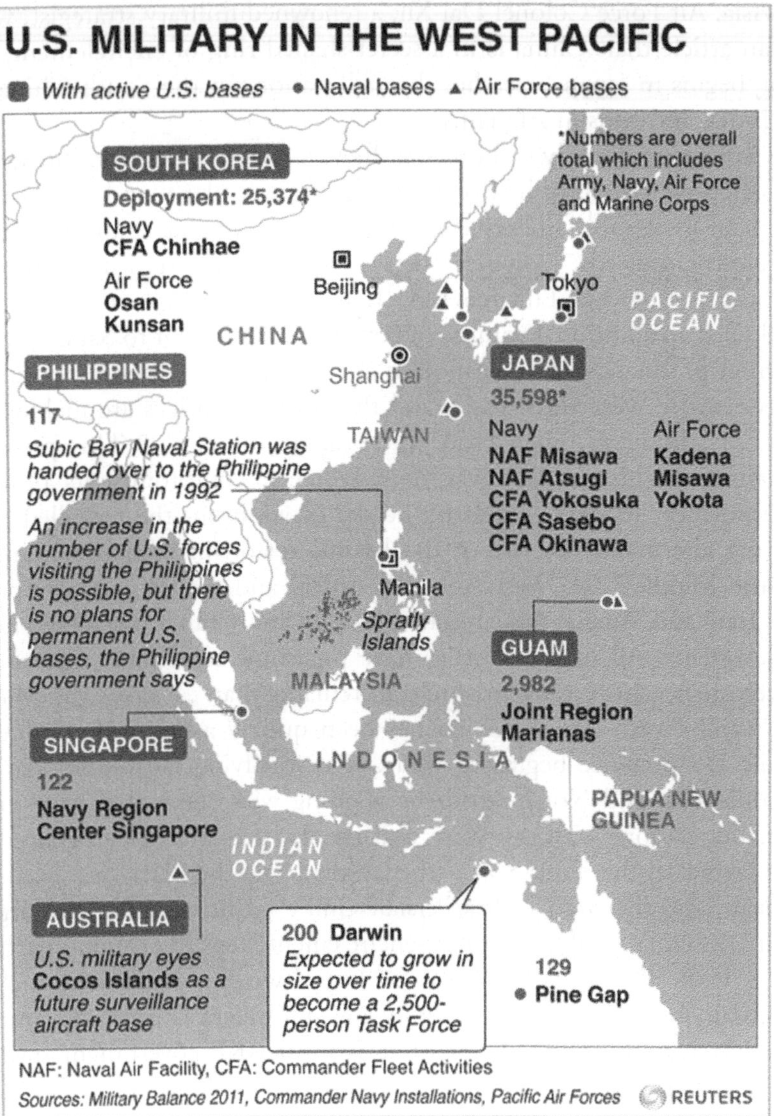

Fig. 3.3 US naval bases and installations in the West Pacific (*Source* Reuters)

in Asia. Air Force Colonel Dai Xu, a renowned military strategist, wrote in an article that 'China is in a secret-shaped ring of encirclement. The ring begins in Japan, stretches through nations in the South China Sea to India, and ends in Afghanistan'.[15]

China relies heavily still on coal. Its vast coal reserves, combined with an extensive system of hydro and nuclear power plants and the development of Renewable Energy Resources, offer a credible solution to cover its electricity needs. However, the issue of transportation fuel remains a critical issue. As Gal Luft remarks, 'If all of China is to 'rise', by 2025, it will need an additional volume of petroleum equivalent to Saudi Arabia's current production. But turmoil in the Middle East, mounting tension between the West and Russia, and the fall in oil prices are making the task of adding so much crude increasingly difficult'.[16] This probably means that China will be ever more willing to compromise its 'peaceful rise' to meet its energy security needs. In addition, the so-called Arab Spring aftermath, the rise of the Islamic terrorism in many countries of the Middle East, West Asia and Africa, and the sanctions imposed on Iran and Russia pose negative effects for China's overseas energy investments. All of these issues have put some of China's top foreign investment at risk. Dr Christina Lin remarks that '…across the Middle East and North Africa, China has been quietly asserting its influence while US presence begins to wane. It is fortifying its foothold in the Middle East and North Africa by aligning with states that have problematic relations with the West, and are also geo-strategically placed on the littoral of the 'Four Seas'—the Caspian Sea, the Black Sea, Mediterranean, and the Arabian Sea/Persian Gulf'.[17] Christina Lin also argues that '…China is taking counter-encirclement steps, by forging partnerships with key pivotal countries with anti-western sentiments such as Pakistan, Iran, Turkey and recently Egypt. Ankara is especially important given its EU/NATO ties while Islamist AKP government steers it

---

[15]Christina Lin, China's Strategic Shift towards the Region of the Four Seas: The Middle Kingdom Arrives in the Middle East, (Berlin: ISPSW Strategy Series, Issue No. 226, 2013), 16.

[16]Gal Luft, Strategic Implications of Chinese Energy Policy, (Washington, DC: The American Interest www.the-american-interest.com, Vol. 10, Number 4, 2015), 1–2.

[17]Christina Lin, 1.

towards a 'look east' orientation and, secondly, having one foot in NATO and another in Shanghai Cooperation Organization (SCO). As such, it is an important partnering pole in western Eurasia for China to protect its influence and counter-balance the US and the West'. Moreover, 'China is now seeking Ankara's cooperation and reciprocity in respecting Xinjiang as China's territory in exchange for supporting Turkey's stance on the Kurds in Turkey'.[18]

3.3.2 Beijing's Middle East Policy

Beijing's interests in the Middle East relate to China's ever-growing need for energy flows that can sustain the county's economic growth. As Christina Lin explains, 'As the US pivots toward Asia, China will naturally seek strategic depth in areas that were once dominated by Washington and its western allies'. Beijing also tries to further its influence via coalitions of non-western countries such as BRICS or the Non-Aligned Movement, especially the Shanghai Cooperation Organization (SCO), achieving to counter US and NATO influence in Central Eurasia. In particular, as G. Kemp and R. E. Harkavy explain, 'China is aligning with key countries' of geostrategic significance 'that have problematic relations with the United States and the West –Iran, Syria and Turkey- and lie on the littoral of the four seas: Caspian, Black, Mediterranean and Arabian/Persian Gulf, in the 'strategic energy ellipse''.[19] The 'Strategic Energy Ellipse', seen here in Fig. 3.4, includes over 70% of the world's proven energy reserves.

---

[18]Ibid., 17.

[19]Geoffrey Kemp and Robert E. Harkavy, Strategic Geography and the Challenging Middle East (Washington, DC: Carnegie Endowment and Brooking Institution Press), 1997, and Christof van Agt, The Importance of Eurasia in the Global Energy Security Agenda, (The Hague: Clingendael International Energy Programme, CIEP), 2010.

For more on China–SCO interaction in Central Eurasia see Thrassy N. Marketos, China's Energy Geopolitics: The Shanghai Cooperation Organization and Central Asia, (London: ed. Rutledge, Taylor & Francis Books Ltd), 2009.

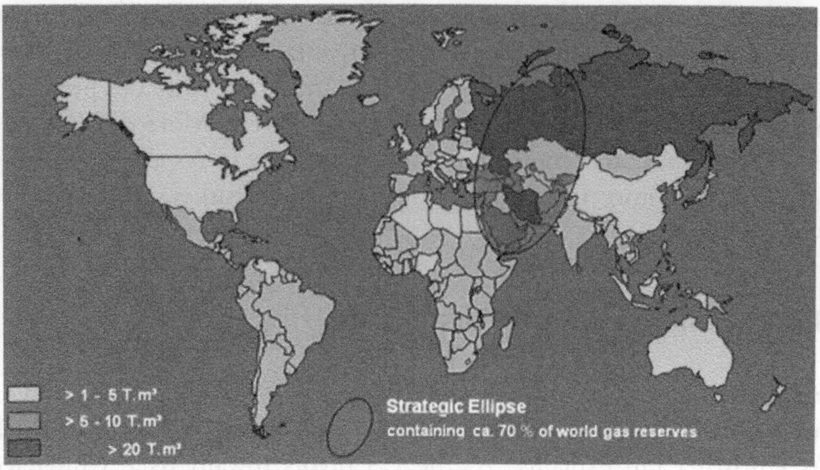

Fig. 3.4 Strategic Energy Ellipse (*Source* T. Boon von Ochssee)

3.3.3 The 'Four Seas' Strategy

Tony Badran argues that Syria's Bashar al Assad first promoted the concept of a 'Four Seas Strategy' to transform his country into a trade hub, and the Ankara-Damascus-Tehran triangle would become the nucleus of an approach that aims to include Iraq and the Caucasus in a geographical continuum linking the Four Seas. Assad peddled the idea to Turkey's President Abdullah Gul in 2009 that was approved by Iran's Supreme Leader Ali Khamenei the same year. He underscored that '…Once we link these four seas, we become the compulsory intersection of the whole world in investment, transport and more'.[20] N. Mozes adds that this would be 'A nucleus that will soon be augmented by Lebanon, Jordan and Iraq'.[21] Ch. Lin adds that 'Syria would then act as an access point for EU countries seeking to enter markets in the Arab world and Western Asia'. Ch. Lin stipulates that 'Although at present there is a division within the Ankara-Damascus-Teheran nucleus

[20]Tony Badran, A Syria in Minor Key, (Washington, DC: Foundation for Defense of Democracies, 2010).

[21]N. Mozes, Syria Regains Pivotal Regional, International Role—The Triumph of the Course of Resistance, (Washington DC: MEMRI Report no. 583, 2010), 5.

over the Syrian crisis, China has nonetheless maintained strong diplomatic ties with all three, wielding its United Nation's Security Council power to shield Iran's nuclear program, water down sanctions, protect Syria's Assad regime, and upgrade military ties/strategic partnership with NATO member Turkey'.[22]

3.3.4 The Pan-Asian 'Energy Club' Analogy in the Eastern Mediterranean

China's size, the pace of its growth, its regional and global aspirations and its complicated relations with its neighbours, beg for a more managed approach to its energy security, one that is part and parcel of a regional energy security architecture for the entire region. Today, energy security, particularly as it relates to oil and gas, is Asia's potential unifying factor, just as concern about competition over coal and steel was Europe's unifying factor post-World War II (European Coal and Steel Community-1951). A Pan-Asian 'energy club' could address the deeply destabilising disputes in the South and East China Seas. Since the US maintains a close relationship with most Asian countries, America should assume the role of 'community organiser'.[23] Analogically, a Pan-East-Mediterranean 'energy club', based on the Eastern Mediterranean Gas Forum (EMGF), including Turkey as well, and organised under the auspices of the US and major European countries, would greatly promote the security environment in EM and beyond. But in addition, such a scheme would serve the purposes of the EU Energy Union prerogatives, providing exceptional service to the European Union's need to diversify its market from Russian gas and an overall bust to EU Mediterranean security policy approach. Around the Mediterranean, Beijing is acquiring stakes in shipping and logistics companies by expanding ports in Greece (Piraeus), France (Marseille container terminal), Spain (Barcelona), as well as rail, air terminals, and fibreoptic networks in Portugal, Italy (air terminal), Lebanon (Tripoli port), Israel (rail connecting Tel Aviv-Eilat

22Christina Lin, China's Persian Gulf Strategy: Israel and the Nuclearizing Iran, (Washington DC: China Brief www.jamestown.org, Vol. 9, Issue 21, 2009).

23Gal Luft, Ibid., 5.

and Eilat-Ashdod and Haifa ports) and Egypt (COSCO has 20% of a container port in Port Said). Across the Suez Canal in the Red Sea, China is enlarging Port Sudan which gives Beijing the ability to deliver maritime shipments (whether civilian or military) to Sudan, East Africa and the Horn of Africa region. Near the Persian Gulf, China has taken operational control of Pakistan's Gwadar Port.[24] China also has vast investments in Cairo's hydrocarbon industry. Furthermore, it has taken advantage of other economic opportunities in the Suez Canal Zone, further enhancing its stance in controlling and securing its worldwide influence on strategic trade and communications chokepoints.

3.3.5 China's and Turkey's Maritime Expansionist Strategies Analogy

Cairo is pursuing important projects to develop its offshore reserves, also in cooperation with Cyprus and Israel. Cairo's plans clash, however, with Turkey's, and this is problematic considering that Egypt and Israel have expressed firmed willingness to defend their right to exploit EM energy reserves. Ankara has adopted an aggressive stance towards Cyprus and has not hesitated to employ the military to protect its interests, like preventing Saipem 12000, the drilling ship owned by Italy's company ENI, from conducting authorised gas exploration activities off the Cypriot coast. As such, Turkey's behaviour resembles that of China's in the South China Sea. Similar to what happens there, Ankara is alienating Greece, Cyprus, Israel and Egypt in the area, thus pushing them into closer energy and defence anti-Turkish alignment. Along with Italy, these countries have reached an agreement to build the EastMed pipeline to Europe. Egypt hopes to become part of the deal by connecting its giant Zohr gas field with this pipeline. But like the Iran-Iraq-Syria pipeline, similar projects bypass Turkey. Ankara's actions to assert its claims over the EM are therefore as much a way to access its hydrocarbon deposits as well as an effort to stop the construction of pipelines that would circumvent its territory and therefore reduce its economic, political and strategic

[24]Nassos Mihalakas, Part II: Chinease Investments in Europe-A Year in Review, (Washington, DC: Foreign Policy, www.foreignpolicy.com, 2011).

leverage.[25] Ankara has elevated its relationship with Beijing to one of strategic partnerships and signed agreements to build a silk railway, increase bilateral trade and upgrade military relations. With EU membership stalling, Turkey is looking increasingly east to China and sees it can play a mediating role between China's SCO and NATO. China itself sees Turkey as an important node on its New Silk Road and more importantly, a potential source of advanced NATO technologies. Both Beijing's and Ankara's interests correlate indeed. Or, as Alessandro Gagaridis stipulates, 'Ankara's behaviour (in EM) resembles that of China in the South China Sea (SCS)…The similarities between the two maritime regions are stunning: in both areas there is a vast zone of primary importance for international trade and rich in hydrocarbons upon which a major state (China in the SCS, Turkey in the EM) advances its claims by adopting an assertive diplomatic-military stance, thus causing neighbouring powers to react by strengthening their cooperation. Akin to what China does in the SCS, Turkey may develop its maritime law enforcement units, create naval militias, set up logistic/military facilities (…) and conduct intimidating manoeuvres to assert its influence over the area, to harvest its resources, and to avoid being bypassed by the construction of new pipelines. Again, similarly to what happens in the SCS, this increases the risk that an incident may escalate to a broader armed conflict. And considering the growing energy demands in the area and beyond, this is not an unrealistic scenario over the long term'.[26]

3.4 Conclusions

The analytical process followed in this chapter has revealed a distinct similarity between China's approach in the SCS and Turkey's attitude in the EM, respectively. Both these mindsets are applied in areas that are extremely important for SLOC's and energy transfer channels, particularly for Beijing's energy procurement from the Persian Gulf. Besides

---

[25] Alessandro Gagaridis, Ibid., 3.

[26] Ibid., 4.

the above, it should be noted that for Ankara, such attitude is generated also by post-Ottoman Turkey's abrupt rediscovery of the geostrategic importance of the sea during the last decade. Ankara has thus skilfully rushed into geographical representations, showing-off the 'Blue Homeland' (Mavi Vatan) and claiming vital sea areas from traditionally sea-born and structured nations like Greece and Cyprus. In relation to that, in needs to be noted that until the—fatal for Asia Minor's Hellenic populations—war between Greece and Turkey in 1922—when Greece was let to be drifted by its allies into an expansionist policy—Greece's economic and civilizational epicentre was located in the centre of the Aegean Sea and in trading—since millennia—in the waters of the greater EM. Turkish allegations in the EM, similar to the ones proclaimed by China in the SCS, demonstrate both countries' newly found self-esteem in an international system paving from the unipolar to the semi-unipolar structure. Without any doubt, peace in the EM, as in the SCS, can be maintained only through a 'Pan-East Mediterranean Energy Club', that will be including an International Law of the Sea-observing Turkey; an analogue to the SCS's 'Pan-Asian Energy Club'. Only then, is there hope that a military escalation in both regions can be averted.

In the EM particularly, where a web of conflicting legal claims to the recently discovered gas fields is merging with old and new conflicts, constructing a more functional EU-Turkey relationship is a dire necessity, so that the bloc protects its core interests in a wide array of issues such as migration, energy and Middle East security. In order to avoid China and Russia leading further inroads in the region, the EU Energy Union policy should focus on a more constructive geostrategic engagement with Turkey, one that highlights both parties' shared interests in trade, energy and regional security. European Union member states should in full harmony cooperate with the US, so that the Eastern Mediterranean Gas Forum (EMGF) does not exclude Turkey. In order to alleviate Ankara's fears that, this slowly emerging, political, economic and security-interests structure will challenge its regional position, the EU should host technical-level discussions related to gas exploitation, with the goal that both Greek and Turkish communities in Cyprus can benefit from the island's gas reserves. This process could be associated with a moratorium on gas exploration in Cyprus's exclusive economic zone, while Turkey would have to draw back its drilling ships and navy from the area. European countries should also avoid interfering in the—destabilising regional conflict—between Turkey and the UAE. The EU needs to provide Turkey with incentives to negotiate. Proposing a Turkish access to EMGF, would not only ease exploration and development grievances between the EU and Turkey, but also help improve relations between Turkey and Egypt, that are essential for a breakthrough in the proxy-war-torn Libyan state.

References

Al-Ahram Weekly, Issue No. 699.

Aviad, Rubin and Ehud, Eiran. *Regional Maritime Security in the Eastern Mediterranean: Expectations and Reality*. University of Haifa, 2019.

Badran, Tony. A Syria in Minor Key: Foundation for Defense of Democracies, 2010.

Bellodi, Leonardo. Il Mare dei Destini Incrociati (The Sea of Crossed Destinies): LIMES, www.limesonline.com, 2020.

Canali, Laura. Mediterranean Sea chock points: LIMES—www.limesonline.com, 2020.

Cyril, Widdershoven and Rob,van der Stel. *Cyprus Energy Future Depends on Ankara: Verocy EastMed Insights,* www.verocy.com for www.financialmirror.com, 2019.

Downs Erica. *China's National Oil Companies Return to the World Stage: Navigating Anticorruption, Low Oil Prices, and the Belt and Road Initiative*. The National Bureau of Asian Research, NBR Special Report#68, 2017.

Friedman George. US Naval Bases and Installations in Asia: Geopolitical Futures—www.geopoliticalfutures.com.

Gagaridis, Alessandro. *Rising Tensions in the Eastern Mediterranean: Another South China Sea?* Situation Reports, Geopolitical Monitor (www.geopoliticalmonitor.com), 2018.

Giuli, Marco. Egypt at the Center of Gas Geopolitics in the Eastern Mediterranean: Limes, www.limesonline.com, 2018.

Hegazy, Adam. Eastern Mediterranean Gas Forum countries (EMGF): 337259/ CC BY-SA: www.en.wikipedia.org.

Lin, Christina. *China's Strategic Shift towards the Region of the Four Seas: The Middle Kingdom Arrives in the Middle East*: ISPSW Strategy Series, Issue No. 226, 2013.

Lin, Christina. *China's Persian Gulf Strategy: Israel and a Nuclearising Iran*: China Brief, Vol. 9, Issue 21, 2009.

Luft, Gal. *Strategic Implications of Chinese Energy Policy: The American Interest*, Number 4, 2015.

Kemp, Geoffrey and Harkavy, Robert E. *Strategic Geography and the Challenging Middle East*: Carnegie Endowment and Brooking Institution Press, 1997.

Marketos, Thrassy. China's targeting on Central Asia's Geopolitical Linking to South Asia: The Iranian Factor in the Evolving Nexus of Energy, Security and Maritime Power in Eurasia Geopolitics: *Central Asia and the Caucasus Journal*, http://www.ca-org/online, 2012.

Marketos, Thrassy. *China's Energy Geopolitics: The Shanghai Cooperation Organization and Central Asia*: Routledge, Taylor & Francis Books Ltd, 2009.

Marketos, Thrassy. Turbulence in Europe's Energy Security Map (in Greek): *Foreign Affairs* Hellenic Edition, www.foreignaffairs.com, 2015.

Marketos, Thrassy. Το Γεωπολιτικό Επίκεντρο του 21ου Αιώνα (The Geopolitical Epicenter of the 21st Century): *Foreign Affairs* Hellenic Edition, 61, 2019–2020.

Mathioulakis, Michalis. *Russian Energy-Related Buffer Zones in Europe*: (https://www.linkedin.com/pulse/russian-energy-related-buffer-zones-europe-michalis-mathioulakis/, 2020.

Mihalakas, Nassos. *Part II: Chinease Investments in Europe-A Year in Review*: Foreign Policy, www.foreignpolicy.com, 2011.

Mozes, N. *Syria Regains Pivotal Regional, International Role—The Triumph of the Course of Resistance*: MEMRI Report no. 583, 2010.

Reuters. U.S. military in the West Pacific. https://blogs.thomsonreuters.com/answerson/us-military-in-the-west-pacific-graphic/.

Santoro Daniele. Ankara (Ri)discovers the Sea: Limes, www.limesonline.com, 2020.

Shiping, Tang. Lixianganquanhuanjingyuxinshijizhonguo da zhuanlue (Ideal Security Environment and China's Grand Strategy in the New Century): MEMRI Strategy and Management, no. 6, 2000.

Tzimitras, Harry. Eastern Mediterranean Hydrocarbons Prospects: In Need of a Pragmatic Geopolitical Assessment. Instituto Affari Internazionali, 2020.

Van Agt, Christof. The Importance of Eurasia in the Global Energy Security Agenda: Clingendael International Energy Programme, CIEP, 2010.

von Ochssee, T. Boon. *Strategic Energy Ellipse. Mackinder and Spykman and the New World Energy Order*: Clingendael International Energy Programme, CIEP, 2007, 3 https://exploringgeopolitics.org/publication_boon_von_ochssee_timothy_mackinder_and_spykman_and_the_new_world_energy_order/.

Widdershoven, Cyril and van der Stel, Rob. *Cyprus Energy Future Depends on Ankara*: Verocy EastMed Insights www.verocy.com for www.financialmirror.com, 2019.

Widdershoven, Cyril and van der Stel, Rob. Egypt Offshore Energy. Verocy, 2020.

Widdershoven, Cyril and van der Stel, Rob. Libya Conference's Shaky Results could Force Europe to Back EMGF, Verocy, verocy.com for www.financial mirror.com, 2020.

Widdershoven, Cyril and van der Stel, Rob. Libya's Haftar Moving towards a Military Victory as Europe Pushes for Peace, Verocy www.verocy.com for Al-Arabiya, 2020.

Zambelis, Chris. *Down the River Nile: China Gains Influence in Egypt*: China Brief, Vol. 5, Issue 22, 2005.

Zicheng, Ye. *Inside China's Grand Strategy: The Perspective from the People's Republic*: ed. Steven I. Levine and Guili Liu, The University Press of Kentucky, 2011.

4

Strategic Elements of the Energy Union and the Role of Regulation in Energy Security

Michalis Mathioulakis

4.1 Introduction

Among the five dimensions of the Energy Union, energy security holds a primary position and a crucial role in the overall success of the Commission's objectives in the energy sector. To achieve its goals, the EU needs to build an integrated European energy system, where various forms of energy flows freely within each country and between the Member States, and energy is produced, procured, transported, distributed, traded, stored and consumed, under the fundamental principles of the free market, competition, efficient use of resources and the effective application of the regulatory framework. Most aspects of these goals relate directly or indirectly to elements of energy security, making it one of the crucial connecting elements within the Energy Union framework. As will be analysed further in this chapter, energy security includes a broad spectrum of definitions spanning from ensuring the infrastructure that

M. Mathioulakis (✉)
Greek Energy Forum, Athens, Greece

M. Mathioulakis (ed.), *Aspects of the Energy Union*,
Energy, Climate and the Environment,
https://doi.org/10.1007/978-3-030-55981-6_4

supports energy demand to efficiently responding to abrupt discrepancies in energy supply, and from securing the availability of energy resources to providing affordable prices for energy products. The International Energy Agency, (IEA) defines energy security as the uninterrupted availability of energy sources at an affordable price.[1] This, rather simplistic but inclusive definition, incorporates almost all of the elements that consist an energy system regarding the way that energy is produced, procured, transported, distributed, traded, stored and consumed, all of which are a subject of the Energy Union policy in the EU. Energy security is, therefore, a structural element of the Energy Union and as such, it is going to be examined in this chapter. For the European Union, where more than 55% of its energy needs are covered through imports,[2] energy security is heavily related to the control over, and diversification of, energy resources and energy infrastructure and therefore influenced drastically by EU's external relations and its foreign and security policy. The structural gravitas of energy in all aspects of economic and social life, inevitably makes extended energy dependency a critical issue, raising concerns of political vulnerability on a national and EU level, thus creating links between energy and national security. Control over energy resources or major energy infrastructure is therefore often considered to carry strategic value to the parties involved, whether it being the Member States or EU as a whole. However, the mere size of an energy project or the volume of energy flows alone, do not constitute sufficient requirements for them to be characterised as 'strategic'. There are specific elements that attribute strategic value to an energy project. This chapter identifies these elements and provides a comprehensive link between energy and national security. However crucial energy dependence is, regarding national security and political vulnerability, EU's foreign and security policy does not directly include it among its objectives, neither possess direct institutional or legislative tools to protect

[1]"Energy Security—Areas of Work", IEA, Accessed January 29, 2020. https://www.iea.org/areas-of-work/energy-security.

[2]"Energy Production and Imports", Eurostat, Accessed January 29, 2020. https://ec.europa.eu/eurostat/statistics-explained/index.php/Energy_production_and_imports.

against it.[3] External energy relations of the EU with third parties are therefore addressed either through EU's participation in international treaties like the Energy Charter Treaty (ECT) and international organisations like the Energy Community—both discussed extensively in other chapters of this book—or by the use of regulatory instruments that, while initially built to address 'internal' issues within the EU borders, possess elements of 'externality' with a direct effect in EU's energy relations with third parties. This chapter thus analyses also the mechanisms under which two principal regulatory instruments, the provision for 'Third Party Access' to energy networks and for 'Ownership Unbundling' of energy infrastructure, are being utilised to deliver the Energy Union's goals for energy security within the EU.

4.2 The Context of Energy Security

The significance of energy security spawns from the broad spectrum of influence energy holds in all aspects of economic activity. Far from just transportation and heating, energy determines the ways human societies are structured, the intensity of production, urbanisation, communications, global trade, quality of the environment and relations between countries. Energy security is primarily linked to concerns regarding the dependable acquisition of energy supply, the regional concentration of energy resources and the effects of their strategic management. The context of energy security has been broadening throughout the years alongside the reasons that carry the capacity to generate disruptions in the energy supply chain. Political unrest like the Russia-Ukraine natural gas disputes in 2006–2009, instability in the Middle East affecting the global oil supply, natural disasters like the earthquake and tsunami in Japan in 2011 and power system failures like the 2003 blackout in the northeast US, raise concerns over the vulnerability of economic systems and human activity and wellbeing due to energy-related disruptions. This led to an expansion of the energy security context to include the

[3] "Foreign and Security Policy", European Union, August 6, 2019. https://europa.eu/european-union/topics/foreign-security-policy_en.

resilience of energy infrastructure, the structure and interconnectivity of energy networks and the composition of the energy mix; the ways energy sources are combined to support economic and social activity. Furthermore, the above concerns raised the issue of the volatility of energy costs that can cause essential discrepancies in social welfare and competitiveness of the economy, thus bringing the issue of affordability of energy into the context of energy security. Rising and volatile energy costs raised in-turn the significance of energy savings adding the notion of energy efficiency into the ever-growing palette of energy security objectives. Finally, current concerns over the effects of global warming, air pollution and protection of the environment, generated a wave of demand for structural changes in the energy mix that affects all the above-mentioned aspects of energy security. The extended diversity of issues and concerns included in the context of energy security stipulates its essential role in human life but also the difficulty to identify, combine and structure coherent and effective policies to address its broad and often conflicting objectives. Attempting to develop a simple, coherent and inclusive description of energy security, Professor's Benjamin Sovacool definition offers a suitable solution, concluding that energy security relates to four distinct dimensions that include the availability, affordability, efficiency and environmental stewardship of energy.[4]

4.2.1 The Need for EU Energy Security

The importance of a coherent energy security strategy of the EU derives from the Union's growing dependency on energy imports. Despite regressing rates of energy demand, the steep decline in primary energy production in the EU led to an increase of its 'energy dependency rate', the percentage of net energy imports to gross available energy, to more than 55% in 2017.[5] Production of primary energy in the EU declined

[4]Brown, Marilyn A., Yu Wang, Benjamin K. Sovacool, and Anthony Louis D'Agostino, "Forty Years of Energy Security Trends: A Comparative Assessment of 22 Industrialized Countries", *Energy Research & Social Science* 4 (2014): 64–77. https://doi.org/10.1016/j.erss.2014.08.008.

[5]"Energy, transport and environment statistics", Eurostat, Accessed January 29, 2020. https://ec.europa.eu/eurostat/documents/3217494/10165279/KS-DK-19-001-EN-N.pdf/76651a29-b817-eed4-f9f2-92bf692e1ed9.

by 12.1% in the last decade summing up to 758 million tonnes of oil equivalent (Mtoe) primarily due to depleted production of natural gas fields and the rendering of coal fields as uneconomical for further exploitation. Primary energy produced within the European borders diversifies through a variety of energy sources, with renewable energy leading with a 29.9% share, nuclear energy 27.8%, coal 16.4%, natural gas 13.6% and crude oil with an 8.8% share of the EU total production.[6] Furthermore, the EU's robust resolve to battle the effects of global warming inevitably raises concerns over its energy security regarding the rate that higher renewable energy production can replace lower nuclear and coal production. Together with concerns over the stability of the European power-production system during this transitional period, the trend towards depleted nuclear (and coal) production raises the need for natural gas imports. This, in turn, raises the EU's dependence and vulnerability from third countries, primarily from Russia. According to the latest published official data from Eurostat, Russia accounts for 38.9% of the EU's imports of coal, 33.7% of its crude oil imports and 38.7% of total EU imports of natural gas.[7] Severe dependency from Russian energy sources leads to excess vulnerability of EU's energy systems while simultaneously exposing the Member States to political pressure from the Kremlin. The 2006–2009 Russia-Ukraine conflict and the discrepancy of gas flow towards the EU has been the catalyst to showcase such vulnerabilities. Volatility over gas prices and uncertainty over gas supply has also been manifested during the 2014 annexation of Crimea as well as the 2019 legal disputes between Russian Gazprom and Ukrainian Naftogaz that ended with the signing of a new accord between the two companies only a few days before the expiration of the previous one on the 31st of December 2019, thus keeping the flow of natural gas to Europe intact.[8] Growing concerns on EU's dependency on energy imports led the European Commission to release in early 2014

[6]"Energy production and imports", Eurostat, Accessed January 29, 2020. https://ec.europa.eu/eurostat/statistics-explained/index.php/Energy_production_and_imports.

[7]Eurostat, "Energy, transport and environment statistics".

[8]"Russia, Ukraine Sign Gas Transit Deal Ahead of Deadline", Deutsche Welle, Accessed January 30, 2020. https://www.dw.com/en/russia-ukraine-sign-gas-transit-deal-ahead-of-deadline/a-51841576.

its Energy Security Strategy, aiming to secure stability and abundancy of energy supply while simultaneously examining the effects of discrepancies from Russian gas imports to the European energy system.[9] The objectives and stress-test results included in the Commission's Energy Security Strategy led to the formulation of a coherent framework strategy in 2015 for a resilient Energy Union with a forward-looking climate change policy.[10] The birth of the Energy Union, from which the current Commission's European Green Deal originates, can, therefore, be traced back to the fundamental objectives and necessities of the energy security context. This realisation grants energy security analysis a central role for the proper design, structure and development of all energy and environmental policies within the European regulatory and political framework.

4.2.2 The Externality of Energy Security Policy

The powerful influence of high energy dependency rates in the formation of energy security policies reveals an element of 'externality' in the latter. When planning to address energy security concerns on a national (i.e. the EU Member States), or supranational (i.e. the European Commission) level, the applied policies and measures often generate results in third countries with the potential to damage their national interests. Furthermore, when the supply of energy resources and their strategic management from producing countries (i.e. limiting oil production or withholding gas flows) is used to yield political advantages in non-energy-related disputes, energy security policies are inevitably strengthened with 'externality' tools to deter the forthcoming adverse effects. For Europe, Russian dominance over the overall supply of energy has been a major concern, with the Kremlin being accused of utilising its energy flows to support its political objectives. A 2018 report from the

[9] "European Energy Security Strategy", European Commission, May 28, 2014. https://eur-lex.europa.eu/legal-content/EN/ALL/?uri=CELEX:52014DC0330.

[10] "A Framework Strategy for a Resilient Energy Union with a Forward-Looking Climate Change Policy", European Commission, February 25, 2015. https://eur-lex.europa.eu/legal-content/EN/ALL/?uri=CELEX:52015DC0080.

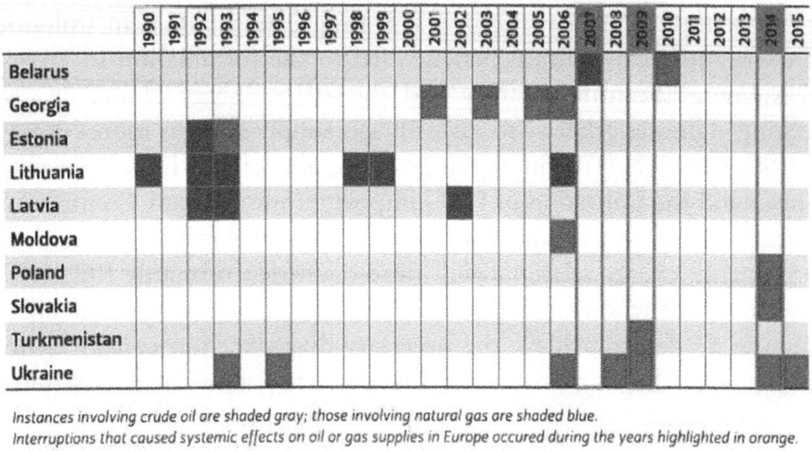

Fig. 4.1 Known or probable politically-driven energy supply or price manipulations by Russia (1990–2015) (*Source* Baker institute for public policy, Rice University)

Directorate-General for External Policies of the Union to the European Parliament's Committee on Foreign Affairs identifies 20 cases between 2003 and 2019 where it is 'probable' that Russia used its energy leverage, in and outside of Europe, for political purposes.[11] Similarly, a 2017 brief by the Baker Institute for Public Policy of Rice University (US), identified 28 events between 1990 and 2015 where 'Russia used price and physical volume manipulation of crude oil or natural gas supplies -often amid political tensions- to pressure consumers located in Central and Eastern Europe and the former Soviet countries'.[12] While the countries and years involved are shown in Fig. 4.1, the brief stipulates that 'the 'true' number of energy supply disruptions Russian entities have caused in and around the former Soviet zone is likely significantly higher than the number reported in this study'.

[11] "Energy as a Tool of Foreign Policy of Authoritarian States, in Particular Russia", European Parliament, April 2018. https://www.europarl.europa.eu/RegData/etudes/STUD/2018/603868/EXPO_STU(2018)603868_EN.pdf.

[12] Gabriel Collins, "Russia's Use of the 'Energy Weapon' in Europe", *Issue Brief* no. 07.18.17. Rice University's Baker Institute for Public Policy, Houston, Texas, USA (2017). https://www.bakerinstitute.org/media/files/files/ac785a2b/BI-Brief-071817-CES_Russia1.pdf.

As mentioned earlier, the overwhelming cases of Russian utilisation of energy to yield political benefits, led to the formulation in 2014 of EU's Energy Security Strategy and the 2015 Energy Union Strategy. Seeking—among others—to diversify gas supply and gas routes through the Southern Gas Corridor, the gas potential of the Eastern Mediterranean and the utilisation of LNG import terminals, both Commission's strategies carry a significant 'externality' footprint since they are designed to influence energy relations with third countries; primarily Russia. The crucial role of this 'externality' element, renders the subsequent infrastructure needed to achieve the necessary diversification equally critical. Since 2013, the European Commission draws a list of energy infrastructure projects named 'Projects of Common Interest' (PCIs) that have the capacity to link the energy systems of EU Member States and are expected to help the EU achieve its energy policy and climate objectives. The significance of such infrastructure and the—usually—high level of investment and political commitment necessary for their materialisation often leads to the notion that they are of strategic importance to the parties involved. However, since the mere size of an energy project is not related to its potentially strategic value, the issue needs to be further analysed and identified.

4.3 Strategy, Security and Energy

Among the current analysis' goals, lies the necessity to identify the elements linking control over energy flows with national security goals. Projects involving infrastructure on energy production or energy networks are often characterised as 'strategic'. However, no typology exists to identify the elements such a description would include. While, almost intuitively, most major national or transnational energy projects involving vast capital investment are characterised as 'strategic', it seems that attributing such a quality based solely on the economic aspect of a project is an incomplete way to approach the issue.

4.3.1 Defining the Concept of Strategic Objectives

For state actors, strategy refers to their ability to utilise all available means and resources (military, economic, diplomatic, etc.) in order to achieve policy-set objectives.[13] Although more complex, the same principle applies to the European Union. The EU is a *sui generis* international actor with unique characteristics and powers. Although the Union is not a state actor, it does, however, possess several supranational powers over its Member States that constitute it much more than an international organisation. The supremacy of EU Law over national legislation and the extensive influence over specific policy areas such as trade, competition and energy, reinforce EU's ability to act as a single international actor, while, on the other hand, its limited influence in areas like defence and foreign policy, deprive its ability to respond with a unified voice. However, the mere size of EU's global trade capacity and the *sui generis* elements within its structure, allows us to consider that, to a great extent, most aspects of state actors' behaviour in the international system can also be attributed to the European Union. The definition of strategy as the ability to utilise all available means and resources to achieve policy-set objectives creates a direct link between the objectives of strategy and the threats state actors face in their effort to survive in the international system. It is therefore deemed necessary to identify those security threats that lead to the formation of strategic objectives and research the links between them and energy flows. The issue of securitisation has been widely discussed by Barry Buzan.[14] The distinguished scholar offers a broad perspective on security that—besides the traditional military aspect—includes Political, Military, Economic, Societal and Ecological threats. According to Buzan, 'Security is taken to be about the pursuit

[13]Athanassios G. Platias and Constantinos Koliopoulos, *Thucydides on Strategy: Athenian and Spartan Grand Strategies in the Peloponesian War and Their Relevance Today*, pp. 1–21, London: Hurst, 2010.

[14]Barry Buzan, *People, States & Fear: An Agenda for International Security Studies in the Post-Cold War Era*. Colchester: ECPR Press, 2016, pp. 104–126.

of freedom from threat and the ability of states and societies to maintain their independent identity'.[15]

The wide range of elements that constitute a state leads to an equally varied palette of threats. The strategic objectives of state actors are thus directly linked to the security threats they are designed to address. Strategic objectives related to military threats can be diverse, spanning from deterring military aggression to counteracting foreign influence over the state's international legal rights. Political threats, on the other hand, refer primarily to the internal structure of the state in terms of the ideas, cultures, institutions and governing principles that constitute a state and justify its monopoly of access to force within its borders. Lastly, economic threats can be identified as a means to an end rather than a direct threat to national security. The imposition of extreme trade tariffs, economic sanctions, product embargoes or even the rapid expansion of foreign economic influence over a state's existing trade allies and partners, are potential threats to a state's national economy and therefore to its ability to effectively utilise its political and military resources for its national security. The diversity of political and economic threats formulate equally diverse strategic objectives spanning from deterring foreign influence on a state's institutions and governing principles to deterring investments in critical infrastructure by a competing state actor. As presented earlier, aim of this chapter is to identify the elements linking control over energy flows with national security goals. The above approach established that this would entail detecting the link between energy flows, security threats and strategic objectives. Placing security above and beyond the strict, direct framework of military threats, offers the necessary basis for energy and Energy Security to be perceived as part of a broader security framework. The elements of 'externality' in Energy Security carrying the ability to generate shifts in a state's overall levels of power, create the conditions for the 'securitisation' of energy resources and energy flows.

[15] Barry Buzan, "New Patterns of Global Security in the Twenty-first Century", *International Affairs* 67.3 (1991), pp. 432–433.

4.3.2 Linking Strategic and Energy Objectives

According to Kenneth Waltz, one of the fathers of modern International Relations theory, anarchy in the international system, the concept that the system functions under the absence of a central authority, leads state actors to self-help in order to ensure their survival within the system.[16] Under the self-help principle, state actors are forced to combine all factors of production in order to maximise their national interests and ensure their survival in the international system. In this effort, a state's power factors (military, economic, technological, etc.) cannot be managed separately but ought to be thought and handled as a whole that produces the total power necessary for survival. Under Waltz's approach, states seeking to establish their own survival, strive to achieve their goals through internal efforts (internal balancing) and external efforts (external balancing). Under this framework, it is essential to examine the ways and preconditions under which control over energy resources and infrastructure can function towards a county's efforts to maximise its national interests and ensure its survival in the international system. Control over energy resources is directly linked to the concept of Energy Security. Daniel Yergin, the distinguished energy expert and economic historian, identifies ten principles of energy security, namely the diversification of sources and routes, the country's position in the global energy market, its security margin, the functioning efficiency of the energy markets, the formation of relationships with exporting nations, the level of cooperative relations among importing nations, the quality of information regarding the state's energy options, the level of robustness of the domestic industry, the level of research & development in the energy field and energy security's contribution in planning for disruptions.[17] The vast majority of these principles carry elements of 'externality' meaning that they influence, or are influenced by, each state's position in the international system and the balance of power with other states. Out of these ten principles, diversification, relations and cooperation with exporting

[16]Kenneth Waltz, "Theory of International Politics". Boston: McGraw-Hill, 1979.

[17]Daniel Yergin, "Energy Security and Markets" in J.H. Kalicki and D.L. Goldwyn (eds.), *Energy & Security: Strategies For A World In Transition.* Washington, DC: Woodrow Wilson Center Press, 2013, pp. 74–80.

and importing nations, respectively, as well as the element of planning for disruptions, can be identified as crucial regarding the strategic weight they carry for a country's national security.

4.3.3 Defining the Strategic Value of Energy Projects

Combining the components that define a state actor's behaviour for survival in the international system with the principals of energy security, we can identify the elements of an energy project that carry strategic value. As security threats formulate strategic objectives, these are directly linked to a state's effort for survival in the international system. In turn, survival in the international system can be achieved through internal and external balancing. This includes, on the one hand, the strengthening of state power factors and its ability to project and use power, and on the other hand, strengthening its system or network of alliances. Given the above line of thought, energy projects and energy flows can potentially offer strategic value when they contribute to a state's internal and external efforts, thus assisting in achieving its strategic objectives built to address the security threats it faces. Consequently, the strategic framework under which energy projects can be examined needs to incorporate an examination of three distinct but interconnected elements and the degree in which the project contributes to each one. These include the project's contribution in maximising the international actor's levels of power, its ability to generate significant change in the balance of power within the system, and its potential to enhance the international actor's self-help efforts through internal or external balancing.

Analysing Fig. 4.2, at least one of three conditions need to be met. (a) Primarily, in order to be attributed with strategic properties, an energy project needs to contribute in a *significant* way in maximising the country's levels of power in the international system. This could be achieved either by significantly raising its economic impact in the local economy, thus offering the state with the potential to transform it to hard power, or by increasing its ability to project power, and therefore its impact, in the economies of other state actors within the system. (b) An energy project needs to generate a substantial *change* in the balance of

	Strategic element	Energy-project function
1	Power Levels	Contribution in maximisation
2	Balance of Power	Generation of significant change
3	Self-Help potential	Enhancement of Internal or External Balancing

Fig. 4.2 The Strategic Framework of Energy Projects (SFEP)

power within the system in order to generate strategic value for the state actor it evolves. If the rise in—economic or political—power deriving from an energy project has the potential to change the balance of power, then it has contributed meaningfully in the state's efforts to improve its position in the system. (c) An energy project carries strategic value if it leads to the overall enhancement of a country's ability to increase its *self-help* efforts either through internal balancing, such as more robust economic capacity, more stable political system, raising hard power abilities or through external balancing, such as building new alliances or strengthening its role in existing ones. Summarising the above three elements, it can be defined that an energy project has strategic value when it contributes in a significant way to the promotion of a state's security goals for survival within the international system.

4.4 Tools of the Energy Union with an External Dimension

As analysed earlier, the Energy Union strategy consists of measures and tools designed to deliver secure, sustainable, competitive and affordable energy to EU consumers, households and businesses. However, the Energy Union framework includes regulatory instruments that, while initially built to address internal issues within the EU borders, possess elements of 'externality' with a direct effect in EU's energy relations with third parties. This includes the introduction of 'Projects of Common Interest' with the capacity to link the energy systems of EU Member States, as well as two principal regulatory instruments, the provision for

'Third Party Access' to energy networks and for 'Ownership Unbundling' of energy infrastructure. These three elements are being utilised to deliver the Energy Union's goals for energy security within the EU and serve a double goal; to enhance EU's resilience and autonomy while advancing its efforts to protect the environment. The regulatory framework built to address issues related to the energy sector in the European Union, faces a distinct challenge related to the very nature of energy. While the core objective of the Energy Union aligns with the fundamental scope of the European Union itself regarding the establishment of a common market functioning under competition, the nature of energy makes it difficult to establish free market and competition conditions when it comes to the transportation, distribution and storage of energy, in particular electricity and natural—and other forms of—gas. Besides the apparent practical, technical and economic issues, the development of parallel and competing electricity and gas networks would go against a fundamental objective of the Energy Union, liked to the efficient use of resources. Simply put, building multiple, parallel electricity and gas networks, (cables and pipelines) in the European cities and countryside to create free competition, not only degrades living standards and is uneconomical but also wastes valuable resources. This distinctive element means that while production, supply and consumption of energy can function under free-market competition rules, energy networks remain under a primarily monopolistic framework. In order to counteract and minimise the negative effects of monopoly, the EU is utilising its regulatory arsenal to ensure that the ownership and function of energy networks do not reduce the overall effectiveness of free market and competition framework.

The aforementioned regulatory instruments, providing 'Third Party Access' to energy networks and 'Ownership Unbundling' of energy infrastructure, are specifically designed to address this issue effectively. Both tools developed alongside the European energy regulatory framework consisting of four 'Energy Packages'. The 'First Energy Package' aiming to liberalise the energy market was adopted in 1996 regarding electricity and expanded in 1998 for gas, while the 'Second Energy Package' was adopted in 2003 and the 'Third Energy Package' in

2009.[18] These were followed by a fourth regulatory initiative, the 'Clean Energy for all Europeans package' concluded early in 2019 and marking a significant step towards the implementation of the Energy Union strategy.[19] Provisions for 'Third Party Access' stipulate that 'Transmission System Operators' (TSOs), and 'Distribution System Operators' (DSOs) controlling electricity and gas networks, as well as Storage System Operators' (SSOs) controlling storage facilities, are required to provide energy companies with non-discriminatory access to their infrastructure. Under this provision, TSOs, DSOs and SSOs are obliged to offer the same service to different users under identical contractual conditions.[20] Respectively, provisions for 'Unbundling' refer to the separation of energy supply and energy generation activities from the operation of—primarily—transmission networks. The provision refers primarily to ownership unbundling where supply or production companies are not allowed to hold a majority share or interfere in the work of a Transmission System Operator. The purpose of this regulatory tool is to avoid distortion in the function of competition created in a case where a company that produces or sells energy would also operate a network thus having an incentive to obstruct its competitors' access to the infrastructure.[21] The 'externality' element in these two regulatory tools lies on the effects they have on the construction or acquisition of energy infrastructure in the EU by companies from third countries. As will be further analysed below, if for example Gazprom, the Russian state-owned gas production company, wants to construct and operate a gas pipeline in the EU, it needs to comply with 'Third Party Access' rules, so it needs to offer access to its pipeline to gas from other producers, i.e. Norway,

---

[18] "Internal Energy Market", European Parliament, Accessed February 2, 2020, https://www.europarl.europa.eu/factsheets/en/sheet/45/internal-energy-market.

[19] "Clean Energy for all Europeans Package", European Commission, Accessed February 2, 2020. https://ec.europa.eu/energy/topics/energy-strategy/clean-energy-all-europeans_en.

[20] "Access to Infrastructure and Exemptions", European Commission, Accessed February 2, 2020. https://ec.europa.eu/energy/topics/markets-and-consumers/wholesale-market/access-infrastructure-and-exemptions_en?redir=1.

[21] "Third Energy Package", European Commission, Accessed February 2, 2020. https://ec.europa.eu/energy/topics/markets-and-consumers/market-legislation/third-energy-package_en#unbundling.

the North Sea, Azerbaijan, Eastern Mediterranean, US LNG, etc. Similarly, regarding 'Ownership Unbundling', if for example State Grid, the state-owned electricity utility monopoly of China, owns coal, nuclear or gas-fired power plant in the EU, it cannot also directly acquire the electricity network of the country. In addition to the above, a direct provision with 'externality' elements exist in two Directives of the European Parliament and of the Council, part of the 2019 'Third Energy Package' concerning common rules for the internal market. In both their Article 11, Directives 2009/72/EC, the 'Electricity Directive',[22] and 2009/73/EC, the 'Gas Directive',[23] stipulate that in order to grant approval for the acquisition of control of specific energy infrastructure by 'a person or persons from a third country or third countries' in a Member State, the Commission is obligated to take into consideration if such control will put at risk the security of energy supply to the EU or puts at risk the Member State's security of energy supply or the security of energy supply of another Member State. Concerns over the risks associated with the control of critical energy infrastructure by non-EU actors inevitably extend to PCIs since they often involve gas or electricity interconnectors built to link energy networks between the Member States. The—often—strategic characteristics of such PCIs render the aforementioned 'externality' elements of the regulatory framework essential for the energy-related security objectives of the European Union, particularly in regions critical to EU's energy security policy, like South-Eastern Europe and the Eastern Mediterranean.

[22]"Directive 2009/72/EC of the European Parliament and of the Council concerning common rules for the internal market in electricity", European Parliament, July 13, 2009. https://eur-lex.europa.eu/legal-content/EN/ALL/?uri=CELEX%3A32009L0072.

[23]"Directive 2009/73/EC of the European Parliament and of the Council Concerning Common Rules for the Internal Market in Natural Gas", European Parliament, July 13, 2009. https://eur-lex.europa.eu/legal-content/en/ALL/?uri=CELEX%3A32009L0073.

4.4.1 PCIs with Strategic Value in SE Europe and Eastern Mediterranean

In the last quarter of 2019, the European Commission released its 4th list of PCIs containing a total of 149 projects across the EU.[24] The overwhelming majority of projects (100 projects) are on electricity transmission and storage, in comparison to 32 projects on gas, six projects about smart grid deployment, six oil-related projects and five cross-border carbon dioxide networks. It is worth mentioning that on a regional level, South-Eastern Europe and the Eastern Mediterranean involve proportionally a larger percentage of gas rather than electricity projects. When examining the various PCIs under the Strategic Framework of Energy Projects (SFEP) presented earlier in this chapter, the following projects can be identified to carry strategic value both for the Member States they involve and the European Union. It needs, however, to be noticed that most of these projects have been long-planned and proposed, so regardless of being supported by the EU, their final construction cannot be taken for granted. Out of the—few—electricity PCIs in SE Europe and Eastern Mediterranean, the one that carries significant strategic value is the proposed power line connecting Israel with Cyprus, Greece and Italy to the rest of EU. The proposed project, known as 'EuroAsia Interconnector' identified as an EU 'electricity highway' is planned to connect the national electricity grids of Israel, Cyprus and Greece through a 1208 km subsea DC cable with a 2GW capacity. For the two EU Member States involved, the strategic value of the project is evident: The project significantly raises Cyprus' power levels by offering control over significant electricity flows to the EU while at the same time is enhancing its self-help potential both via internal efforts (by generating significant economic revenues) as well as by external efforts (by forging common interests thus strengthening the alliance with Israel). The project's strategic value for Greece derives

[24]"Annex VII, Union List of Projects of Common Interest Referred to in Article 3(4)", European Commission, October 31, 2019. https://ec.europa.eu/energy/sites/ener/files/c_2019_7772_1_annex.pdf.

primarily by its strategic effects on Cyprus. Strengthening Cyprus' position in the regional subsystem of Eastern Mediterranean assists Greece's efforts to deter Turkish aggression, thus improving its overall levels of power. An element of imperative strategic importance for Greece would be the effect that the construction of the interconnection can have on Greece's efforts to deter Turkish aggression regarding the Exclusive Economic Zone (EEZ). The project also contributes to Greece's external self-help efforts by enhancing the alliance perspectives with Israel. For the European Commission, the EuroAsia Interconnector project offers the opportunity to strengthen the supply of electricity in the EU and connect one isolated Member State with the main European grid. However important, this element does not offer significant strategic value to the EU since the diversification of countries of origin for electricity is not linked to the EU's vulnerability to political pressure from third countries.

Regarding natural gas projects, PCIs with strategic value to the EU correspond to the potential to diversify the EU's natural gas imports and lower its dependency from Russian gas. Projects that under the SFEP analysis carry such strategic capacity for the EU include the 'Krk LNG Terminal', the 'Alexandroupolis FSRU', and the 'TAP', 'IGB', 'IBS' and 'EastMed' gas pipelines. Following, are some necessary data on each project that help justify their strategic value to the EU, as described above. (a) The Krk LNG Terminal in Croatia is a 2.6 billion cubic metres per year (bcm/y) floating LNG regasification and import terminal project, under construction at the Krk Island located in the northern Adriatic Sea, Croatia. The terminal can enhance the diversification of gas supply in South-Eastern Europe, providing landlocked countries with alternative sources of natural gas. (b) Part of the broader 'Southern Gas Corridor', TAP is a gas pipeline with a ten bcm capacity, bringing Azeri gas via Georgia and Turkey and connecting Greece to Italy via Albania and the Adriatic Sea. The pipeline is expected to go into full operation by the end of 2020 assisting EU's diversification efforts and lowering its dependence from Russian gas. (c) The Alexandroupolis FSRU is a project in the northeast of Greece, consisting of an offshore Floating, Storage and Re-gasification Unit of LNG. It connects via a subsea and an onshore gas transmission pipeline with the Greek National Natural Gas System, the TAP pipeline and gas interconnections to Bulgaria. With a capacity of 5 bcm/y, the project offers the potential of LNG imports from the US, Qatar, Egypt and other sources of non-Russian natural gas. (d) The IGB pipeline is a 182 km-long natural gas Interconnector between Greece and Bulgaria that can bring natural gas from the TAP pipeline, the Greek national natural gas system and the Alexandroupolis FSRU, to Bulgaria and thus further on to the Balkans and the rest of SE Europe. (e) The IBS pipeline is a 170 km-long natural gas Interconnector between Bulgaria and Serbia with a capacity up to 1,8 bcm/y and the potential to connect to the IGB pipeline and generate diversification of gas supply in the Western Balkan region. (f) The EastMed pipeline, unlike the previously mentioned PCIs, is a project that meets significant technical, economic and political challenges. Designed to bring up to 11 bcm/y of natural gas from the newly discovered Eastern Mediterranean Israeli and Cypriot fields the pipeline has the potential to further-assist EU's diversification efforts. Combined, this network of gas pipelines and LNG terminals in SE Europe and Eastern Mediterranean, identified as PCIs by the Commission, carry the capacity to bring more than 25 bcm of non-Russian natural gas to various regions of the EU. Representing more than 12% of total EU gas imports, these projects indeed carry strategic value by lowering EU's vulnerability to Russian energy-related political influence, thus generating significant change in the balance of power in Europe.

4.4.2 Third Party Access

Besides the 'externality' effect of PCIs approved and supported by the Commission with the obligation to take into consideration EU's security of energy supply, one of the significant—indirect—instruments with 'externality' properties is the provision for Third Party Access (TPA). As presented earlier in this chapter, Member States need to adapt and apply a regulatory framework that ensures access by third parties to the gas and electricity transmission and distribution networks. Although directed towards the internal market, TPA provisions proved to have a substantial effect in EU's energy relations with third countries. In particular, TPA provisions have been applied in—both successful and unsuccessful—efforts to deter or reduce the influence of the Russian natural gas giant of Gazprom, who is the primary source of Russian gas to Europe. In one of the most recent occasions, a decision in September 2019 by the Court of Justice of the European Union (CJEU) has ruled that Gazprom could not use 100% of the capacity of the OPAL pipeline, a Gazprom-owned onshore pipeline running across the eastern part of Germany to the Czech Republic as an extension to the, also Gazprom-owned, Nord Stream 1 natural gas pipeline.[25]

In October 2016, a Commission decision, following a request by the German Regulatory Authority, had led to the exemption of the 36 bcm capacity OPAL gas-pipeline from TPA provisions. Poland's energy company PGNiG appealed, and the CJEU's concluded in 2019 that the Commission's 2016 decision was adopted in breach of the principle of energy solidarity. In a similar situation, Gazprom is facing difficulties regarding its Nord Stream 2 pipeline. The Nord Stream 2 pipeline is designed to carry 55 bcm/y of natural gas from the Russian port of Ust-Luga in the Leningrad region, across the Baltic Sea to Germany in the Greifswald area close to the exit point of Nord Stream 1. Following the CJEU's Judgment in 'Case T-883/16 Poland v. Commission', (the 'OPAL case') Gazprom is expected to face difficulties for Nord Stream 2 to comply with the examination of the security of supply criterion stipulated in Article 11 of the 2009 Gas Directive, or the competition and market-liberalisation provisions in Article 36 of the same Directive. Finally, the Russian company should face similar issues on a third major gas pipeline project, the 2nd line of the TurkStream pipeline. The TurkStream pipeline starts from Russkaya compressor station near Anapa in Russia's Krasnodar region, crossing the Black Sea to the receiving terminal at Kıyıköy in Turkey. It consists of two lines with a capacity of 15 bcm each. The first line is already in operation, delivering 15 bcm of gas to Turkey. The second line is designed to run from Turkey to

[25]"Press Release n°107/19, The General Court Annuls the Commission Decision Approving the Modification of the Exemption Regime for the Operation of the OPAL Gas PIPELINE", General Court of the European Union, September 10, 2019. https://curia.europa.eu/jcms/upload/docs/application/pdf/2019-09/cp190107en.pdf.

Bulgaria, across Serbia to Hungary and Slovakia. The 2nd line of TurkStream is also subject to the EU regulatory framework once it enters EU territory in Bulgaria. The Bulgarian Regulatory Authority is obligated to apply EU energy liberalisation legislation in the light of the OPAL ruling.[26] The above cases stipulate how a regulatory framework built to address internal issues of the European energy market can carry 'externality' elements that affect the EU's relations with third countries.

4.4.3 Ownership Unbundling

Similarly to TPA, the provisions of Ownership unbundling carry 'externality' elements that can affect relations with third countries. Pursuant to Unbundling provisions, System Operators of energy networks are subject to unbundling requirements which oblige the Member States to ensure the separation between generation, distribution, transmission and supply activities. Unbundling requirements determine that the independence of the TSOs must be assessed through a certification process conducted by each Member States' National Regulatory Authority (NRA). Unbundling provisions' 'externality' effect relates particularly in Chinese investments in the European energy sector. Over the last decade, Chinese FDI regarding privatisation or acquisition of energy infrastructure in production or networks has been advancing at an increasing rate. Primarily part of China's Belt and Road Initiative, such investments involve Chinese State-owned Enterprises. With EU and China holding top global trade shares and competing on the global trade and technology scene, such flow of investments into critical energy infrastructure and power generation has raised considerable concern over their potential security implications on a national and supranational level. Currently, Chinese state-owned companies hold shares in the French company nuclear power production company Areva, the British Hinkley Point nuclear power station, the Greek electricity TSO ADMIE, and various

[26] "The 'Principle of Solidarity': OPAL, Nord Stream, and the Shadow Over Gazprom", Atlantic Council, October 17, 2019. https://www.atlanticcouncil.org/blogs/energysource/the-principle-of-solidarity-opal-nord-stream-and-the-shadow-over-gazprom/.

Renewable Energy infrastructure in Spain, Portugal, Germany, the Czech Republic, Poland and others.

Furthermore, state-owned Chinese corporations own shares and finance existing and planned coal-fired power plants in European non-EU countries, including Bosnia and Herzegovina, Kosovo and Serbia. In March 2019, the European Commission in its Strategic Outlook in EU-China relations underlined that: '…China is constructing coal-fired power stations in many countries; this undermines the global goals of the Paris Agreement.'[27] Focusing in the Western Balkans and the EU's neighbourhood, the EU has a crucial interest in strengthening the resilience of countries involved to secure compliance to EU values, norms and standards, notably in key areas such as the rule of law, public procurement, environment, energy, infrastructure and competition. To this end, Ownership unbundling provisions offer protection on the over-exposure of energy infrastructure to Chinese influence since state-owned Chinese companies with a share in power production infrastructure cannot invest in transmission and distribution networks in the same country while the ones with shares in TSOs and DSOs cannot acquire shares in power production facilities. However limited, being an indirect tool, Ownership Unbundling, therefore, offers significant assistance in EU's efforts to maintain control from third countries to critical energy infrastructure within its borders and the EU neighbourhood area, thus enhancing the overall objectives of the Energy Union strategy.

4.5 Conclusions

Energy security is a cornerstone for any successful policy applied in the energy sector. Within its broad spectrum of objectives, securing the availability, affordability, efficiency, and environmental impact of energy incorporates most aspects of the EU's energy policy. The Energy Union is derived through packages of regulatory and policy initiatives designed to address growing concerns about the EU's dependency on energy imports

[27]"EU-China—A Strategic Outlook", European Commission, March 12, 2019. https://ec.europa.eu/commission/sites/beta-political/files/communication-eu-china-a-strategic-outlook.pdf.

and the economic and political vulnerabilities related to it. Such concerns have been focusing on the resilience of EU's energy systems from discrepancies of supply, its ability to incorporate the merits of free market and competition in the European energy sector, and its resolve in protecting the environment. Striving for energy security and the resilience of the European energy system generates a distinctive effect on the EU's relations to third countries. It is therefore linked to security issues on a national and supranational level, involving both the Member States and the European Commission. Examining the fundamental elements of the relationship between energy and security, this chapter identified the framework and conditions under which energy infrastructure acquires strategic value for the stakeholders it involves.

To this end, three fundamental conditions have been identified, where at least one of them needs to be present to attribute strategic value to energy infrastructure. These include the infrastructure's contribution in maximising the international actor's levels of power, its ability to generate significant change in the balance of power within the system, and its potential to enhance the international actor's self-help efforts through internal or external balancing. Furthermore, this chapter identified the tools of the European regulatory and policy framework that, while designed to address internal issues, incorporate an 'externality' effect, thus influencing EU's relations with third countries in the energy sector. Such tools include the identification and support for Projects of Common Interest with the capacity to link the energy systems of EU Member States, and the provisions for Third Party Access to energy networks and Ownership Unbundling of energy infrastructure. These three elements are being utilised to deliver the Energy Union's goals for energy security within the EU and serve a double goal; to enhance EU's resilience and autonomy while advancing its efforts to protect the environment. Striving for energy security has been the driving force behind the structure of a coherent, resilient and efficient Energy Union, determining both the internal development of the European energy sector and its relations with third countries. The successful application of energy security policies will, therefore, determine the overall success of the Energy Union strategy and its impact on social, economic and political developments in the European Union.

References

Atlantic Council. "The 'Principle of Solidarity': OPAL, Nord Stream, and the Shadow over Gazprom". October 17, 2019. https://www.atlanticcouncil.org/blogs/energysource/the-principle-of-solidarity-opal-nord-stream-and-the-shadow-over-gazprom/.

Brown, Marilyn A., Yu Wang, Benjamin K. Sovacool and Anthony Louis D'Agostino. "Forty Years of Energy Security Trends: A Comparative Assessment of 22 Industrialized Countries." *Energy Research & Social Science* 4 (2014): 64–77. https://doi.org/10.1016/j.erss.2014.08.008.

Buzan, Barry. *People, States & Fear: An Agenda for International Security Studies in the Post-Cold War Era*. Colchester: ECPR Press, 2016. pp. 104–126.

Buzan, Barry. "New Patterns of Global Security in the Twenty-first Century." *International Affairs*, 67.3 (1991), pp. 432–433.

Collins, Gabriel. "Russia's Use of the 'Energy Weapon' in Europe". *Issue Brief* no. 07.18.17. Rice University's Baker Institute for Public Policy, Houston, Texas, USA (2017). https://www.bakerinstitute.org/media/files/files/ac785a2b/BI-Brief-071817-CES_Russia1.pdf.

Deutsche, Welle. "Russia, Ukraine Sign Gas Transit Deal Ahead of Deadline". Accessed January 30, 2020. https://www.dw.com/en/russia-ukraine-sign-gas-transit-deal-ahead-of-deadline/a-51841576.

European Commission. "Clean Energy for all Europeans Package". Accessed February 2, 2020. https://ec.europa.eu/energy/topics/energy-strategy/clean-energy-all-europeans_en.

European Commission. "Access to Infrastructure and Exemptions". Accessed February 2, 2020. https://ec.europa.eu/energy/topics/markets-and-consumers/wholesale-market/access-infrastructure-and-exemptions_en?redir=1.

European Commission. "Third Energy Package". Accessed February 2, 2020. https://ec.europa.eu/energy/topics/markets-and-consumers/market-legislation/third-energy-package_en#unbundling.

European Commission. "EU-China A Strategic Outlook". March 12, 2019. https://ec.europa.eu/commission/sites/beta-political/files/communication-eu-china-a-strategic-outlook.pdf.

European Commission. "European Energy Security Strategy". May 28, 2014. https://eur-lex.europa.eu/legal-content/EN/ALL/?uri=CELEX:52014DC0330.

European Commission. "A Framework Strategy for a Resilient Energy Union with a Forward-Looking Climate Change Policy". February

25, 2015. https://eur-lex.europa.eu/legal-content/EN/ALL/?uri=CELEX:52015DC0080.
European Commission. "Annex VII, Union List of Projects of Common Interest Referred to in Article 3(4)". October 31, 2019. https://ec.europa.eu/energy/sites/ener/files/c_2019_7772_1_annex.pdf.
European Parliament. "Energy as a Tool of Foreign Policy of Authoritarian States, in Particular Russia". April 2018. https://www.europarl.europa.eu/RegData/etudes/STUD/2018/603868/EXPO_STU(2018)603868_EN.pdf.
European Parliament. "Internal Energy Market". Accessed February 2, 2020. https://www.europarl.europa.eu/factsheets/en/sheet/45/internal-energy-market.
European Parliament. "Directive 2009/72/EC of the European Parliament and of the Council concerning common rules for the internal market in electricity". July 13, 2009. https://eur-lex.europa.eu/legal-content/EN/ALL/?uri=CELEX%3A32009L0072.
European Parliament. "Directive 2009/73/EC of the European Parliament and of the Council Concerning Common Rules for the Internal Market in Natural Gas". July 13, 2009. https://eur-lex.europa.eu/legal-content/EN/ALL/?uri=CELEX%3A32009L0073.
Eurostat. "Energy Production and Imports." Accessed January 29, 2020. https://ec.europa.eu/eurostat/statistics-explained/index.php/Energy_production_and_imports.
Eurostat. "Energy, Transport and Environment Statistics". Accessed January 29, 2020. https://ec.europa.eu/eurostat/documents/3217494/10165279/KS-DK-19-001-EN-N.pdf/76651a29-b817-eed4-f9f2-92bf692e1ed9.
European Union. "Foreign and Security Policy." August 6, 2019. https://europa.eu/european-union/topics/foreign-security-policy_en.
General Court of the European Union. "Press Release n°107/19, The General Court Annuls the Commission Decision Approving the Modification of the Exemption Regime for the Operation of the OPAL Gas Pipeline". September 10, 2019. https://curia.europa.eu/jcms/upload/docs/application/pdf/2019-09/cp190107en.pdf.
IEA. "Energy Security—Areas of Work." Accessed January 29, 2020. https://www.iea.org/areas-of-work/energy-security.
Platias, Athanassios and Constantinos Koliopoulos. *Thucydides on Strategy: Athenian and Spartan Grand Strategies in the Peloponesian War and Their Relevance Today*, pp. 1–21, London: Hurst, 2010.

Waltz, Kenneth. *Theory of International Politics.* Boston: McGraw-Hill, 1979.
Yergin, Daniel. "Energy Security and Markets" in J.H. Kalicki and D.L. Goldwyn (eds.), *Energy & Security: Strategies for a World in Transition*, Washington, DC: Woodrow Wilson Center Press, 2013, pp. 74–80.

Part II

Regional and Regulatory Aspects of the Energy Union

5

Geopolitical Challenges and Cooperation in the European Energy Sector: The Case of SE Europe and the Western Balkan Six Initiative

Sotirios Manolkidis

5.1 Introductory Remarks

The legal basis for the formation of a European internal energy market (IEM) is provided by the EU Treaty. The ultimate goal is to create well-functioning pan-European electricity and gas markets, putting fair access and a high level of consumer protection in the centre, while guaranteeing adequate levels of generation and interconnection capacity across the continent. Over the years, the EU secondary legislation (Regulations, Directives and Recommendations) has built the road from a historically defined monopolistic regime, to a well-functioning free market. The first steps towards the European Internal Energy Market were actually set in the 1970s; however, it was the environmental issues of the late 80s and early 90s that highlighted the importance of developing a coordinated

S. Manolkidis (✉)
Hellenic Regulatory Authority for Energy (RAE), International Hellenic University, Thermi, Greece
e-mail: s.manolkidis@rae.gr; sotirios@manolkidis.gr

M. Mathioulakis (ed.), *Aspects of the Energy Union*, Energy, Climate and the Environment,
https://doi.org/10.1007/978-3-030-55981-6_5

European policy on energy. This policy leads to various electricity and gas, market liberalization waves. In 1996, the first European Directive concerning the liberalization of the electricity market was adopted. The primary purpose of Directive 96/92/EC was to allow the entrance of private entities into the market, secure account transparency and further promote the avoidance of market domination by securing third party access to the electricity grid at a market-determined price. A similar initiative was taken for gas leading to Directive 98/30/EC, two years later. These two directives, known as the 'First Energy Package', initiated the European energy market liberalization process. In 2003, the 'Second Energy Package' was adopted, including a second wave of Directives on the electricity (and gas) markets (Directive 2003/54/EC and Directive 2003/55/EC). The second electricity Directive introduced two significant changes in the market; it allows suppliers to enter markets in a different Member State from the one in which they generate power, and enables consumers to freely choose their suppliers without territorial restrictions (with the notable exemption of Non-Interconnected Islands). One of the most important initiatives as of the 2000s, is the Energy Action Plan, dating from March 2007. Furthermore, the plan identifies three significant issues to be consequently addressed; sustainability, the security of supply and competitiveness. In 2009, the 'Third Energy Package' was adopted. The Directive 2009/72/EC provides the framework for the generation of electricity, the transmission system operation, the distribution system operation, the unbundling and transparency of the accounts, the access to the system, the role of the national regulatory authorities in the operation of the sector, the operation of the retail market under an energy Clearing House, etc. Furthermore, EC Regulation 714/2009 was issued, stipulating the conditions for accessing the network on cross-border exchanges in electricity. The Regulation aims at setting fair rules on cross-border exchanges in electricity. This package completed the integration of the European electricity and gas markets. The main characteristics of the EU energy legal framework are Ownership Unbundling (separation of generation and retail activities from transmission activities), the obligation for all Member States to establish a National Regulatory Authority (NRA) and the installation of an Agency for the Cooperation of Energy Regulators (ACER), a forum for National Regulators to cooperate at the EU level. For quite the same reasons, Directive 2009/73/EC was issued, concerning common rules for the internal market on natural gas. The Directive also provides the rules for the transmission and storage of Liquefied Natural Gas (LNG), the organization of the operators of the sector, the rules for the distribution and supply, the unbundling and the transparency of accounts, the organization of accessing the system, the role of the national regulatory authorities in the sector and the principal rules of the operation of the retail market. Furthermore, Regulation 715/2009 was issued, on conditions for accessing the natural gas transmission networks. The Regulation aimed at setting non-discriminatory rules for accessing natural gas transmission systems, LNG facilities and storage and at facilitating the emergence of the well-functioning and transparent wholesale market, taking into consideration the particular characteristics of the national and regional markets. The 2009 Directives and Regulations mentioned above constitute the so-called 'Third Energy Package' also adopted by the Energy Community under the "acquis communautaire" rule.

5.2 Balkans: A Region Beyond Turmoil: Energy Growth and Sustainability Within the Energy Community

The Western Balkans, composed of Albania, Bosnia & Herzegovina, Croatia, North Macedonia, Montenegro, Serbia and Kosovo, is a complex region facing significant energy challenges. The conflicts over the break-up of the former Yugoslavia in the 1990s damaged much of the energy infrastructure. They compounded to a clear challenge of providing reliable energy supply and addressing energy poverty in all seven counties. Of the aforementioned countries only Croatia is, since 2013, an EU Member State. For the region of the Western Balkans any reform effort will be focused on, and will depend upon, the advancement of the Energy Community Treaty, a regulatory and market framework to which the entire region (and beyond) has now subscribed.[1] The Treaty aims at creating an integrated regional market for electricity and gas, compatible with, and complementary to, the European Union's internal energy market. Coal (mostly lignite) still dominates the primary energy supply in the Western Balkan region, accounting, in 2017, for 37% of Total Primary Energy Supply.[2] Lack of reliable electricity supply is a serious obstacle to economic development and investment in the region. Furthermore, a reliable natural gas distribution network is crucial, as the Western Balkan Region is strategically located between the resource-rich regions of the Caspian basin (and the Middle East) and the consuming areas of Western and Central Europe. At present, Russia is the dominant supplier of natural gas to the region. The establishment of a financially attractive environment to secure investment for future energy infrastructure projects appears to be a difficult process. Political instability, corruption, and access to the financial markets are the main obstacles to doing business within the WB6 countries. The EU has been strategically involved in the development of a Western Balkan energy sector and, for the last twenty or so years, remains the largest source of financing in the region. In addressing all relevant key challenges of transitioning to a regime compatible with EU transparent and open energy markets and further developing infrastructure in their neighbouring WB6 countries, a central role is attributed to Greece, Bulgaria, Croatia and Romania, which are Member States of the European Union. The progressive implementation of the EU energy *acquis* and the availability of market financing will determine future developments in the Western Balkans. Until today, the EU has predominantly financed energy infrastructure projects in hydropower and other renewable energy sources (RES), as well as, natural gas and electricity transmission networks, whereas China has predominantly financed lignite mining and large-scale power generation.

Currently, national energy portfolios in the Western Balkans rely almost exclusively on lignite and large-scale hydro (Albania) to generate

---

[1] See, *inter alia*, "The Energy Community: Legal Framework, 4th Edition", Energy Community Secretariat, 2018. file:///C:/Users/user/Downloads/ECS_LF4_2018.pdf.

[2] Semkoua, A., et al. (2019). "Integrating Energy Markets in the Wider Europe: The Eastern Dimension" *European Journal of Sustainable Development* 8(4).

electricity, and national energy strategies do not envisage a significant decrease in lignite use before 2030. These national strategies consider the future implementation of the EU emissions trading scheme (ETS) but justify continued investment in lignite power generation with the fact that carbon emissions would effectively be offset by extensive RES utilization. However, as new lignite power generation assets may become stranded considering future emissions targets, and wind power and hydropower are seasonally variable, offshore wind parks might also be the answer in future RES investment. Nonetheless, they are limited to the Adriatic Sea. Furthermore, one should note the focus on lignite and hydro notwithstanding several countries in the region intend to increase electricity generation from non-hydro renewable sources, as well from natural gas. These initiatives correspond to policy measures presented by National Renewable Energy Action Plans, the adoption of which is mandatory under EU energy law and, eventually, will be embraced by all the Energy Community countries. Natural gas is currently not extensively used in the broader region for electricity production, and its role in the near future remains uncertain, with future demand limited by uncertainty in transmission infrastructure development and in ETS pricing. Apparent exceptions are Croatia and Greece, which are currently developing FSRU stations for their supply with the, cheaper, LNG alternative. Taking into consideration the anticipated increase in carbon emissions pricing through the ETS and the EU's long-term goal of establishing a transparent and competitive gas market across the EU and the Energy Community countries, it is most likely that these conditions will be met in the near future.

5.3 The Western Balkans (WB6) Energy Profile and Future

The total power generation capacity across the Western Balkans region is 17.6 GW, of which lignite capacity accounted for 48%, followed by hydropower (46%), gas (4%) and fuel oil (2%). Although new forms of RES (wind power, photovoltaics and hybrid) are beginning to break through, they still form a negligible percentage of the total capacity. In May 2019, the Energy Community Secretariat issued the Electricity Monitoring Report for the Western Balkans.[3] The document echoes some positive developments of the WB6 Initiative and identifies the pace of reforms that should be undertaken by the Western Balkans countries. These include the establishment of a harmonized legal framework in compliance with the Third Energy Package and the formation of a joint Power Exchange. Furthermore, the WB6 Initiative Electricity Monitoring Report outlines some positive steps taken to form Day-Ahead markets in Montenegro and North Macedonia. The effective launch of the Serbian Day-Ahead market is an essential development for the electricity market in SEE. The SEEPEX is the first organized marketplace in the region that provides high-level standards in terms of trading and clearing infrastructures. Athens has yet to launch its energy exchange (HEnEx), envisaged for September 2020. The Hellenic Energy Exchange (HEnEx) will provide access to new liquid energy markets and products. It will also support regional integration by enabling the -ever necessary- market coupling with Greece's neighbours, that initially includes Italy and Bulgaria (both EU Member States). In its 2019 Report, the Energy Community Secretariat highlighted that regarding the latest developments on the electricity market, "Montenegrin electricity power exchange (in Montenegro) is foreseen to go-live in the fourth quarter of 2020; meanwhile North Macedonia has made some progress in setting the national electricity market operator, which is crucial when it comes to its day-ahead market".[4] Also, some notable achievements mentioned in the report consist in establishing the power exchange (APEX) in Albania, finalizing procedures for selecting a universal and last-resort supplier in North Macedonia-recently also launched in Montenegro-and certifying Transmission System Operators (TSOs) in Kosovo and North Macedonia. All the above developments are necessary and in conformity with EU law.

[3] "WB6 Electricity Monitoring Report", Energy Community Secretariat, May 2019. https://www.energy-community.org/dam/jcr:6fcc56dd-4dcc-4986-8171-bb867d10f8fe/EnC_WB6_EL_05_2019.pdf.

[4] Ibid.

5.4 Natural Gas in the Western Balkans

As of 2018, the Western Balkan region consumed annually over 3 billion cubic metres (bcm) of natural gas. The market is considered as relatively small and plans to gasify the region may bring a considerable increase in gas consumption. As of today, neither Montenegro nor Kosovo have natural gas sources. Moreover, both countries do not dispose of gas transport infrastructures and are not involved in gas trade exchange with their neighbours. Total natural gas production in Albania, in 2017, reached almost 50 million cubic metres. Parallel to that, Bosnia & Herzegovina and North Macedonia have no natural gas sources, thus leading the two countries to import Russian-sourced natural gas. Similarly, Serbia's gas resources are insignificant, forcing the country to import most of its needed supplies.[5] Oil and gas originating from the Russian Federation dominates the Western Balkans import portfolio, while several oil refineries in Romania, Bulgaria, Serbia and Bosnia, all Member States of the Energy Community, are controlled by Russian energy companies. Serbia is making efforts to diversify its gas supplies by building a new gas interconnection with Bulgaria. The Serbian-Bulgarian route, whose project was conceived in 2018, once completed, will become, combined with the IGB interconnector between Greece and Bulgaria, part of a significant energy initiative aimed at connecting most South-eastern European countries. Furthermore, the Trans Adriatic Pipeline (TAP) carrying, from 2021 onwards, gas from Azerbaijan across Turkey, Greece and the Adriatic Sea before arriving in Southern Italy, is expected to significantly alter gas supply to the entire region. The TAP pipeline is scheduled to be completed in late 2020. The pipeline's initial capacity is estimated at 10 billion cubic metres (bcm), of which eight bcm of gas were contracted by Italy. The Trans Adriatic Pipeline is planned to be then extended further to the Western Balkans as part of the IAP (Ionian Adriatic Pipeline) project and could supply gas to the region's states after an increase in its capacity.

[5]"Towards a Balkan Gas Hub: The interplay between pipeline gas, LNG and renewable energy in South East Europe", Oxford Institute for Energy Studies. Accessed March 14, 2020. https://www.oxfordenergy.org/wpcms/wp-content/uploads/2017/02/Towards-a-Balkan-gas-hub-NG-115.pdf.

5.5 What About the Environment? Is the Answer Blowing in the Wind?

Regarding the transposition of the environmental EU *acquis*, the most critical issue relates to Bosnia and Herzegovina's failure to respect the Environmental Impact Assessment Directive and the Sulphur in Fuels Directive and the failure of Serbia to comply with the latter. By 2019, the WB6 countries have begun implementing the Directive on the limitation of emissions of certain pollutants into the air from large consumption plants.[6] While Albania is the only WB6 country which does not need to take any further measures, Serbia and Montenegro have started, since 2019, to implement the provisions of the Directive. Serbia has announced that it will shut down eight coal-fired power plant units owned by its Public Power Utility (Elektroprivreda Srbije), which do not comply with the requirements of the Large Combustion Plants Directive, by the year 2024. The implementation of this Directive alone will result in significant positive impacts on human health and the overall environment but will also require an answer to the most "burning" question in the region; what comes after coal for the WB6 Initiative countries? Similar to Albania, considering the fact that it has the most substantial amount of cost-efficient hydro potential in the Western Balkans, Bosnia & Herzegovina can expand its own RES share in the energy mix, as well as sell renewable energy to the rest of the Energy Community countries in the region.

Despite the existence of adequate grid connections between Bosnia & Herzegovina, Serbia, Montenegro and Croatia (which is an EU Member State), Bosnia & Herzegovina appears to suffer from a poor legislative framework and weak, restricted support mechanisms. On the contrary, Serbia has a significant, unexploited potential in renewable energy resources that carry the potential to cover about half of its domestic needs. Although only large wind and some solar parks would actually attract investments by the EU, the disproportion between

[6]"Directive 2001/80/EC on the limitation of emissions of certain pollutants into the air from large combustion plants", European Parliament, October 23, 2001. https://eur-lex.europa.eu/legal-content/EN/TXT/?uri=CELEX%3A32001L0080.

respective RES potential and current applications on the one hand and the potential feed-in-tariff (FIT) capacity on the other, pave the way for, at least, considering Serbia as a cooperation partner, in the framework of the RES Directive. The profiles of the seven countries outlined above, cultivate, therefore, the need to study and explore the potential benefits and risks related to the implementation of Cooperation Mechanisms between the EU and WB6 countries, regarding joint renewable energy projects which will provide not only a green source of readily available energy but will facilitate, as well, power exchange over the regional grid. Furthermore, very little investment and priority are being given to the increase of energy efficiency in energy consumption even in the Balkan EU Member States, i.e. Greece, Bulgaria, Croatia and Romania. Moreover, there are still many constraints (political, technological, financial, legislative, educational, etc.) to the development of renewable energy sources (RES) in order to contribute to the regional energy supply, distribution and security, outside of the EU. Regional cooperation is thus essential for increasing prosperity and economic growth of South-eastern Europe (SEE), in order to make better, faster progress towards future European integration, especially for the WB6 countries in question.

5.6 The Need for Institutional Reform Within the Energy Community

The reform of the Energy Community Treaty represents a golden opportunity to meet these weighty challenges with a cohesive approach and a sense of common purpose within the Western Balkan (WB6) Initiative six countries. The Energy Community needs to be strengthened by the ongoing commitment of its members to an enhanced legal framework and the EU's long-term goals such as decarbonization (by 2050) and environmental protection. Parallel to the above, there is also an urgent need to address systemic failures, e.g. dysfunctional governance and structural flaws in decision-making, lack of transparency, as well as high-level corruption among the signatory states, especially in the

Balkans. As one element of its work, the Energy Community Secretariat has recently made greater efforts to include the views of Civil Society in improving the functioning of the Energy Community's institutions. The EU's own position on this matter supports such a move, stating, in a Communication from the Commission, that "an empowered civil society is a crucial component of any democratic system and is an asset in itself. It represents and fosters pluralism and can contribute to more effective policies, equitable and sustainable development and inclusive growth…".[7] Any additional reform initiatives should express the growing demand for transparent and accountable governance within the Energy Community.

A critical issue that needs to be addressed is whether the existing dispute settlement mechanism should further be strengthened by establishing a Tribunal, to enhance the current (rather ineffectual) administrative mechanism. If one agrees on the necessity of an enhanced dispute settlement mechanism, any such Tribunal should be founded upon the principles of Impartiality, Comprehensiveness and Openness. Regarding impartiality, the majority of judges should come from countries outside the region and represent the highest level of competence. The Tribunal could meet in chambers of three Judges. Concerning comprehensiveness, the whole legal framework -meaning all parts of the Energy Community Treaty and the *acquis communautaire*- should be treated with the same level of importance. Finally, regarding openness, there is the need for access to justice by private entities and not only by representatives of the Member States of the Energy Community. A Tribunal, if established, should ensure that Contracting Parties comply with the Energy Community Charter and the EU *acquis* within its infringement procedures regarding both acts and omissions of the Energy Community Member States. The Energy Community needs to send a strong and clear message to its members on the application of Europe's future

[7]"The roots of democracy and sustainable development: Europe's engagement with Civil Society in external relations", European Commission, September 12, 2012. https://europa.eu/capacity4dev/policy-forum-development/documents/eu-communication-roots-democracy-and-sustainable-development-europes-engagement-civil.

goals.[8] Setting defined goals and ambitions with equal rules that ensure sustainable market development will guarantee that all members of the Energy Community compete on a level playing field. This is of particular significance for the Western Balkan states due to the current substantial discrepancies among them. The Energy Community Secretariat and ACER mutually benefit from close cooperation since 2016. This includes the participation of the Secretariat in ACER's gas and electricity working groups, joint workshops on the CACM Regulation and inclusion of Energy Community contracting parties in its annual market monitoring report. This cooperation contributes to the alignment of regulatory rules for the benefit of creating a pan-European energy market. An amendment to the Energy Community Administrative Arrangement[9] in 2019, gives the Secretariat the option to take part in the Market Integrity and Transparency Working Group of ACER. This allows the Secretariat to closely follow EU regulators' discussions on REMIT and thus support the Energy Community contracting parties' Regulators in implementing the REMIT Regulation in alignment with EU developments. The amended Administrative Arrangement aligns with the Contracting Parties' obligation to transpose the REMIT Regulation by the end of 2020. Moreover, it needs to be stressed that there is a necessity for a regulatory framework that ensures the creation of a single energy market encompassing both the EU and the Energy Community *acquis*. Creating a situation where the regulatory framework establishes different obligatory rules only for some of its members falls on a conflict in principle. It creates a high-risk situation that might disturb the path towards a level playing field or even result in electricity and gas market distortions within the Energy Community, and, consequently, the Western Balkan states. Furthermore, the Energy Community should not offer its support to any new generation projects unless the host country is on track to meet

[8]"A Roadmap for moving to a competitive low carbon economy in 2050", European Commission, March 8, 2011. https://eur-lex.europa.eu/LexUriServ/LexUriServ.do?uri=COM:2011:0112:FIN:EN:PDF.

[9]"Administrative Arrangement between ACER and Energy Community on Working Modalities in the Areas of Gas and Electricity", Energy Community, October 7, 2019. https://www.energy-community.org/dam/jcr:447ef1a3-b14e-400b-9193-9c8392fe7cef/Administrative%20Agreement%20between%20ACER%20and%20Energy%20Community%20Secretariat_consolidated%20version.pdf.

its energy efficiency targets in-line with EU legislation and long-term EU policies (the 2050 targets) and which are based on realistic demand scenarios.[10]

5.7 Concluding Remarks

Forthcoming economic diversification and the ever-growing relations with actors other than Russia, present both great opportunities and challenges to the Energy Community Member States. Such a challenge includes deepening economic ties with the European Union, but also with China and Turkey.[11] In recent years several Energy Community Member States have been working to deepen their economic relationship with the EU. These changes in energy economics will have long-term geopolitical consequences at the expense of Russia's vital interests. On the other hand, the US, under the Trump administration, has chosen to invest in the Balkans through their NATO allies and partners, i.e. Greece, Bulgaria, Albania, North Macedonia and Croatia. If the Energy Community is to make a positive environmental and social impact in the Balkan region, it needs to move its decision-makers' thinking beyond the mindset of thermal power capacity and large-scale power generating installations. Large-scale centralized energy generation limits energy planning towards energy transmission within the regional energy markets, which not only will cost billions of Euros in required investments but also dictates the perpetuation of an inefficient, centralized, large-scale, inherently based on fossil fuel and/or nuclear power paradigm, in which transmission losses can go far beyond a, technically acceptable, fifteen per cent (15%) level. While the aim of interconnecting regional energy systems certainly facilitates trade and decreases inefficiencies, to some extent, it should not be considered an optimum model by

[10]Małgorzata Smolak, Barbara Skardzińska, Aleksandra Bujaroska, Dragana Mileusnić, and Garret, Tankosić-Kelly, "Recommendations Regarding the Reform of the Energy Community Treaty", June 2014, http://www.caneurope.org/docman/south-east-europe/2833-recommendations-for-the-reform-of-the-energy-community-treaty.

[11]Adam Balcer, "A real game changer in the region", *New Eastern Europe*, Issue 1–2, January 28, 2020, https://neweasterneurope.eu/2020/01/28/a-real-game-changer-in-the-region/.

itself. Taking the transmission losses and their environmental, social and investment costs into account, they may significantly reduce the gains from investing in, across borders, regional transmission and interconnection models, when compared to a decentralized energy system. A framework for small-scale installations, located where they are most needed, and where the local resources are available, coupled with increased energy efficiency and with limited, but necessary, investment in the energy transmission and distribution with neighbouring regions, might be the answer. Decentralized energy production also provides significant employment opportunities, as does improving energy efficiency. This switch cannot possibly be triggered by liberalized market forces alone and calls for policy changes and new regulatory frameworks to be put soon in place by the Member States of the Energy Community. Additionally, as many needless barriers persist to the widespread adoption of renewable energy within the Energy Community and in particular the WB6 states, the Energy Community Secretariat needs to intensify its efforts to assist signatory states in removing barriers to the expansion of renewable (small and large-scale) energy production, while respecting current EU environmental requirements and future common goals.

References

Balcer, Adam. "A Real Game Changer in the Region", *New Eastern Europe*, Issue 1–2, January 28, 2020. https://neweasterneurope.eu/2020/01/28/a-real-game-changer-in-the-region/.

Energy Community Secretariat. "The Energy Community: Legal Framework, 4th Edition", 2018. file:///C:/Users/user/Downloads/ECS_LF4_2018.pdf.

Energy Community Secretariat. "WB6 Electricity Monitoring Report", May 2019. https://www.energy-community.org/dam/jcr:6fcc56dd-4dcc-4986-8171-bb867d10f8fe/EnC_WB6_EL_05_2019.pdf.

European Parliament. "Directive 2001/80/EC on the Limitation of Emissions of Certain Pollutants into the Air from Large Combustion Plants", October 23, 2001. https://eur-lex.europa.eu/legal-content/EN/TXT/?uri=CELEX%3A32001L0080.

European Commission. "The Roots of Democracy and Sustainable Development: Europe's Engagement with Civil Society in External Relations", Communication from the Commission to the European Parliament, the Council, the European Economic and Social Committee and the Committee of the Regions, September 12, 2012. https://europa.eu/capacity4dev/policy-forum-development/documents/eu-communication-roots-democracy-and-sustainable-development-europes-engagement-civil.

European Commission. "A Roadmap for Moving to a Competitive Low Carbon Economy in 2050", Communication from the Commission to the European Parliament, the Council, the European Economic and Social Committee and the Committee of the Regions", March 8, 2011. https://eur-lex.europa.eu/LexUriServ/LexUriServ.do?uri=COM:2011:0112:FIN:EN:PDF.

Energy Community. "Administrative Arrangement between ACER and Energy Community on Working Modalities in the Areas of Gas and Electricity", October 7, 2019. https://www.energy-community.org/dam/jcr:447ef1a3-b14e-400b-9193-9c8392fe7cef/Administrative%20Agreement%20between%20ACER%20and%20Energy%20Community%20Secretariat_consolidated%20version.pdf.

Oxford Institute for Energy Studies. "Towards a Balkan Gas Hub: The Interplay between Pipeline Gas, LNG and Renewable Energy in South East Europe", accessed March 14, 2020. https://www.oxfordenergy.org/wpcms/wp-content/uploads/2017/02/Towards-a-Balkan-gas-hub-NG-115.pdf.

Semkoua, A. et al. (2019). "Integrating Energy Markets in the Wider Europe: The Eastern Dimension." *European Journal of Sustainable Development* 8(4).

Smolak, Małgorzata, Barbara Skardzińska, Aleksandra Bujaroska, Dragana Mileusnić, and Garret, Tankosić-Kelly, "Recommendations Regarding the Reform of the Energy Community Treaty", June 2014. http://www.caneurope.org/docman/south-east-europe/2833-recommendations-for-the-reform-of-the-energy-community-treaty.

6

Integrating Energy Markets in the Wider Europe

Panayotis Glavinis, Anna Konstantinidou, Afroditi Semkou, Elias Andreadis, and Elias Kolovos

6.1 Introduction

The EU has constituted an Energy Union by absorbing the energy markets of the EU Member States, which now operate within the single European market. The Energy Union has its own rules and institutions, where independent energy regulators in the Member States are coordinated by the European Commission to regulate national energy markets effectively. It is built upon a uniform set of rules, largely of EU origin, that ensures a level playing field for energy players in the EU. The internal European energy market has a long way to go before forming a genuine single market, subject to the same pan-European operating rules. This is because the European energy market is still primarily divided into national energy markets, as national energy networks are not fully interconnected. This feature also affects the regulation of the single European

P. Glavinis (✉) · A. Konstantinidou · A. Semkou · E. Andreadis · E. Kolovos
University of Thessaloniki, Thessaloniki, Greece
e-mail: glavinis@law.auth.gr

M. Mathioulakis (ed.), *Aspects of the Energy Union*,
Energy, Climate and the Environment,
https://doi.org/10.1007/978-3-030-55981-6_6

energy market, which cannot reach the highest degree of integration, unless the pan-European energy networks are interconnected in such a way that the national energy markets of the EU Member States will function as communicating vessels, where the same level of fluid will indicate the absolute harmonization of their operating rules. The EU is energy-dependent on sources located in third countries where primary energy is produced and transported to Europe, where it is consumed. These countries are geographically situated in the area that we call Wider Europe. The markets in these countries are not governed by EU law. For the security of its supply and the smooth and predictable operation of its energy market, the EU has an interest in extending the *acquis communautaire*, in other words, the European energy *acquis*, to non-member states neighbouring it. From the above derives the necessity to interconnect the countries of the wider Europe by subjecting the regional energy markets on both sides of the European frontiers to the same operating rules to ensure an uninterrupted flow of sufficient energy resources to Europe at affordable prices. The EU possesses three main tools to foster the integration of the energy markets in the wider Europe: The Energy Community, the TEN-E strategy, and the Energy Charter Treaty.

6.2 The Need to Harmonize Laws Across Borders and the Energy Community

Energy law is "the allocation of rights and duties concerning the exploitation of all energy resources between individuals, between individuals and the government, between governments and between states".[1] As a result, since energy trade has overcome national borders, internationalization of the energy markets is of paramount importance and can be considered; either as regards to the influence that international law has on energy markets or as internationalization of national laws.[2] This

[1] A. Bradbrook, Energy Law as an Academic Discipline, *Journal of Energy and Natural Resources Law*, 14, 1996, 194.

[2] K. Talus, *Research Handbook on International Energy Law*, Research Handbooks in International Law, 2014.

international aspect has expanded to all fields of energy law, from exploration and generation to transportation, downstream energy markets, national sovereignty and as a consequence, to its regulatory framework. The international and cross-border essence of the energy industry is indisputable. Since the occurring problems are many when it comes to conflicting rules, absence of rules or regulatory instruments that don't relate from one jurisdiction to another,[3] a degree of harmonization of national regulatory regimes is highly required. Even the most hardcore national energy-related areas, such as government contracts or production licencing are significantly influenced by international law, the practices followed are common and the rules applied are uniform regardless of geographical location.

EU Law is a hybrid system, somewhere between national and international law. It has grown dramatically over the last twenty years. First, the electricity and gas directives in 1996 and 1998, then the second energy package in 2003 and the third energy package in 2009, turned the EU energy *acquis* to a key element of liberalization, through exporting its energy laws and policies. The dependency of the EU on imported energy is a fact; the limitation of fossil resources with the simultaneous consumption growth is another fact. Natural gas comes from Norway and Russia; oil comes from the Middle East, Nigeria and Norway and coal comes from Colombia and the United States. This extent of dependency is problematic when it comes to possible disturbances affecting the availability of energy, such as political circumstances in the exporting country, conflicts between countries, etc. The case of Russia and Ukraine, for example, was critical for the EU to realize that the risks relating to energy dependency should be minimized and security of supply must be enhanced by all means.[4] Energy Security is the constant and uninterrupted availability of affordable energy[5] that can be achieved through various methods, some of which are the diversification of the energy mix, the decentralization of energy supplies, the improvement of energy

[3]T. Wälde, "Why OGEL-an Oil-Gas-Energy Law Intelligence Service?", 1 OGEL (2003).

[4]K. Talus, *Introduction to EU Energy Law.* First Edition, Oxford University Press, 2016.

[5]K. Talus, "Security of Supply: An Increasingly Political Notion", in Bram Delvaux, Michael Hunt, Kim Talus (eds.) *EU Energy Law and Policy Issues* (Euroconfidentiel, 2008), pp. 123–149.

efficiency, the multiplication of energy sources and the adaptability of the markets that is accomplished by upgrading international cooperation. Since the energy markets are international, the external relations among the countries are of paramount importance and should be the cornerstone of the energy policy. The EU considers that its global energy interests are better served when third countries incorporate the operating rules of its internal market. This provides a framework of stability and predictability to the flow of energy resources from outside Europe into the European Union. To this end, the EU has signed a plethora of bilateral agreements reflecting the geopolitical approach to energy security. The most interesting one is the Treaty on the Energy Community also discussed in other chapters of this book. This Treaty entered into force in 2006 for a period of ten years that was further extended until 2026. It established a community among the EU countries on the one hand and the Albania, Bulgaria, Bosnia and Herzegovina, Croatia, North Macedonia, Montenegro, Romania, Serbia and Kosovo on the other hand, to meet specific targets. To begin with, what was first needed was the establishment of a stable regulatory framework, attractive for new investments not only in power generation but also in networks, ensuring equal access to the energy supply to all parties. Furthermore, product markets constituted a big geographic area that had to be coordinated in the context of a single regulatory space for trade, with an eye to the development of network energy market competition. This space should provide an investment-friendly environment, capable of intensifying the security of supply, by promoting connections to the Caspian, North African and Middle Eastern gas reserves, or utilization of regional energy resources.

Finally, all the aforementioned should take place in an environmentally-friendly climate, fostering the use of renewable energy. With regard to its structure, the Energy Community follows an institutional system that is similar to that of the EU[6]; the Ministerial Council, for the provision of policy guidelines and the adoption of procedural acts, the Regulatory Board for the issuance of recommendations when

[6]M. Hunt and R. Karova, "The Energy Acquis under the Energy Community Treaty and the Integration of South East European Electricity Markets: An Uneasy Relationship?", in B. Delvaux, M. Hunt, and K. Talus (eds.), *EU Energy Law and Policy Issues* (Euroconfidentiel, 2010), pp. 51–86.

it comes to cross-border affairs and the Secretariat based in Vienna, which administratively supports the other two institutions and monitors the obligations of the parties.[7] Legally binding decisions can be taken, and non-legally-binding recommendations can be issued, whereas the measures needed are based either on the Commission's or a contracting party's proposal. The basic idea of the Energy Community was to export the EU energy *acquis* to all the parties of the Treaty, which practically means that the energy regulation, the internal market rules and competition standards are all implemented by the contracting states. Given that the ECT has not succeeded its target to become a bridge between East and West, the EU's focus regarding its external energy policy has switched to the Energy Community, that is nowadays expanding towards the Caspian Sea and developing its content by implementing the framework of the Third Energy Regulatory Package, in line with the ultimate target of security supply.

6.3 TEN-E Strategy and Projects of Common Interest

Energy infrastructure is of the utmost importance for the EU. The EU is highly reliant on third countries to secure the necessary energy resources to meet its energy targets. Therefore, the security of the energy supply itself is crucial for the energy future of the EU, which of course entails the necessity of further improving, developing and enhancing the energy infrastructure of EU countries. To note, the development of sustainable infrastructure that will also meet the environmental targets of the EU is also a crucial challenge for the EU. Taking a step towards this direction, the Union has already undertaken proper actions especially by adopting Regulation (EU) 347/2013 of the European Parliament and the Council on guidelines for Trans-European energy infrastructure, effectively adopting the Trans-European Networks for Energy (TEN-E) Strategy to strengthen further and integrate the internal market and

[7] Report from the Commission to the European Parliament and the Council under Art. 7 of Decision 2006/500/EC (COM(2011)105 final (10 March 2011)).

to enhance the security of the energy supply and diversify its energy suppliers. The Regulation: "(a) addresses the identification of projects of common interest necessary to implement priority corridors and areas falling under the energy infrastructure categories in electricity, gas, oil, and carbon dioxide set out in Annex II ('energy infra-structure categories'); (b) facilitates the timely implementation of projects of common interest by streamlining, coordinating more closely, and accelerating permit granting processes and by enhancing public participation; (c) provides rules and guidance for the cross-border allocation of costs and risk-related incentives for projects of common interest; (d) determines the conditions for eligibility of projects of common interest for Union financial assistance".[8] The prompt implementation of the TEN-E strategy is crucial for the EU to accomplish its energy targets, reduce production and operation costs and render the wholesale prices more competitive while furthering the potential for development in the energy sector and enhancing the energy security. For this purpose, the EU identified a framework within which specific energy infrastructure projects would get specific incentives and benefits. These projects are called Projects of Common Interest (PCIs).

But how does an energy project qualify as a PCI? Article 4 of the Regulation stipulates three criteria: "(a) the project is necessary for at least one of the energy infra-structure priority corridors and areas; (b) the potential overall benefits of the project, assessed according to the respective specific criteria in paragraph 2, outweigh its costs, including in the longer term; and (c) the project meets any of the following criteria: (i) involves at least two Member States by directly crossing the border of two or more Member States; (ii) is located on the territory of one Member State and has a significant cross-border impact as set out in Annex IV.1; (iii) crosses the border of at least one Member State and a European Economic Area country". If the project fulfils the criteria mentioned above, then it should also fall under one of the nine priority corridors as they are laid out in Annex I of the regulation or under one of the three priority thematics regarding electricity, gas, and oil corridors.[9]

---

[8] See article 1, para. 2.

[9] https://eur-lex.europa.eu/legal-content/EN/TXT/?uri=CELEX%3A32013R0347.

In particular, the Regulation stipulates that regarding priority electricity corridors: "(1) Northern Seas offshore grid (NSOG): integrated offshore electricity grid development and the related interconnectors in the North Sea, the Irish Sea, the English Channel, the Baltic Sea and neighbouring waters to transport electricity from renewable offshore energy sources to centres of consumption and storage and to increase cross-border electricity exchange. (2) North-South electricity interconnections in Western Europe (NSI West Electricity): interconnections between Member States of the region and with the Mediterranean area including the Iberian Peninsula, notably to integrate electricity from renewable energy sources and reinforce internal grid infrastructures to foster market integration in the region. (3) North-South electricity interconnections in Central Eastern and South-Eastern Europe (NSI East Electricity): interconnections and internal lines in North-South and East-West directions to complete the internal market and integrate generation from renewable energy sources. (4) Baltic Energy Market Interconnection Plan in electricity (BEMIP Electricity): interconnections between Member States in the Baltic region and reinforcing internal grid infrastructures accordingly, to end isolation of the Baltic States and to foster market integration inter alia by working towards the integration of renewable energy in the region". As regards to the priority gas corridors, the Regulation stipulates that: "(5) North-South gas interconnections in Western Europe (NSI West Gas): gas infrastructure for North-South gas flows in Western Europe to further diversify routes of supply and for increasing short-term gas deliverability. (6) North-South gas interconnections in Central Eastern and South-Eastern Europe (NSI East Gas): gas infrastructure for regional connections between and in the Baltic Sea region, the Adriatic and Aegean Seas, the Eastern Mediterranean Sea and the Black Sea, and for enhancing diversification and security of gas supply. (7) Southern Gas Corridor (SGC): infrastructure for the transmission of gas from the Caspian Basin, Central Asia, the Middle East and the Eastern Mediterranean Basin to the Union to enhance diversification of gas supply. (8) Baltic Energy Market Interconnection Plan in gas (BEMIP Gas): gas infrastructure to end the isolation of the three Baltic States and Finland and their dependency on a single supplier, to reinforce internal grid infrastructures accordingly, and to increase diversification and security of supplies in the Baltic Sea region." Finally, regarding the priority oil corridors, the Regulation requires that: "(9) Oil supply connections in Central Eastern Europe (OSC): interoperability of the oil pipeline network in Central Eastern Europe to increase the security of supply and reduce environmental risks".

The three priority thematic areas provided in the Regulation are smart grids deployment, electricity highways and cross-border carbon dioxide networks. The TEN-E strategy encapsulates the EU's targets for decarbonization, further development of renewable energy resources and the development of electricity storage facilities. Once it has been decided that a project has satisfied all the necessary prerequisites, it is included in the biannual list that contains all the PCIs. Every two years since 2013, the EU, after proper consultations, issues a list of PCIs. The 2017 list contains 173 projects, of which 106 are for transport and storage of electricity, four for the deployment of smart grids, 53 natural gas, six oil and four cross-border carbon dioxide networks. Specific incentives and benefits are applicable to the PCIs, which acquire "Priority status". This means that PCIs benefit from faster planning and licencing procedures enhanced regulatory terms, lower administrative costs, increased participation of the public through consultations and increased visibility towards potential investors. The list is accompanied by a crucial technical document, which contains more information regarding each project separately and more specifically its location, its type and the technology used for its implementation, as well as the relevant dates for the start of its implementation. In addition, the PCIs have access to the "Connecting Europe Facility" (CEF) amounting up to 5.35 billion euro, a fund dedicated to speeding the energy projects and attracting potential investors. CEF will enjoy an even larger budget for the 2021–2027 periods amounting up to 42.3 billion euro. PCIs receive strategic and political guidance from the High-Level Groups, the Regional Groups and the Expert Groups, which correspond to the aforementioned priority corridors or thematic areas.

6.3.1 The Crucial Projects

To integrate the energy markets of the wider Europe into the EU energy market, we need further prompt and targeted interventions on the part of the EU on this very crucial sector. Unfortunately, of the 173 projects included in the 2017 PCI list, only 12 have a wider Europe cross-border impact. These projects, however, are not minor or insignificant. They are an essential starting point on which the EU can build upon to realize its energy and environmental targets, but also the integration of the wider Europe energy Markets, which, for the most part, are its future energy suppliers. Regarding electricity, these include: Nordlink (Germany—Norway), North Sea Link (UK—Norway), Ice Link (UK—Iceland), Green connector (Italy—Switzerland), Interconnection between Airolo and Baggio (Switzerland—Italy), Euroasia Interconnector (Israel—Cyprus—Greece), Mid Continental East Corridor (Romania—Serbia and Italy—Montenegro) and ElMed (Italy—Tunisia). In natural gas, they include Passo Gries (Italy—Switzerland), IBS (Bulgaria—Serbia), the Trans-Caspian Gas Pipeline (TCP), the South Caucasus Pipeline Future Expansion (SCPFX), Trans Anatolian Natural Gas Pipeline (TANAP) (Turkmenistan—Azerbaijan—Georgia—Turkey—Greece) and the EastMed (Israel—Cyprus—Greece). In order to implement these PCIs across borders, considerable investments are needed from the private sector, in addition to any grants that the EU is prepared to allocate, especially at the design phase. The Energy Charter Treaty (ECT), also discussed in other chapters of this book, is the main instrument available today to regulate cross-border Foreign Direct Investments.

6.4 The Regulation of Cross-Border Investments Under the Energy Charter Treaty

The fundamental objective of the ECT was to create a level playing field for energy investments across the geographical scope of the Treaty to minimize the non-commercial risks associated with the investments made in the energy sector.[10] Western European states wanted equal opportunities for access to Eastern and Central European markets while Eastern and Central European states sought to protect the domestic investors and control their economic policies. This dimension is also reflected in the relevant provisions of the Charter.[11] EU's core initiative behind the ECT was the protection of the existing East-West energy flows, the promotion of the anticipated West-East investment flows, as well as the aforementioned export of its legal framework to as many countries possible. However, although the ECT was envisaged to bring together and combine all the actors and forces of the global energy industry, there are still crucial shortcomings, since the main gas suppliers of the EU such as Russia[12] and Norway haven't ratified it, whereas Algeria and the producers from the Middle East are just observers in the ECT.[13] It is the consuming or transit countries that are the ones showing eagerness to adopt the Treaty rather than the producing states. On the other hand, there is indeed a big gap between the ECT and the EU energy *acquis*. Since the ECT was negotiated by the time of the First EU Energy Directives, it didn't include liberalization notions such as the mandatory Third-Party Access rule. As a consequence, the ECT provides minimum liberalization standards compared to the EU. Furthermore, Russia would never accept the EU energy *acquis*.

6.4.1 Investment Promotion Rules

To regulate investments, the Charter distinguishes the pre-investment from the post-investment stage. Providing access to foreign investments is the core of the promotion rules contained in a treaty promoting and protecting such investments. Promoting cross-border investments

[10] "The Energy Charter Treaty and Related Documents. A Legal Framework for International Energy Cooperation", Energy Charter Secretariat, 2004, p. 14.

[11] C.P. Speed and T.W. Walde, "Will the Energy Charter Treaty help international energy investors?", UNCTAD/ITE/IIT, Vol. 5, No. 3, 1996, p. 8.

[12] D. Doeh, A. Popov, and S. Nappert, "Russia and the Energy Charter Treaty: Common Interests or Irreconcileable Differences?", 5(2)OGEL, 2007.

[13] C. Bamberger and T. Wälde, "The Energy Charter Treaty", in M. Roggenkamp et al. (eds.), *Energy Law in Europe*, Oxford University Press, 2008.

through market-opening increases investment flows but can also affect the capacity of domestic industry negatively while limiting the host state's regulatory capacity.[14] Under International Law, a state is not obliged to accept any investment in its territory. The economic dimension of state sovereignty enables the state to decide whether to open its borders to foreign investors or to exclude certain areas of economic activity and to determine the conditions to be met for the admission of an investment.[15] However, once an investor is accepted in a host country, the state's obligations vis-à-vis his investment are binding.[16] On the contrary, measures that regulate the entry of foreign investment remain with the host state to take, while protecting investments is not an option.[17] At the pre-investment stage, the ECT seems to follow a restrictive approach, reducing the possibility of integrating energy markets by opening them to foreign investments. The principles of the Charter are those of open markets, such as free access to business and liberalization of trade in energy products.[18] Thus, the ultimate goal of the Charter is to create a liberalized energy market. However, this objective is difficult to achieve without the corresponding liberalization of investments. In the context of accepting an investment, the most common pre-investment rules set by the host state are export quotas, supplies from local producers, as well as the requirement for a local partner to participate in the investment scheme. These measures, which aim at maximizing the benefits for the host country and controlling investments, are clearly in conflict with the intended liberalization of investments and may therefore adversely

[14] UNCTAD, Series on International Investment Policies for Development, "Investment Promotion Provisions in International Investment Agreements", UN, New York and Geneva, 2008, pp. 7–9.

[15] R. Dozler and C. Scheuer, *Principles of International Investment Law*, Oxford University Press, 2008, pp. 79–80.

[16] A. Konoplyanik and T. Wälde, "Energy Charter Treaty and its Role in International Energy", *Journal of Energy and Natural Resources Law*, Vol. 24, No. 4, 2006, p. 33.

[17] Th. Papanastasiou, "Protecting Investments in the Hydrocarbons Sector", in *Hydrocarbons Law*, ed. Nick. Farantouris and T. Kosmidis, Law Library, 2015, p. 344.

[18] K. Iliopoulos, "The International Law of Energy", European State, 2009, pp. 599–600.

affect potential investment flows.[19] In the present circumstances, where the amendment process of the ECT has been launched, it seems difficult to include pre-establishment phase arrangements in its regulatory field through binding commitments. For the EU, it is not a matter of priority to regulate issues concerning the pre-establishment stage.[20] It does not support the inclusion of investment input in the provisions of the Charter and in the event of a positive outcome of the negotiations, it will oppose to the inclusion of issues that will arise at this stage in the dispute settlement system.[21] Turkey is on the same line that the pre-investment phase should remain outside the provisions of the Treaty. Similarly, Georgia is of the view that the Charter should not cover the pre-investment phase. At the same time, however, Georgia does not oppose the adoption of soft law provisions on issues relating to the pre-establishment stage.[22]

6.4.2 Investment Protection Rules

In accordance with Article 10 (1) of the Charter, "each Contracting Party shall, (…), encourage and create stable, equitable, favourable and transparent conditions for Investors of other Contracting Parties to make Investments in its Area. Such conditions shall include a commitment to accord at all times to Investments of Investors of other Contracting Parties fair and equitable treatment. Such Investments shall also enjoy the most constant protection, and security and no Contracting Party shall in any way impair by unreasonable or discriminatory measures their management, maintenance, use, enjoyment or disposal. In no

19"Sharing Pre-Investment Rules in the Energy Sector", A discussion paper for the Energy Charter Expert Meeting removing pre-investment barriers in energy, mobilizing energy investments towards universal access and energy transition, Brussels, 14 June 2016, International Energy Charter, p. 3.

20European Commission, Recommendation for a Council Decision authorizing the entering into negotiations on the modernization of the Energy Charter Treaty, Brussels, 14 June 2019, COM(2019) 231, final, p. 2.

21Council of the European Union, Negotiating Directives for the Modernisation of the Energy Charter Treaty—Adoption, Brussels, 2 July 2019, 10745/19, p. 5.

22Energy Charter Secretariat, Room Document 1 Suggesting Policy Options Regarding the Modernisation of the ECT, 11 July 2019, p. 5.

case shall such Investments be accorded treatment less favourable than that required by international law, including treaty obligations. Each Contracting Party shall observe any obligations it has entered into with an Investor or an Investment of an Investor of any other Contracting Party". With regard to the first sentence of the above provision, the question is whether it includes an independent standard of treatment or merely it acts as an introduction to the other treatment measures of the article.[23] The jurisprudence of the arbitral tribunals appears divided. In *AES v. Kazakhstan*, the tribunal held that the first sentence of Article 10 (1) is not an autonomous standard of protection. Instead, it has an introductory and programmatic character.[24] The court reached the same conclusion in the case of *Isulux v. Spain*, where, according to the tribunal, the first sentence does not incorporate an independent protection measure.[25] In contrast to the case of *Energolliance v. Moldova*, the tribunal treated the provisions of the first sentence as an independent model of treatment.[26] Beyond the different case-law approaches, the creation of a stable investment environment, as defined in the first sentence, is closely linked to the principle of fair and equitable treatment. This results both from the grammatical wording of the article and its semantic sequence (the use of the term *such as* combining the first with the second sentence shows undoubtedly the relevant interlinking) and the teleological interpretation of the terms. Among the methods of interpretation set out in Article 31 (1) of the Vienna Convention on the Law of Treaties is one which emphasizes the purpose and the object of the Treaty. Article 2 of the Charter provides under the heading "Purpose of the Treaty" that it "establishes a legal framework in order to promote long-term cooperation in the energy field, based on complementarities and mutual benefits". Thus, the Treaty seeks to enhance the stability

[23]C. Verburg, "Modernising the Energy Charter Treaty: An Opportunity to Enhance Legal Certainty in Investor-State Dispute Settlement", Journal of World Investment & Trade, 20 (2019), p. 430.

[24]AES Corporation and Tau Power B .Vv. Republic of Kazakhstan, ICSID Case No. ARB/10/16, Award, 1 November 2003, par. 383.

[25]Isolux Infrastructure Netherlands B.V v. Spain, SCC V2013/153, Award, 12 July 2016, par. 764.

[26]Energoalliance LTD v. The Republic of Moldova, UNCITRAL Arbitration, Arbitral Award, 23 October 2013, par. 356.

required to achieve the intended long-term cooperation.[27] However, the context in which the investment operates and develops is subject to change as it is adapted to new circumstances.[28]

6.4.2.1 Fair and Equitable Treatment

This is an absolute standard, as it is not defined concerning any other level of protection, contrary to the national treatment and the MFN treatment, which are relevant standards of treatment insofar as they are determined by the host State's treatment of domestic investors and foreign investors, respectively. The FET principle has a dominant position in the law of protection of foreign investment. The tribunal's view in *Petrobart v. Kyrgyz Republic* is typical. Article 10 (1) is intended to safeguard the FET as a whole.[29] The tribunal seems to regard the FET clause as a general principle incorporating all the other treatment measures referred to in the article.[30] There is no single and universally accepted definition of the meaning and the content of the principle of FET.[31] The vague and ambiguous expression of "fair" and "equitable" is not defined in the Charter. Therefore, the content of the principle is subject to a variety of interpretations.[32] The arbitral tribunals have concentrated their efforts on attributing the substantive content of the principle which is not equated with the principle of *ex aequo et bono,* but it is a rule of law instead.

---

[27]Eiser Infrastructure Limited and Energia Solar Luxembourg S.A.R.I and Kingdom of Spain, ICSID Case No. ARB/13/36, Award, 4 May 2017, par. 378.

[28]AES Summitt Generation Limited and AES—Tisza Eromukft. v. Republic of Hungary, ICSID Case No. ARB/07/2, Award, 23 September 2010, par. 9.3.29.

[29]Petrobart Limited v, The Kyrgyz Republic, Arbitral Award rendered in Stockholm on 29 March 2005, in Arbitration No. 126/2003 of the Arbitration Institute of the Stockholm Chamber of Commerce, p. 76.

[30]C. Schreuer, Selected standards of treatment available under the Energy Charter Treaty, at *Investment Protection and the Energy Charter Treaty*, G. Coop and C. Ribeiro Editors, Arbitration institute of the Stockholm Chamber of Commerce, Juris Net LLC, 2008, p. 65.

[31]S. Schill, *The Multilateralization of International Investment Law*, Oxford University Press, 2009, p. 262.

[32]J. Salacuse, *The Law of Investment Treaties*, Oxford University Press, 2010, pp. 131–132.

Clarifying the constituent elements is an ongoing process where the arbitral tribunals have a crucial role to play. According to an arbitrator, the term is deliberately vague to give arbitrators a quasi-legislative power to formulate the rules necessary to achieve the object and purpose of the treaty in a particular dispute.[33] The decisions of the arbitral tribunals have developed the essential criteria that constitute the concept of FET. It relates to the existence of a stable and transparent legal framework, legitimate expectations, extortion of pressure, denial of justice,[34] legal proceedings, discrimination and manifestly abusive behaviour of the host State.[35] The concept and definition of FET are included in the issues for the amendment of the Charter. The EU is in favour of a clear definition of the principle, based on a closed list of elements that will include those formed by the jurisprudence of the arbitral tribunals referred above. Azerbaijan agrees to add a list of actions that would be considered as a breach to the principle. At the same time, it proposes that the FET principle should be based on the minimum standard of treatment of aliens in accordance with international law. Georgia agrees that the concept of FET should be clarified, but with reference to the minimum standards of international customary law. Thus, any inclusion of a list that defines the concept of authority should be in accordance with the minimum standard of treatment under customary international law.[36] The views of Azerbaijan and Georgia reflect the wider issue of whether the FET principle corresponds to a minimum level of individual rights, as expressed by the minimum standard set by international customary law, or whether it extends beyond customary international law.[37] According to Schreuer, the FET principle is a broader standard of treatment. For example, the measure of transparency extends beyond the traditional international minimum standard of the treatment of aliens.[38] On the contrary, the

---

[33] UNCTAD Series on Issues in International Investment Agreements II, Fair and Equitable Treatment, United Nations New York and Geneva, 2012, pp. 61–62.

[34] Mamidoil Jet Oil Greek Petroleum Products Societe S.A. v. Republic of Albania, ICSID Case No. ARB/11/24, Award 30 March 2015, par. 598–675, 691–735, 742–749, 764–771.

[35] UNCTAD, Ibid., p. 62.

[36] Energy Charter Secretariat, Brussels, 6 October 2019, Decision on the Energy Charter Conference Adoption by Correspondence—Policy Options for Modernization of the ECT, p. 7.

[37] K. Stefanou, C. Gortsos, *International Economic Law*, Law Library, 2005, p. 133.

[38] C. Schreuer, Ibid., p. 88.

approach of Azerbaijan and Georgia significantly limits its scope. At the same time, it contradicts the jurisprudence of the arbitral tribunals, which has expanded and broadened the content of the relevant principle.

6.4.2.2 Legitimate Expectations

Undoubtedly, legitimate expectations are one of the primary expressions of FET. There are two basic approaches to define their content. According to the broad approach, the foreign investor expects the host state to behave in a consistent, clear and transparent manner without any ambiguity, in order for the investor to know in advance both the laws governing the investment and the objectives of the host state so as to plan his investment.[39] This approach leads to a rigid view of the concept of legitimate expectations, which, in essence, undermines the regulatory power of the host state. The investor should be protected from unfair changes, but at the same time, the state should have a reasonable degree of regulatory flexibility to achieve its goals and adapt to the changing conditions. The requirement for fairness does not mean that the legal framework should remain unchanged. What is meant is that any subsequent changes should be made in a fair, consistent and predictable manner, taking the circumstances of the investment into account.[40] In the narrow approach, some elements should be taken into account: first, the specific assurances or commitments made by the host State where the foreign investor has relied on them. Second, the fact that the foreign investor should be aware of the general regulatory framework of the host State. Third, the strike of a balance between the legitimate expectations and the regulatory power of the host state.[41] The legitimate expectations have been analyzed in four recent arbitral awards rendered in cases involving the amendment of the renewable energy legal framework in Spain and Italy. In *Charanne v. Spain*, the tribunal rules that the

[39]Tecnic as MedioambientalesTecmeds v. The United Mexican States, ISCID Case No. ARB (AF)/00/2, Award 29 May 2003, par. 154.

[40]ElectrabelS.A v. Hungary, ICSID CASE NO. ARB/07/19, Award 25 November 2015, par. 7.77.

[41]UNCTAD, Ibid., p. 68.

host State's commitments should be made either based on a stabilization clause or through any statement by the state in relation to a particular investor that the existing regulatory framework will not change. Converting a regulatory framework to a specific commitment because of the limited number of people concerned would significantly limit the power of the state to regulate the economy to achieve the public interest. The tribunal held that the decrees at issue did not contain specific commitments to the investors.[42] In *Novenergia v. Spain*, the arbitral tribunal ruled that the commitments and assurances do not need to be specific. If a regulatory framework is designed to attract investments, investors may have a legitimate expectation that this regulatory framework will be stable.[43] On the other hand, *Blusun v. Italy* reaffirmed the need for specific commitments by the State to keep the legal framework unchanged, which was not the case here.[44] It further distinguished between laws that create rights and obligations, contractual commitments, and expectations arising from a relationship. These expectations, even legitimate, are elements that must be taken into account when applying other rules, but they cannot be assimilated to rules.[45] Promises arising from a regulatory regime for the promotion of the investments cannot create legitimate expectations. In *Eiser v. Spain*, the tribunal was in line with the basic premise that in the absence of specific commitments to investors, the State would not change its legal framework. The State may proceed with changes to achieve the public interest. At the same time, however, it clarified that the regulatory power of the State should not undermine the obligations arising from the FET principle. The FET under the Charter protects the investors against the fundamental changes in the regulatory framework when this is done without taking into account the conditions under which the existing investments

---

[42]Charanne B.V., Construction Investments S.A.R.L. v. The Kingdom of Spain, Final Award 21 January 2016, Arbitration No.: 062/2012, par. 490 and 493.

[43]Novenergia II—Energy & Environment (SCA) (Grand Duchy of Luxembourg), SICAR v. The Kingdom of Spain, SCC Arbitration (2015/063), Arbitration Institute of the Stockholm Chamber of Commerce, Final Arbitral Award 15 February 2018, par. 546 and 548.

[44]Blusun S.A., Jean Pierre Lecorcier and Michael Stein v. Italian Republic, ICSID Case No.ARB/14/3, Award 27 December 2016, par. 319(5) and 374.

[45]Ibid., par. 371.

were made.[46] The FET obligation means that a regulatory regime cannot be radically modified in relation to existing investments in the way of damaging these investments.[47] In Eiser, the tribunal held that the State measures were substantially different from those in *Charanne*, which had far less drastic consequences.[48] The element of the non-fundamental change in the regulatory framework was also accepted in *Charanne*. According to the tribunal, even where there are no specific commitments, the investor may have legitimate expectations, which in the event of a change in the regulatory framework will not be unreasonably disproportionate or contrary to the public interest. The proportionality test applies when changes are not unnecessary and do not amount to a sudden and unpredictable elimination of the essential features of the existing regulatory framework.[49]

6.4.2.3 Expropriation

According to Article 13(1) of the Charter, investments may not be expropriated or subject to measures equivalent to the expropriation unless the expropriation is (1) in the public interest, (2) it is not discriminatory, (3) the legal process is followed and (4) it is accompanied by a prompt, adequate and effective compensation. The use of the term "measures" refers not only to direct but also to indirect modes of expropriation. Direct expropriation consists of the mandatory transfer of a property right of an investor to the State by one or more acts of the domestic law. In the case of the indirect expropriation, the ownership remains intact. The state measures do not entail the transfer of the investment to the state, but they restrict the enjoyment of the investment or substantially limit the investor's control over his assets. Direct expropriations are now rare. The most usual form of expropriation is the indirect one. Compensation is the most challenging issue concerning expropriation cases. The

---

[46]Eiser Infrastructure Limited and Energia Solar Luxembourg S.A.R.I and Kingdom of Spain, ICSID Case No. ARB/13/36, Award, 4 May 2017, par. 362–363.

[47]Ibid., par. 382.

[48]Ibid., par. 368.

[49]Charanne v. Spain, ibid., par. 514 and 517.

Charter uses the terms “prompt, adequate and effective”. These terms usually refer to the concept of “full compensation”. Compensation is prompt when it is paid to the beneficiary without delay; effective when paid in the currency in which the investment funds were imported in the host country, provided that the currency in question is freely convertible, otherwise in the currency chosen by the investor used by the International Monetary Fund. Compensation is adequate when it corresponds to the fair market value of the investment. The Charter accepts the concept of fair market value by stating that the compensation for the expropriation will correspond to the fair market value of the investment expropriated as existed before the expropriation was known.

6.4.2.4 Regulatory Measures and Indirect Expropriation

States should be able to act in the broader public interest by taking measures such as to protect the environment, modify existing tax regimes, change tariff levels and take other similar arrangements. These reasonable state measures would not be taken if an investor who would be adversely affected could claim compensation.[50] In essence, the right of the state to legislate contradicts the strict and rigid nature of foreign investment protection.[51] The right of the state to exercise its regulatory power often violates the right of foreign investors to the peaceful enjoyment of their investments.[52] In *Vatenfell v. Germany*, Vattenfall's project to build a coal-fired power plant in the city of Hamburg in 2004 had sparked a lot of controversy. Environmental and political groups argued that the plant would be much larger than what was needed to meet Hamburg's energy needs and thus would unduly burden the environment. Despite the public opposition, the Hamburg authorities entered into a provisional agreement with Vattenfall to build the plant, which included some environmental restrictions on the plant's impact

---

[50] Marvin Feldman v. United Mexican States, ICSID Case No. ARB(AF)/99/1, Award 16 December 2002, par. 103.

[51] M. Sornarajah, *Resistance and Change in the International Law in Foreign Investment*, Cambridge University Press, 2015, p. 214.

[52] P. Glavinis, *International Economic Law*, Sakkoulas Publications, 2009, p. 638.

on the Elbe River. However, the terms of the agreement were subject to final authorization. Subsequently, the Hamburg Urban Development and Environment Authority issued a preliminary construction permit allowing Vattenfall to proceed with certain aspects of the construction. Final approval was granted in September 2008 and included additional restrictions on the impact of the power plant on the Elbe River. These additional measures were at the heart of the dispute. Hamburg, for its part, argued that the terms of the licence were necessary in order to comply with EU law and the restrictions imposed on all industries in the Elbe River region. At the same time, they were in line with the EU Water Directive, which requires all EU Member States to ensure certain levels of water quality. According to Vattenfall, the relevant conditions were making the plant unworkable beyond what was originally agreed. By imposing stricter conditions at the final permit, Germany had violated the ECT.[53] The ECT refers to indirect expropriation but makes no mention of regulatory measures. The main difference between regulatory measures and indirect expropriation is that the former does not imply an obligation to pay compensation. So, the crucial point is to delineate the boundary between regulatory measures and indirect expropriation.[54] To this effect, three main approaches have been developed. The first is known as the "solo effect doctrine", according to which the primary criterion for ascertaining expropriation is the impact of the measures. The degree of state intervention is important when considering whether an expropriation occurred or not.[55] Simple intervention by the state is not enough. It should be so restrictive that the investor would be deprived of the control over his investment.[56] The second, known as "policy powers doctrine", focuses on the purpose and nature of the measures.[57] According to a third approach, both the impact of the

[53]N. Bernasconi, Background paper on Vattenfall v. Germany arbitration, International Institute for Sustainable Development, 2009, pp. 1–2.

[54]A. Reinisch, Expropriation at *The Oxford Handbook of International Law*, P. Muchlinski. F. Ortino and C. Schreuer, Oxford University Press, 2008, p. 426.

[55]U. Kriebaum, "Regulatory Takings: Balancing the Interests of the Investor and the State", The Journal of World Investment and Trade», Vol. 8, No. 5, October 2007, p. 723.

[56]P. Glavinis, ibid., p. 639.

[57]M. Brunetti, "Indirect Expropriation in International Law", Forum Du Droit International, Vol. 5, No. 3, August 2003, p. 151.

measures and their character must be taken into account. At the same time, a balance must be struck between the legitimate expectations of the investor and the right of the state to take action for the common benefit.[58] A careful examination of the relevant case-law shows that it is in line with the third approach as a prerequisite for establishing regulatory expropriation: (i) the extent to which investor rights are affected; (ii) the purpose and nature of government measures; and (iii) reasonable investor expectations.[59] The definition of the indirect expropriation and the concept of regulatory measures are among the issues which are mostly discussed in the amendment process of the ECT, where the EU proposes to safeguard the right of the state to exercise its regulatory power to serve the public interest, such as protecting the health and the environment. At the same time, it clarifies that the provisions of the investment protection cannot be regarded as a commitment by the State not to modify its legislation in the future if this has a negative impact on investor's expectations regarding investment profits. Only regulatory measures that are manifestly excessive in the light of their objectives should qualify for an indirect expropriation. Indirect expropriation is established only when the investor is substantially alienated from the fundamental features of his investment. The mere fact that the government measures increase the cost of the investor does not amount to indirect expropriation. The tribunal must conduct a detailed and careful examination prior to determining whether an indirect expropriation has occurred.[60] According to Turkey, the right to regulate should be included in both the preamble to the Treaty and in a separate article specifying it with a non-stabilization clause. Regarding the indirect expropriation, its content could be clarified by establishing criteria referring mainly to the economic impact and the nature of the measures.[61] At this point, Turkey

[58] T. Gazzini "Drawing the Line between Non-Compensable Regulatory Powers and Indirect Expropriation of Foreign Investment—An Economic Analysis of Law Perspective", Manchester *Journal of International Economic Law*, Vol. 7, Issue 3, 2010, p. 42.

[59] OECD Working Papers on International Investment, 2004/04, ""Indirect Expropriation" and the "Right to Regulate" in International Investment Law", p. 22.

[60] Energy Charter Secretariat, Brussels, 6 October 2019, Decision on the Energy Charter Conference Adoption by Correspondence—Policy Options for Modernization of the ECT, pp. 15 and 22.

[61] Ibid., pp. 16 and 23.

differs from the established case-law. Georgia proposes to include separate provisions on regulatory measures that will underline the power of the Contracting States to exercise their legislative and regulatory powers in a non-discriminatory, non-arbitrary and proportionate manner in the pursuit of the public interest. At the same time, it does not oppose to include such a provision in the preamble of the Charter. On the issue of the indirect expropriation, Georgia understands the need to define it by taking into account the prevailing tendency of arbitral tribunals and international investment agreements to avoid conceptual misunderstandings.[62] Finally, Azerbaijan is of the view that a balance between investor rights and the regulatory authority of the state should be preserved. Non-discriminatory measures aimed at serving the public interest such as environmental protection, public health, the safety and the labour rights, should not fall under the scope of indirect expropriation.[63]

6.5 Conclusions

The EU is far from getting autonomous on primary energy resources any time soon. It will be for a long time dependent on imports from energy-producing countries laying outside Europe. This is the main reason why the EU promotes market access and free competition in the energy-producing countries and encourages new investments in order to secure an adequate and uninterrupted supply of energy resources at affordable prices from these countries to Europe. The Energy Charter Treaty has secured the effective protection of European investors in many countries of the wider Europe. Its amendment process has been launched, but nobody may predict what will be the outcome. The integration of regional energy markets extending across border with third countries depends on other factors as well. In addition to providing market access to European foreign direct investment in the energy sector, these countries need to harmonize their operating rules in line with the acquis communautaire. To achieve this goal, the EU uses various regulatory and

[62] Ibid., pp. 15 and 22.

[63] Ibid., pp. 15 and 22.

action tools in order to impose the required degree of harmonization of the operating conditions in these markets. The Energy Community is the most significant one. Moreover, the development of cross-border energy-transportation infrastructure and interconnection networks is a prerequisite for the integration of the energy markets in the wider Europe. Energy markets cannot be integrated unless they are technically interconnected. To this end, the TEN-E Strategy is the primary tool of action adopted by the EU. However, regional energy markets operate within the existing relations between the EU and third countries. These relations are developed in the framework of various partnership and association agreements entered into between the EU and its member states, on the one hand, and third countries concerned, on the other. The implementation of these agreements is affected by the geopolitical conditions prevailing in the regions surrounding Europe. To what extent these conditions allow the integration of the energy markets in the wider Europe and how effective the EU can be in the exercise of energy diplomacy remains to be seen. Unless the EU develops a genuine foreign energy policy and the EU member states allow for the development of such a policy at EU level, no effective regional integration of the energy markets in the wider Europe may occur in the near future.

References

Books

T. Dimitroff and T. Edwards (consulting eds.), *More on Risk and Energy Infrastructure: Value Chains, Stakeholders and Black Swans*, Globe Law & Business, 2017.

R. Dozler and C. Scheuer, *Principles of International Investment Law*, Oxford University Press, 2008.

P. Glavinis, *International Economic Law*, Sakkoulas Publications, 2009.

W. E. Hughes, *Fundamentals of International Oil & Gas Law*, Pennwellbooks, 2016.

T. Hunter (ed.), *Regulation of the Upstream Petroleum Sector*, Edward Elgar, 2015.

R. King (ed.), *Arbitration in the International Energy Industry*, Globe Law & Business, 2019.
P. Martin, B. Kramer, K. Hall, and A. Ritchie, *The Law of Oil and Gas: Cases and Materials,* Foundation Press, 10th ed., 2016.
M. Peeters, *Renewable Energy Law in the EU*, Edward Elgar, 2014.
Reinisch, *Standards of Investment Protection,* Oxford University Press, 2008.
P. Roberts (ed.), *Oil & Gas Contracts,* Sweet & Maxwell, 2016.
T. Roe, M. Happold (eds.), *Settlement of Investment Disputes under the Energy Charter Treaty,* edited in consultation with James Dingemans, Cambridge University Press, 2011.
M. Roggenkamp, C. Redgwell, A. Ronne, and I. del Guayo (eds.), *Energy Law in Europe: National, EU and International Regulation,* Oxford University Press, 3rd ed. 2016.
N. Saint-Paul, *Summers Oil and Gas,* Thomson Reuters, 3rd edition.
J. Salacuse. *The Law of Investment Treaties,* Oxford University Press, 2010.
S. Schill, *The Multilateralization of International Investment Law*, Oxford University Press, 2009.
M. Sorharajah, *Resistance and Change in the International Law on Foreign Investment*, Cambridge, 2015.
K. Talus, *Introduction to EU Energy Law*, Oxford University Press, 2016.
K. Talus (ed.), *Research Handbook on International Energy Law*, Edward Elgar, 2015.

Articles

N. Bernasconi, Background paper on Vattenfall v. Germany arbitration, International Institute for Sustainable Development, 2009.
M. Brunetti, "Indirect Expropriation in International Law", Forum du droit international, Vol. 5, No. 3, August 2003.
T. Gazzini, "Drawing the Line between Non-Compensable Regulatory Powers and Indirect Expropriation of Foreign Investment—An Economic Analysis of Law Perspective", *Manchester Journal of International Economic Law*, Vol. 7, Issue 3, 2010.
Konoplyanik and T. Wälde, "Energy Charter Treaty and its Role in International Energy", *Journal of Energy and Natural Resources Law*, Vol. 24, No. 4, 2006.

U. Kriebaum, "Regulatory Takings: Balancing the Interests of the Investor and the State", *The Journal of World Investment and Trade*, Vol. 8, No. 5, October 2007.

Th. Papanastasiou, "Protecting Investments in the Hydrocarbons Sector", in *Hydrocarbons Law*, eds. Nick. Farantouris and T. Kosmidis, Law Library, 2015.

A. Reinisch, Expropriation at the *Oxford Handbook of International Law*, P. Muchlinski. F. Ortino and C. Schreuer (eds.). Oxford University Press, 2008.

C.P. Speed and T.W. Walde, "Will the Energy Charter Treaty Help International Energy Investors?", UNCTAD/ITE/IIT, Vol. 5, No. 3, 1996.

Schreuer, Selected standards of treatment available under the Energy Charter Treaty, at *Investment Protection and the Energy Charter Treaty*, G. Coop and C. Ribeiro Editors, Arbitration Institute of the Stockholm Chamber of Commerce, JurisNetLLC, 2008.

Verburg, "Modernising the Energy Charter Treaty: An Opportunity to Enhance Legal Certainty in Investor-State Dispute Settlement", *Journal of World Investment &Trade*, 20 (2019).

7

Soft Measures for Energy Market Reform in the Western Balkans

Maria Bozoudi

7.1 Introduction

Apart from the fact that it lies at the Eurasian energy crossroads, Southeast Europe has always been viewed as integral to European infrastructure networks. In their 2001 paper on regional transport and energy infrastructure, European Commission services note that "the development of transport and energy networks in the South-East Europe region, connected and compatible with the corresponding European internal networks and those of the neighbouring candidate countries, is an important means of improving links within the region and integrating the countries of the area into the political and economic mainstream of Europe".[1] With regard to energy networks, in particular, the European

[1] European Commission. *Transport and Energy Infrastructure*, 3.

M. Bozoudi (✉)
American College of Thessaloniki, Thessaloniki, Greece
e-mail: mbozoudi@act.edu

M. Mathioulakis (ed.), *Aspects of the Energy Union*,
Energy, Climate and the Environment,
https://doi.org/10.1007/978-3-030-55981-6_7

Commission considered that "the development of energy infrastructures should… ensure security of supply, and the necessary interconnection between the region and the neighbouring systems (for instance that with the European electrical system UCTE). It should also develop interconnections of regional interest within the South-eastern European region".[2] Going beyond infrastructure networks, the 2015 Energy Union Strategy underlines the importance of competition and the proper functioning of the retail market for consumer costs and highlights the need for a market design that offers enough predictability and effective incentives for investment in the energy sector. It also focuses on improving connectivity and increasing energy security.[3] For the past quarter of a century, policy and decision-makers across the region and in Brussels have been challenged to address the need of attracting investment, implementing effective institutional reform of the energy sector in the Balkans, and harmonizing local markets to the European body of law governing the energy sector. These challenges are the result of years of infrastructure neglect during the communist era, as well as of market and institutional distortions created at the time. The impact of Yugoslav wars on physical infrastructure and the painful process which led to the birth of new states in the Western part of the peninsula is to be considered, too. The first comprehensive approach to resolving this puzzle was driven by the Stability Pact for South-eastern Europe,[4] which was launched in 1999 as a joint endeavour and mechanism of the international community for regional stability and security, under the auspices of the Organization for Security and Cooperation in Europe (OSCE), led by the EU. Infrastructure rehabilitation lays at the core of early reconstruction efforts. As seen in Fig. 7.1, more than $12 bn was pledged during

[2]European Commission. *Transport and Energy Infrastructure*, 18.

[3]European Commission. *A Framework Strategy*, 3.

[4]More about the Stability Pact for South-eastern Europe in Busek & Kuehne, *From Stabilization to Integration.*

Pledges (in USD)	1999	2000	2001	Total
Total	6,027bn	6,629bn	6,656bn	19,312bn
Of which				
Reconstruction and economic development	2,681bn	4,683bn	4,853bn	12,199bn
Humanitarian Assistance	1,344bn	405mn	440mn	2,190bn

Fig. 7.1 Pledges for economic assistance to South-Eastern Europe 1999–2001 (*Source* IMF—Finance & Development, vol. 39, no. 1)

the three regional donors' conferences[5] led by the regional High-level Infrastructure Steering Group.[6]

In 2002, under Special Coordinator and former Austrian Vice-Chancellor, Dr Erhard Busek, the Stability Pact took up the task of streamlining processes and initiatives that funds had been pledged for, with the aim of turning them into concrete and practical deliverables for the region—and donors, too. One of the objectives brought forward was strengthening regional energy co-operation in the field of electric power, based on the European Commission's 2001 paper on "Energy and Transport Infrastructure for Southeast Europe", which encouraged the creation of a modern and efficient energy infrastructure, able to meet demand for electricity, oil and gas in a sustainable way. It also called for the establishment of a South-East Europe Regional Electricity Market and pushed for reforms on the national and regional level, placing emphasis on making energy investments on economic grounds. The Stability Pact thus embarked upon an effort to enhance cooperation among countries on energy affairs, resulting in what is today known as the Energy Community. The Energy Community was launched with the Athens Memorandum of 2002, which outlined the principles and institutional necessities for developing a regional electricity market.[7] Proposed by the European Union and signed by Albania, Bosnia and Herzegovina, Bulgaria, Croatia, Greece, FYR of Macedonia,

[5]Detailed reports on project implementation progress and actual fund disbursements by sector are available in the 2001 and 2003 reports of the joint European Commission/World Bank Office for Southeast Europe (Office for Southeast Europe, *Report*, 2001, and *Report*, 2003).

[6]Joint European Commission/World Bank initiative. The role of the Infrastructure Steering Group was to select priority infrastructure projects, also known as "Quick Start" projects.

[7]More in Busek, "Achievements and Challenges", 199–208.

Romania, Serbia,[8] Montenegro and Turkey, it was revised in 2003 to include natural gas and envisioned a South-East Europe Regional Energy Market (SEEREM) as part of the EU's internal energy market. The following year, the EU worked towards turning the "Athens Process" into a legal agreement, and the Treaty establishing the Energy Community was signed in 2005, including UNMIK on behalf of Kosovo. The Treaty entered into force in 2006 and has been extended until 2026. In other words, the Energy Community is an international organization, aiming to expand the EU internal energy market to South-eastern Europe and beyond. Contracting parties have committed to implement the EU energy acquis (though with a lot of exceptions and loopholes), which requires that an appropriate regulatory framework is in place and that local energy markets are liberalized.

7.2 Then and Now

At the time of the Energy Community conception in the early '00s, EU priorities for the energy sector in the region were based on TEN[9]-Energy guidelines drawn in 1996.[10] Considering Southeast Europe's transit role for Eurasian networks, improving infrastructure quality and network connectivity between the region and the EU were critical objectives. Therefore, when it came to energy security and energy efficiency of local markets, efforts—both European and regional—were aligned and directed towards building infrastructure that would connect a limited number of existing large generators to the major consumption centres, gasifying and diversifying the broadly coal-dominated electricity generation mix of the Balkans, and deregulating the sector, with the aim to create a level-playing field that would attract much-needed investment

[8]Serbia and Montenegro at the time.

[9]Trans-European Networks.

[10]European Commission, *Transport and Energy Infrastructure*. As noted in the paper, "the basic principles and overall objectives in the field of electricity and gas networks have been defined by the TEN-Energy guidelines contained in the decision N° 1254/96/CE of the European Parliament and the Council" (p. 19).

and technological know-how. As already mentioned, in 2001, the European Commission was calling for a series of reforms on the national and regional level that included—among others—progressive liberalization, market opening, unbundling of electricity utilities, developing tariff systems in line with costs and improved rate of revenue collection. It was expected that such reforms would lead to enhanced system efficiency and better use of the existing infrastructure, therefore ensuring the long-term viability of the existing Balkan systems and better integration with the Community.

For the electricity sector, in particular, it was considered that the installed power production capacity within the region (including now EU members Bulgaria, Romania and Croatia) was sufficient to meet its demand. Adequate production capacity, however, was well below installed levels, due to war damage, mismanagement and ineffective market regulation. The premise of the Commission's approach at the time was that no new power production capacity would be required in the short and medium-term if appropriate market liberalization measures were taken. Two decades later, the very same priorities that were diligently drawn using the 1996 TEN-Energy guidelines as their basis seem misplaced. They were put forward as catalytic for the region's integration into European energy systems and markets, yet, by 2020, climate change has radically transformed the European approach to energy security and efficiency. As such, the EU now expects of the region policies, actions and results that are in essence contradictory to the original plan. Pursuing a centralized architecture of energy infrastructure is not in line with the priority for decarbonization, which calls for decentralized electricity generation and the introduction of renewable energy resources into consumption patterns. Furthermore, although gasifying the regional electricity generation mix is still relevant, yet, energy infrastructure needs to support more than fossil fuels. Finally, decarbonization is becoming a more and more regulated process, a reality that contradicts calls for more deregulation of the sector. The point is that EU targets for a sustainable Europe by 2030[11] centre around policies that were not decided with Western Balkans realities in mind. As noted in the introduction of the

[11] European Commission, *Toward a Sustainable Europe*.

Country	Per capita CO_2 emissions from fossil fuels (metric tons equivalent)
Poland	8,76
Bosnia-Herzegovina	7,88
EU-28	6,78
Bulgaria	6,34
GLOBAL	4,97
Romania	4,09
North Macedonia	3,87
Albania	1,64

Fig. 7.2 Per capita carbon dioxide emissions from fossil fuels in Central and Eastern European countries, 2018 (*Source* European Commission—JRC)

2019 report of the Energy Community Secretariat, "the clean energy transition, which occupies the rest of Europe and most of the world, has not been initiated by and for the Energy Community".[12] Unlike the EU, in the Western Balkans, green policies and environmental ideas are still peripheral to the political debate. As demonstrated in Fig. 7.2 below, Western Balkan countries continue to rely heavily on fossil fuels for electricity generation, especially relevant to the size of their market, while little—if any—attention is given to pursuing green targets.

Frustration of the Energy Community Secretariat with pursuit of 2030 targets is expressed in the relevant 2019 progress report: "The planned adoption of 2030 targets on energy efficiency, renewable energy and greenhouse gas emission reduction and the incorporation of the Clean Energy Package did not materialize, and not even a serious debate among decision-makers has taken place".[13] Amid a triple transition from a carbonized to a decarbonized economy, from centralized to decentralized production and distribution, and from a fragmented to an integrated regional market, Western Balkan countries have to deal with increasing electricity demand and limited growth of their energy production capacities. According to the International Energy Agency (IEA), Albania's energy production has been unchanged since the 1990s, even though final electricity consumption has grown by 200% since then.[14] Bosnia's

[12]Energy Community Secretariat, *Implementation Report 2018/2019*, 7.

[13]Ibid.

[14]International Energy Agency, *Albania.*

energy production and electricity consumption have only now reached the '90s levels,[15] while Serbia's energy production is more than 28% less than before the Yugoslav wars, and energy supply is still down 20% compared to the same period.[16] North Macedonia is probably the only economy in the region where energy supply grew by 50%, and electricity consumption by 40% vis-à-vis three decades ago.[17]

7.3 Soft Measures and Investment

Overall, regional energy production is restricted, and energy supply is often problematic because of market distortions. The one dimension of the 2001 EU strategy that remains unchanged is the importance of regulatory reform and institutional fundamentals for the regional market to function. In a push to improve energy connectivity and work towards an integrated regional energy market, the six Western Balkan countries (Albania, Bosnia-Herzegovina, North Macedonia, Montenegro, Kosovo, Serbia) committed to implementing a list of "energy soft measures" in 2015.[18] These soft measures can be considered as the energy equivalent of non-tariff barriers to trade, and comprise actions both on a regional and on a national level. Regional measures focus on strengthening cross-regional institutions and cooperation, whereas national measures aim at removing national obstacles to regional electricity market development. Among others, objectives include the elimination of price regulation and subsidies,[19] unbundling transmission and distribution operators and ensuring the independence of national administrative authorities. According to the latest report on the implementation of soft measures for the electricity market,[20] progress is inconsistent on a national level, and uneven more so among the different countries as presented in Fig. 7.3

[15] International Energy Agency, *Bosnia-Herzegovina.*

[16] International Energy Agency, *Serbia.*

[17] International Energy Agency, *North Macedonia.*

[18] Western Balkans Summit, *Final Declaration.*

[19] In 2017 alone, coal subsidies in the Western Balkans were valued at €1,2 bn. (Vujasinovic & Gallop, *Western Balkan Subsidies*).

[20] Energy Community Secretariat, *Energy Monitoring Report.*

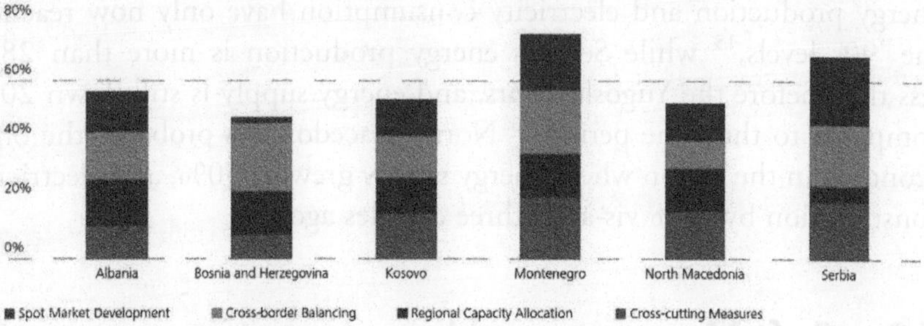

Fig. 7.3 Overall implementation of soft measures in Western Balkan countries (2019) (*Source* Energy Community Secretariat, *WB6 Electricity Monitoring Report*. Vienna [5/2019]: 3)

below. For example, when it comes to spot market development, the process of establishing day-ahead markets continued in Montenegro and North Macedonia in 2018/2019 but stalled in Albania. Bosnia-Herzegovina was identified as the only country to introduce competition among service providers for cross-border balancing, and even though the transmission system operator of Serbia started allocating capacities at its borders with Bulgaria, it still doesn't do so with its other neighbours (i.e. with Bosnia-Herzegovina, Montenegro and North Macedonia).

In effect, inconsistent progress on implementation of soft measures perpetuates market fragmentation, as core legal and institutional fundamentals of an integrated energy market continue to be missing. National energy authorities are neither effectively independent, nor efficient enough to respond to the combined stresses of climate change, increasing energy demand and technological advancements. As a result, coal continues to be subsidized. The sector remains over-regulated, and excessive legislation ends up blocking implementation of meaningful reform. Despite political commitment to align with EU priorities,[21] in the absence of common targets, the level of ambition for transformation is left upon each Contracting Party. In practice, this ambition lies below the European mainstream, and climate planning does not advance fast enough. Soft measures are not only meant to create a functional regional

[21] Energy Community Secretariat, *Energy Community Ministerial Council*.

market harmonized to the European acquis. Framework regulation is key to creating market stability, which can then attract investors. As long as the regional operational environment continues to be incoherent, it will be unattractive for investment. Due to limited investment flows, the regional energy sector remains reliant on ageing conventional infrastructure, and registers muted growth in power generating capacity. Bureaucracy in issuing permits and insufficient availability of information in other than each local language comprise obvious barriers for foreign investors. Obstacles, in particular for wind energy, lead to long project development periods (e.g., Bosnia and Herzegovina, or Serbia) and higher investment risk perception. Even when positive steps have been taken to simplify procedures, such as Serbia's introduction of a one-stop-shop for permits, they are met with doubt by project developers.[22] What is also hampering investment is the lack of a clear policy focus on renewable energy. Coal power remains key to several markets in the whole of Central and Eastern Europe as a source of cheap electricity and electricity security. In the Western Balkans in particular, more than half of total energy consumption relies on locally produced coal and lignite, while the energy mix also includes oil, biomass and waste fuels.[23] Solid fuels—primarily coal and firewood—are used by more than 60% of the Western Balkans populations for heating, and district heating connections exist for only 12% of buildings. Notably lacking from the regional energy scene is natural gas, with close to no non-hydropower renewables capacity additions foreseen until 2028.[24] In the midst of everything, all countries are in the process of revising their respective frameworks for state aid, so as to comply with the new EU guidelines and eventually allow the introduction of effective competition.[25] Still, state financing schemes for the energy sector are not designed with

---

[22] International Renewable Energy Agency (IRENA), *Cost-competitive Renewable Power Generation.* 84.

[23] Sartori, "The energy path of the Western Balkans".

[24] Fitch Solutions, *Europe Power Report*. The region's 6% share of non-hydro-power renewables in total consumption is almost exclusively concentrated in Serbia thanks to the Petroleum Industry of Serbia (NIS), in which the Russian gas giant Gazprom holds a majority stake. Elsewhere in the region, the gasification level is virtually non-existent, as is the level of penetration of non-hydropower renewables.

[25] IRENA, *Cost-competitive Renewable Power Generation.* 85.

renewables in mind. Limited government experience with institutions of a liberalized energy market, combined with the regular adoption of retroactive measures in most of the Western Balkans, create a framework that is not very conducive to investment in the energy sector. In Albania, for example, the renewables support scheme was established without in-depth evaluations of its impact, while the Energy Community Secretariat accuses Bosnia and Herzegovina of breaching state aid rules in granting a guarantee for a Chinese loan to build the Tuzla 7 lignite plant.[26] Kosovo is facing a similar process, as the Secretariat is concerned about the legality of state aid in relation to the Kosova e Re lignite power plant.[27] The importance of these cases lies as much in limited respect for the rule of law, as in showcasing how governments are still pursuing new coal projects without considering the implications—neither on the environment nor on human health. According to the United Nations Environmental Programme, people living in seven Western Balkan cities lose up to 1.3 years of life to air pollution, which causes nearly 5.000 premature deaths in these cities.[28] Air pollution is responsible for at least 15% of premature mortality in North Macedonia (19% in the city of Tetovo alone). The primary sources of particulate matter emissions are thermal power plants that use lignite coal. Public health costs from air pollution from coal-fired power plants in the Western Balkans are estimated at 8.5 EUR billion per year. According to the Energy Community Secretariat's 2019 implementation report, none of the contracting parties which rely on coal in their energy mix are in compliance with

---

[26]Energy Community Secretariat, *Secretariat sends Reasoned Opinion.* "Today, the Secretariat sent a Reasoned Opinion to Bosnia and Herzegovina concerning State aid compliance of a public guarantee granted by the Federation's Ministry of Finance in favor of the Export-Import Bank of China for a loan by the latter to the public utility Elektroprivreda BiH d.d. Sarajevo for the Tuzla 7 project (Case ECS-10/18). The Secretariat is of the opinion that the guarantee for the Tuzla 7 project constitutes State aid in the sense of Article 18(1)(c) of the Treaty, and that the decision by the State Aid Council of Bosnia and Herzegovina coming to the opposite conclusion is wrong". Given that the total sum of the guarantee is not even specified in the Bosnian Government's documents, this guarantee could cost the country dearly.

[27]Energy Community Secretariat. *The Contractual Framework Pertaining to the Kosova e Re Project—The Secretariat's Preliminary Concerns.*

[28]United Nations Environment Programme (UNEP), *Air Pollution and Human Health.*

the emission ceilings that they, themselves, set in their National Emission Reduction Plans (NERPs)[29] with regard to the three pollutants (nitrogen oxides, sulphur dioxide and dust) that are addressed by the Large Combustion Plants Directive (LCPD). In particular, Kosovo failed to meet standards for all three pollutants, while both Bosnia and Herzegovina and North Macedonia exceeded sulphur dioxide and dust limits. In many cases, plants which operate using an "opt-out" clause for their emissions—meaning that emission levels are set by individual directives—may soon have to shut down. One such example is Montenegro's plant in Pljevlja, which will have to be shut down in 2020 unless there is a radical technical upgrade to enable meeting stricter emission limits. Similarly, Bosnia and Herzegovina's Tuzla plant (Block 4) would have to shut down in May 2021, and three plants in Serbia will have exhausted their emission ceiling by December 2023.[30]

In addition to the lack of sophisticated regulations governing procurement and investment in the sector, access to the grid, grid connections and grid capacities to integrate energy from different sources comprise additional challenges. The overall ability of the grid to integrate large renewable energy projects, such as large-scale solar PV and wind energy, often poses limitations (e.g. Kosovo, Croatia). Transmission System Operators (TSOs) tend to be restrictive on integrating renewable energy into the grid, as there is insufficient experience with non-hydro technologies. According to the International Renewable Energy Agency (IRENA),[31] renewable energy producers in Bulgaria are subject to rules limiting generation to specific times of the day and setting an upper limit on the amount of electricity eligible for feed-in support. Taking all these factors into account, it is understandable that financing institutions shy away from projects in the region, making access to capital difficult and expensive. A series of energy-related scandals across all Western Balkan countries is a direct result of ineffective regulation and only adds to investor and financier hesitation.[32] As in all sectors, such incidences

[29]Serbia doesn't even have such a plan in place yet!

[30]Vujasinovic & Gallop, *Western Balkan Subsidies.*

[31]IRENA, *Cost-competitive Renewable Power Generation*, 89.

[32]Just between 2013–2016, there was a case of money laundering surrounding a hydro concession in Albania, another involving energy trading in Bosnia-Herzegovina, and a similar one

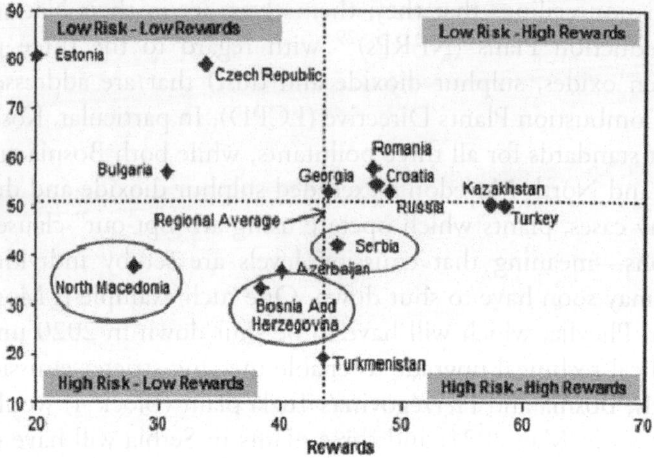

Fig. 7.4 Fitch Solutions Power Sector Risk/Rewards, select transition countries (2020) (*Source* Fitch Solutions, *Europe Power Report Q1 2020*. London [January 2020]: 7)

drive reputable investors away. The region appears unwilling to transition towards renewable energy and away from ageing conventional power generation assets—primarily coal and gas-fired power—which weighs down on power sector opportunities. Lack of market reform results in muted growth in power generating capacity and leads to failure to attract much-needed investment. According to Fitch, in a report whose results are presented in Fig. 7.4 below, countries in the Western Balkans are considered as high-risk and mostly low-reward destinations for investment in the power sector.

7.4 Conclusions

After years of conflict and socio-economic decline during the '90s, Western Balkan countries chose membership to the European Union

on energy trading and bribing in Croatia. There were political campaign financing violations surrounding KEK in Kosovo, an electricity trading cartel was reported in North Macedonia, an electricity theft scandal shook Montenegro, and a mining scandal Serbia.

and Euro-Atlantic institutions as a response to both old and new challenges. Most of them are rooted in past decades of central planning and communist rule, which had resulted in obsolete infrastructure networks, ineffective governance and management systems, as well as social divides running along geographic, ethnic and economic lines. In other words, since the mid-'90s, Western Balkans have undergone a complicated post-communist and post-conflict socio-economic and political transition in parallel with a state-building process. We are witnesses to this ongoing transformation, including all successful, not-so-successful and uncertain steps.

Building the missing physical links with European Union infrastructure networks was identified early on as a high priority by both EU and regional policy and decision-makers. For the biggest part of the '00s, under the Stability Pact for South eastern Europe, the European Union led concerted efforts to promote economic revival, social peace, solid institutional capacities and the rule of law in the region. But, while Western Balkan countries were juggling with fluid domestic socio-political balances and geopolitical realities, the development paradigm they were trying to adjust into started to shift. Southeast Europe became once again caught up in traditional balance-of-power politics and competition—this time including the West, Russia, Turkey and China. The impact from the global financial crisis deprived the region of potential financial resources, giving rise to extremism and populism, while Europe and its institutions became more self-involved. The 2015 refugee crisis only added fuel to the domestic turbulence. In the midst of these developments, the urgency of responding to climate change became a conscious reality for the West and the whole globe, causing a drastic reconsideration of energy resources, energy infrastructure architecture and energy use by end consumers. For these—among many—reasons, progress towards the European acquis and towards meeting modern infrastructure standards has been uneven—also—in the energy sector.

Energy projects in the Western Balkans have been mostly small and medium scale. Still, they have often been delayed due to a mixture of technical difficulties, political preferences for grandiose—instead of marketable—projects, legal complexities, lack of local technical know-how, or financing difficulties. Early attempts to develop a sensible legal frame for public-private partnership schemes (PPPs) were encouraging (eg Serbia, 2003), and driven rigorously by the World Bank, international financial institutions (IFIs) and the OECD.[33] However, it took at least until the establishment of the European Energy Community before it was possible for Western Balkan countries to effectively implement any combination of Build-Own-Operate-Transfer projects. Involvement of IFIs, like the European Bank for Reconstruction and Development or the European Investment Bank, has indeed been catalytic in moving forward with investment in energy infrastructure, and despite all difficulties, infrastructure development has progressed. Yet, building a functional, cohesive and transparent institutional framework for the regional energy market to function has admittedly lagged. Complex articulation of EU priorities on energy security, energy efficiency, energy diversification and, lately, green energy, has not facilitated local efforts, either. As acknowledged by the Energy Community Secretariat, there has been non-uniform progress on institutional fundamentals across the Western Balkans.

In his address to COWEB in 1999, then German Minister of Foreign Affairs, and architect of the Stability Pact for South-eastern Europe, Joschka Fischer, noted that the European Union's response to the Yugoslav crisis had been flawed for a long time, because it had considered Balkan problems as different from those of Europe.[34] As geopolitical competition in the region centres more and more on energy, his message is just as relevant for EU policy makers today as it was back then. Since 2001, the Western Balkans have not taken centre stage in the EU's energy strategy, allowing for the space and distance between the EU and WB-6 to grow. Combined with increasing non-EU influence via cheap loans for coal projects, much-needed private sector involvement for more green energy in the region is discouraged.

---

[33]The OECD's Investment Compact for Southeast Europe (eventually transformed into the organization's current SEE Programme, focusing only on WB-6 countries) had a pivotal role in building public administration capacities for PPPs between 2000–2005.

[34]German Foreign Ministry, Mitteilung für die Presse, No. 1041/99, 11 June 1999. *Speech of the German Foreign Minister, Joschka Fischer, at the Conference of the Foreign Ministers concerning the Stability Pact for South Eastern Europe, Cologne, 10 June 1999.*

In 2018, Western Balkan governments reiterated their commitment to follow the EU's energy and climate footsteps, by setting 2030 targets mirroring those of the EU. The objective was to reduce greenhouse gas emissions, improve energy efficiency and increase the share of renewable sources in their energy mix.[35] A clear roadmap and process to achieve these targets is a *sine qua non* precondition for their meaningful, effective and—eventually—successful pursuit. Climate change makes the need for such roadmaps even more pressing. Yet, as long as the Energy Community is deprived of enforcement mechanisms and monitoring tools to observe progress on soft measures governing regional energy markets, Western Balkan administrations have little incentive to avoid procrastination. The slower Western Balkans move away from coal in the present, the higher the cost of compliance and transition for society in the future. The smaller the steps towards regional energy markets integration, the longer the road towards environmental sustainability, economic growth and membership to the EU. Disregarding or failing to implement soft measures only complicates an already complicated technical, institutional, infrastructural and technological transition for Western Balkan energy markets. It further discourages investors, raises costs, wastes resources, creates market distortions in favour of vested interests, and hurts public trust in the rule of law, democracy, and the European Union.

References

Busek, Erhard. "The Stability Pact for South Eastern Europe: Achievements and Future Challenges". *OSCE Yearbook 2002* (2003): 199–208.

Busek, Erhard. "Assessing the Western Balkans' EU Membership Hopes", in *Europe's World.* Spring 2009.

Busek, Erhard & Björn Kühne (eds.). *From Stabilisation to Integration: The Stability Pact for South Eastern Europe. Vol. 1: Articles, Vol 2: Documents.* Weimar: Boelau Verlag Wien—Koeln, 2010.

Casey, Zoey (ed.). *Chronic Coal Pollution: EU Action in the Western Balkans Will Improve Health and Economies Across Europe.* Brussels: HEAL, CAN Europe,

[35]Energy Community Secretariat. *16th Energy Community Ministerial Council.*

Sandbag and CEE Bankwatch Network, and Europe Beyond Coal. February 2019.

Energy Community Secretariat. *The Contractual Framework Pertaining to the Kosova e Re Project—The Secretariat's Preliminary Concerns.* Vienna, 14 June 2018.

Energy Community Secretariat. *Energy Community Implementation Report 2018/2019.* Vienna, 1 November 2019.

Energy Community Secretariat. *16th Energy Community Ministerial Council Adopts Key Energy Infrastructure Projects, Sets Direction for Adopting 2030 Energy and Climate Targets.* Vienna, 29 November 2018.

Energy Community Secretariat. *Analysis of Direct and Selected Indirect Subsidies to Coal Electricity Production in the Energy Community Contracting Parties.* Vienna, March 2019.

Energy Community Secretariat. *Secretariat sends Reasoned Opinion regarding state guarantee for the Tuzla 7 project in Bosnia and Herzegovina.* Vienna, 15 January 2020.

European Commission. *Transport and Energy Infrastructure in Southeastern Europe*. Brussels, 15 October 2001.

European Commission. *A Framework Strategy for a Resilient Energy Union with a Forward-Looking Climate Change Policy*. Brussels, 2015.

European Commission. *Reflection Paper: Towards a sustainable Europe by 2030.* Brussels, 30 January 2019.

Figure 1. Demekas, Dimitri G., Johannes Herderschee, James McHugh, Saumya Mitra. "Southeastern Europe After the Kosovo Crisis", in *Finance & Development*, March 2002, vol. 39, no. 1.

Figure 2. Muntean, M., D. Guizzardi, E. Schaaf, M. Crippa, E. Solazzo, J.G.J. Olivier, & E. Vignati. *Fossil CO2 Emissions of all World Countries—2018 Report*, EUR 29433 EN, Publications Office of the European Union: Luxembourg, 2018, ISBN 978-92-79-97240-9, https://doi.org/10.2760/30158, JRC113738.

Figure 3. Energy Community Secretariat. *WB6 Electricity Monitoring Report.* Vienna, May 2019.

Figure 4. Fitch Solutions. *Europe Power Report, Q1 2020*. London, January 2020.

Friis, Lykke & Anna Murphy. *Negotiating in a Time of Crisis: The EU's Response to the Military Conflict in Kosovo.* EUI Working Papers RSC 2000/20. Rome: European University Institute, 2000.

German Foreign Ministry. *Speech of the German Foreign Minister, Joschka Fischer, at the Conference of the Foreign Ministers concerning the Stability Pact for South Eastern Europe*, Cologne, 10 June 1999—Mitteilung für die Presse, No. 1041/99, 11 June 1999.

Hancher, Leigh & Antonis Metaxas. *EU Energy Law and Policy: Meeting the Challenges of a Low Carbon Economy*. Portland: Claeys and Casteels, 2018.

International Energy Agency. *Key Energy Statistics Database 2018—Albania*. Paris, 2019.

International Energy Agency. *Key Energy Statistics Database 2018—Bosnia-Herzegovina*. Paris, 2019.

International Energy Agency. *Key Energy Statistics Database 2018—Serbia*. Paris, 2019.

International Energy Agency. *Key Energy Statistics Database 2018—North Macedonia*. Paris, 2019.

International Renewable Energy Agency (IRENA). *Cost-competitive Renewable Power Generation: Potential Across South-East Europe*. Bonn, January 2017.

Office for Southeast Europe. *Report on Activities of the European Commission/World Bank Office for South-East Europe 2001*. Brussels, 2001.

Office for Southeast Europe. *Report on Activities of the European Commission/World Bank Office for South-East Europe 2003*. Brussels, 2003.

Regional Table of the Stability Pact for Southeastern Europe. *Decision of the Meeting and Work Plan*, Brussels, 9/1999.

Regional Table of the Stability Pact for Southeastern Europe. *Agenda for Stability*. Thessaloniki, 8 June 2000.

Regional Table of the Stability Pact for Southeastern Europe. *Extraordinary Meeting Conclusions*. Bucharest, 5 June 2001.

Regional Table of the Stability Pact for Southeastern Europe. *Meeting Conclusions*. Brussels, 27–28 June 2001.

Regional Table of the Stability Pact for Southeastern Europe. *Strategic Framework and Working Table Priorities*. Bucharest, 25–26 October 2001.

Regional Table of the Stability Pact for Southeastern Europe. *Chairman's Conclusions—Setting Priorities for 2003 and beyond*. Thessaloniki, 16 December 2002.

Sartori, Nicolo. "The Energy Path of the Western Balkans", *About Energy*. 13 March 2018 (available online at https://www.aboutenergy.com/en_IT/topics/energy-path-of-western-balkans.shtml, last accessed on 19 January 2020).

United Nations Environment Programme—UNEP. *Air Pollution and Human Health: The Case of the Western Balkans*. Nairobi, 4 June 2019 (available online at https://www.unenvironment.org/news-and-stories/press-release/air-pollution-responsible-one-five-premature-deaths-19-western, last accessed on 19 January 2019).

Vujasinovic, Stevan & Pippa Gallop. "Energy Community: Western Balkan coal subsidies worth over 1.2 billion euros in 2017". *Climate Action Network Europe*: 25 March 2019—(available online at http://www.caneurope.org/publications/press-releases/1747-energy-community-western-balkan-countries-subsidised-coal-with-half-a-billion-euros-in-just-three-years, last accessed on 19 January 2019).

Western Balkans Summit. *Final Declaration by the Chair of the Vienna Western Balkans Summit.* Vienna, 27 August 2015.

8

The Role of the Energy Charter Process in Accelerating the Energy Transition and Ensuring Energy Security in South East Europe in Line with the Energy Union

Gokce Mete Mete and Janessa Goh Pei-Ru

8.1 Introduction

Globally, there is a strong momentum for thinking more innovatively about the pathways for accelerating the production and consumption of low-carbon energy sources (the energy transition). As energy is responsible for two-thirds of global greenhouse gas (GHG) emissions,[1] the energy transition is plays a critical role in the emission reduction efforts of countries to achieve the goals of the 2015 Paris Agreement to limit global temperature rise to well below 2 degrees Celsius

[1] International Energy Agency (IEA), 'Climate Change' (undated) available at https://www.iea.org/topics/climatechange/.

G. M. Mete (✉)
International Energy Charter, Brussels, Belgium
e-mail: gokce.mete@encharter.org

J. G. Pei-Ru
Ministry of Trade and Industry, Singapore, Singapore

M. Mathioulakis (ed.), *Aspects of the Energy Union*,
Energy, Climate and the Environment,
https://doi.org/10.1007/978-3-030-55981-6_8

above pre-industrial levels and to pursue efforts to further limit its rise to 1.5 degrees Celsius.[2] Furthermore, the energy transition supports the achievement of the Sustainable Development Goals (SDG).[3] Aside from SDG 7 of ensuring access to affordable, reliable, sustainable, and modern energy for all, energy is intertwined with other SDGs as it is connected with climate change, food production, increased economic and job opportunities, and access to energy for all.[4] In addition to reducing GHG emissions and facilitating sustainable development, the actions taken by countries in pursuit of the energy transition play a part in improving energy security,[5] which has been defined as reliable and affordable access to all fuels and energy sources.[6] In particular, an increased share of renewables in the energy mix can facilitate the diversification of energy sources, reduce import dependency, and increase decentralisation (such as through solar installations that enable local energy generation).[7] That being said, it is important to keep in mind that the contribution of the energy transition to the enhancement of energy security is tampered by new energy security challenges arising from an increased reliance on renewables, which include cyber security, grid security, and system integration.[8] With energy production and consumption at the heart of modern economies, the challenge for policymakers is to find a balance between the "trilemma" of economic development, energy security, and environmental protection. In July 2014, at the outset of the presidency of the former President of the European Commission, Jean-Claude Juncker, the need for a resilient energy union with a forward-looking climate change policy was identified as a priority of the

[2] Paris Agreement (adopted 5 October 2016, entered into force 4 November 2016) UNTS 54113.

[3] United Nations General Assembly, 'Transforming Our World the 2030 Agenda for Sustainable Development' (21 October 2015) A/RES/70/1.

[4] United Nations, 'Ensure Access to Affordable, Reliable, Sustainable and Modern Energy' (undated) available at https://www.un.org/sustainabledevelopment/energy/.

[5] World Energy Council (WEC), 'World Energy Trilemma Index 2019' (2019) available at https://www.worldenergy.org/assets/downloads/WETrilemma_2019_Full_Report_v4_pages.pdf.

[6] IEA, 'Energy Security' (undated) available at https://www.iea.org/topics/energy-security.

[7] WEC (n 5).

[8] Ibid.

Commission.[9] This priority emerged against the context of the Russian invasion of the Crimea in February 2014 and concerns about heavy reliance on Russian energy by the EU.[10] In 2015, the European Commission presented the Energy Union framework strategy,[11] which was slated to create a new momentum for the transition to a low-carbon, secure and competitive economy.[12] The goal of the Energy Union is to give EU consumers secure, sustainable, competitive and affordable energy,[13] and the Energy Union strategy comprises five dimensions[14]:

- Security, solidarity, and trust:
  Energy sources, suppliers and routes must be diversified and there should be solidarity and trust between Member States that understand their interdependence on the delivery of secure energy.
- A fully integrated internal energy market:
  Energy should flow freely across the EU based on competition and without technical or regulatory barriers and energy islands should be connected.
- Energy efficiency:
  Energy efficiency in the building and transport sectors should be improved to cut emissions, reduce the dependence of the EU on

---

[9]Jean-Claude Juncker, 'A New Start for Europe: My Agenda for Jobs, Growth, Fairness and Democratic Change. Political Guidelines for the Next European Commission' (15 July 2014) available at https://www.eesc.europa.eu/resources/docs/jean-claude-juncker---political-guidelines.pdf.

[10]Donald Tusk, 'Roadmap Towards an Energy Union for Europe' (submitted to the European Commission on 10 April 2014).

[11]European Commission, 'Communication from the Commission to the European Parliament, the Council, the European Economic and Social Committee, the Committee of the Regions and the European Investment Bank, "A Framework Strategy for a Resilient Energy Union with a Forward-Looking Climate Change Policy"'(25 February 2015) available at http://ec.europa.eu/transparency/regdoc/rep/1/2015/EN/1-2015-80-EN-F1-1.PDF (hereinafter 'Energy Union strategy').

[12]European Commission, 'Communication from the Commission, "State of the Energy Union 2015"' (18 November 2015) available at https://eur-lex.europa.eu/legal-content/EN/TXT/?qid=1449767367230&uri=CELEX:52015DC0572.

[13]Energy Union strategy (n 11).

[14]Ibid.; European Commission, 'Energy Union and Climate' available at https://ec.europa.eu/commission/priorities/energy-union-and-climate_en.

energy imports, and drive jobs and growth by leading in energy efficiency technology.

– Climate action—decarbonising the economy:
  This is achieved through an ambitious EU climate policy in furtherance of its commitment to the Paris Agreement and an energy policy to make the EU the number one in renewable energy.
– Research, innovation and competitiveness:
  Supporting research and innovation in low-carbon and clean energy technologies to progress the energy transition and boost competitiveness.

South East European countries, including those still outside the EU, have progressively converged their energy and economic policies with EU strategy through regional efforts.[15] Such efforts are most clearly embodied by the Energy Community, which is an international organisation founded in 2005 for the purpose of extending the internal energy market of the EU to South East Europe and the Black Sea region.[16] It comprises the EU, Albania, Bosnia and Herzegovina, Georgia, the Republic of North Macedonia, and Kosovo,[17] and its Contracting Parties are Moldova, Montenegro, Serbia, and Ukraine.[18] In respect of other South East European countries, countries like Bulgaria, Croatia, Greece, and Romania are both EU Member States and part of the Energy Union. It has been observed however, that there is a distinct divergence between the energy strategies of the EU and South East European countries, which are highly dependent on hydrocarbons (particularly coal), and face a difficult and uncertain path towards the energy transition, due in part to repeated policy failures and electricity grid constraints with respect

[15]International Renewable Energy Agency, 'Europe' (undated) available at https://www.irena.org/europe; OECD, 'Competitiveness in South East Europe: A Policy Outlook 2018' (24 April 2018) available at https://read.oecd-ilibrary.org/development/competitiveness-in-south-east-europe_9789264298576-en#page470 .

[16]European Commission, 'Energy Community' (undated) available at https://ec.europa.eu/energy/en/topics/international-cooperation/international-organisations/energy-community.

[17]In line with UNSCR 1244 and the ICJ Opinion on the Kosovo declaration of independence.

[18]European Commission (n 16).

to renewables.[19] The mobilisation of investment in energy infrastructure, renewables, and energy efficiency plays a key role in facilitating the energy transition and improving energy security in South East Europe.[20] In that respect, almost all South East European countries (with the exception of Serbia, which has observer status and not affiliated Kosovo) are Contracting Parties to the Energy Charter Treaty (ECT),[21] which is a multilateral investment treaty that was signed in 1994 and entered into effect in 1998.[22] The ECT counts the EU and its member states (minus Italy[23]) among its constituency,[24] and has been identified by the EU as a framework for cross-border cooperation between the EU and third countries.[25] The Euro-centric nature of the ECT constituency is founded in the origins of the broader Energy Charter process, which was initiated through the 1991 European Energy Charter—a political declaration on establishing energy cooperation among the states of Eurasia.[26] The governing body for the Energy Charter process is the Energy Charter Conference, which has, since 2010, been working on the modernisation of the ECT.[27] These efforts have resulted in the adoption of the International Energy Charter in 2015—a declaration of political intention that updated the 1991 Energy Charter and expanded the Energy Charter

---

[19] Institute of Energy for SE Europe, 'South East Europe Energy Outlook 2016/17' (May 2017).

[20] OECD (n 15).

[21] The Energy Charter Treaty and Related Documents, 17 December 1994, UNTS 2080, 100 34 I.L.M. 360 (ECT).

[22] International Energy Charter (IEC), 'The Energy Charter Treaty' (updated on 18 February 2019) available at https://www.energycharter.org/process/energy-charter-treaty-1994/energy-charter-treaty.

[23] On 31 December 2014 Italy notified the Depository of its withdrawal from the Energy Charter Treaty (ECT). Under Article 47, paragraph 2 of the ECT, any such withdrawal shall take effect upon the expiry of one year after the date of the receipt of the notification by the Depositary. Therefore, the withdrawal from the ECT by the Italian Republic took effect on the 1st January 2016.

[24] Ibid.

[25] 'European Energy Charter' available at https://eur-lex.europa.eu/legal-content/EN/TXT/?uri=LEGISSUM:l27028.

[26] IEC, 'The European Energy Charter' (2 June 2015) available at https://www.energycharter.org/process/european-energy-charter-1991.

[27] Decision of the Energy Charter Conference, 'Road Map for the Modernisation of the Energy Charter Process' (24 November 2010) available at https://www.energycharter.org/fileadmin/DocumentsMedia/CCDECS/CCDEC201010.pdf.

process to more than 90 signatory states. It has also led to the creation of various tools and soft law instruments, such as the Conflict Resolution Centre[28] and Model Instrument on Investment Dispute Management,[29] which assist signatory states and the industry in avoiding disputes. The potential revision of the text of the ECT however,[30] is still in the process of negotiation.[31] As it stands, the core of the ECT focuses on (i) the protection of foreign energy investments; (ii) the facilitation of energy trade and transit; (iii) providing multiple dispute resolution mechanisms for energy investments; and (iv) promoting energy efficiency and minimising the environmental impact of energy production and use. This chapter will therefore seek to address how the ECT can accelerate the energy transition and ensure energy security in South East Europe, in furtherance of the Energy Union. The structure of this chapter is as follows. Part one first illustrates the commitment of the EU to the energy transition via the Energy Union, followed by the role that the ECT can play in supporting the energy transition, and lastly, a discussion on how the ECT can accelerate the energy transition in South East Europe to support its convergence with EU law and policy. Part two first sets out the energy security concerns of the EU and its Energy Union strategy. It then illustrates how the ECT facilitates secure energy flows among its constituency. Lastly, it discusses how the ECT can enhance cooperation on energy security between South East Europe and the EU. In part three, concluding remarks are made.

---

28 IEC, 'Conflict Resolution Centre' (updated on 16 April 2019) available at https://www.energycharter.org/what-we-do/dispute-settlement/conflict-resolution-centre.

29 Decision of the Energy Charter Conference, 'Adoption by Correspondence—Model Instrument on Management of Investment Disputes' (23 December 2018) available at https://www.energycharter.org/fileadmin/DocumentsMedia/CCDECS/2018/CCDEC201826_-_INV_Adoption_by_correspondence_-_Model_Instrument_on_Management_of_Investment_Disputes.

30 Decision of the Energy Charter Conference, 'Modernisation of the Energy Charter Treaty' (28 November 2017) available at https://www.energycharter.org/fileadmin/DocumentsMedia/CCDECS/2017/CCDEC201723.pdf.

31 Decision of the Energy Charter Conference, 'Modernisation of the Energy Charter, Treaty: Mandate, Procedural Issues and Timeline for Negotiations' (6 November 2019) available at https://www.energycharter.org/fileadmin/DocumentsMedia/CCDECS/2019/CCDEC201910.pdf.

8.2 Energy Transition

8.2.1 The Commitment of the EU to the Energy Transition via the Energy Union Strategy

Following the Energy Union framework strategy presented by the European Commission in 2015, the European Commission presented the Clean energy for all Europeans package (CEP)[32] in 2016, which marked a significant step towards the implementation of the Energy Union strategy.[33] The CEP consists of eight legislative acts on areas such as energy performance in buildings, renewable energy, and energy efficiency.[34] It builds on the EU's progressive targets towards the energy transition and revises relevant legislation to elevate the EU's ambition and commitment towards greening the EU economy. In line with the Energy Union strategy, the EU pursues an ambitious climate policy by setting itself targets for the years 2020, 2030, and 2050 to reduce the GHG emissions of its Member States. The 2020 package contains binding legislation to meet a 20% reduction in GHG emissions from 1990 levels; to obtain 20% of the EU energy mix from renewables; and to improve energy efficiency by 20%. These targets are implemented through directives and regulations such as the Renewable Energy Directive[35] and the Energy Efficiency Directive,[36] both of which have been revised in 2019 with an upward ambition of a 32% share of renewables in the energy mix and a 32.5% improvement in energy efficiency by 2030. The 2050 targets, under the newly established European Green Deal agenda, aim for net-zero GHG emissions by 2050, in line with the

[32] European Commission, 'Clean Energy for All Europeans' (26 July 2019) available at https://op.europa.eu/en/publication-detail/-/publication/b4e46873-7528-11e9-9f05-01aa75ed71a1/language-en?WT.mc_id=Searchresult&WT.ria_c=null&WT.ria_f=3608&WT.ria_ev=search.

[33] Ibid.

[34] Ibid.

[35] Directive 2009/28/EC of the European Parliament and of the Council of 23 April 2009 on the promotion of the use of energy from renewable sources and amending and subsequently repealing Directives 2001/77/EC and 2003/30/EC, OJ L 140.

[36] Directive 2012/27/EU of the European Parliament and of the Council of 25 October 2012 on energy efficiency, amending Directives 2009/125/EC and 2010/30/EU and repealing Directives 2004/8/EC and 2006/32/EC, OJ L 315.

Paris Agreement.[37] As part of the European Green Deal, the Commission has also stated its intention to propose the first European "Climate Law" by March 2020, thus enshrining the 2050 net-zero GHG emission objective in legislation.[38] The EU has also recognised the importance of sustainable investment in driving the energy transition and encourages investments in "green" products. In 2019, the EU co-legislators agreed on the generation of low-carbon and positive-carbon-impact benchmarks to create a unified classification system on what can be deemed an environmentally sustainable economic activity (taxonomy).[39] This aims to regulate disclosure obligations for investors on environmental, social, and governance factors as well as to better discern the carbon footprint of investments.[40] This therefore facilitates investments in "green" products. Such investments are crucial as it is estimated that additional investments in the amount of EUR142–199 billion a year between 2030 and 2050 are required to reach the EU's 2050 goal of a net-zero GHG emissions.[41] As investments from the private sector account for more than 75% of EU investments in clean energy research and innovation, the engagement of the private sector is crucial. Progress towards the implementation of the stated aims of the Energy Union and binding obligations under the CEP are tracked under the State of the Energy Union reports[42] and through the 10-year National Energy and Climate Plans for the period covering 2021–2030, which outlines how Member States will meet the new 2030 targets for renewable energy and for energy efficiency.[43] Energy

[37] European Commission, 'Communication from the Commission to the European Parliament, the European Council, the Council, the European Economic and Social Committee and the Committee of the Regions: The European Green Deal' COM(2019) 640 final (11 December 2019).

[38] Ibid.

[39] Fourth Report, p. 22.

[40] Ibid.

[41] Fourth Report, p. 21.

[42] https://ec.europa.eu/commission/priorities/energy-union-and-climate_en.

[43] Regulation (EU) 2018/1999 of the European Parliament and of the Council of 11 December 2018 on the Governance of the Energy Union and Climate Action, amending Regulations (EC) No 663/2009 and (EC) No 715/2009 of the European Parliament and of the Council, Directives 94/22/EC, 98/70/EC, 2009/31/EC, 2009/73/EC, 2010/31/EU, 2012/27/EU and 2013/30/EU of the European Parliament and of the Council, Council Directives 2009/119/EC

Efficiency is at the forefront of these monitoring reports and the subsequent recommendations of the Commission. For instance, the "Fourth Report on the State of the Energy Union"[44] noted that post-2014, energy consumption began to increase and was, at the time of the report, slightly above the linear trajectory for the 2020 targets.[45] In response, the Commission established a Task Force with Member States to exploit energy efficiency potentials so that increased consumption would not result in a failure to meet the 2020 targets.[46] Hence, this section has illustrated the commitment of the EU to the energy transition via the setting of targets to reduce emissions and increase the share of renewables in its energy mix, as part of its Energy Union strategy. The EU has also taken concrete steps to realise these targets through the CEP and facilitating investments in "green" products, among others. Further, the EU has shown its commitment to the energy transition by continuously tracking the progress of its objectives under the Energy Union through its State of the Energy Union reports and 10-year implementation plans for Member States.

8.2.2 The Role of the ECT in Supporting the Energy Transition

One of the main goals of the ECT is to limit the environmental impacts of energy production and use. The preamble recognises the "increasingly urgent need for measures to protect the environment".[47] The "Agreements of the United Nations Climate Change Conference reached in Cancun on 11 December 2010" are to be considered in the ECT.[48] These sentiments are reflected in the text of the ECT, which

and (EU) 2015/652 and repealing Regulation (EU) No 525/2013 of the European Parliament and of the Council, PE/55/2018/REV/1, OJ L 328.

[44]Fourth Report on the State of the Energy Union (Fourth Report), p. 6 https://ec.europa.eu/commission/sites/beta-political/files/fourth-report-state-of-energy-union-april2019_en_0.pdf.

[45]Fourth Report, p 4.

[46]Ibid.

[47]Such as the United Nations Framework Convention on Climate Change and the Convention on Long-Range Transboundary Air Pollution and its protocols.

[48]Such reference is found by reading the Preamble in conjunction with the Annex.

commits parties "to strive to minimise in an economically efficient manner harmful Environmental Impacts..." in accordance with commitments under international environmental agreements and in pursuit of sustainable development.[49] The ECT's objectives on "Energy efficiency and environmental protection" also imply the promotion of a sustainable energy mix through the use of renewable energy sources and clean technologies, among others. The broader Energy Charter process also facilitates the protection of low-carbon energy investments through soft law instruments such as the In-Depth Review of the Energy Efficiency (IDEER), knowledge-sharing (including country reports), and capacity-building programmes. In 2019, all Energy Charter Secretariat capacity-building programmes and executive training, symposium, and forums focused on the energy transition as well as electrification and regional electricity supply security including cross-border power pools. Further, in the June 2019 meeting of the Industry Advisory Panel (IAP) (the expert consultative body to the Energy Charter Conference), the main theme was "Investment for a low-carbon energy sector".[50] There, one of the two thematic sessions concerned new rules for the low-carbon energy industry and carbon capture utilisation and storage. Indeed, one of the roles of the IAP is to provide policy advice and expertise to the Energy Charter Conference on the promotion of low-carbon investment. The dispute settlement system under the ECT can also promote and protect low-carbon energy investments. Numerous investors have relied on the ECT to bring arbitration proceedings against a number of EU Member States concerning their renewable energy investments in those countries. The ECT therefore plays a role in promoting renewable energy investments as it protects investors against a breach of fair and equitable treatment standards,[51] expropriation,[52] and/or emergency situations.[53] In order to qualify as a protected investment under the ECT,

---

49 Article 19 ECT.

50 Who we Are/Industry Advisory Panel/Meetings/Industry Advisory Panel (4 June 2019) https://www.energycharter.org/who-we-are/industry-advisory-panel/meetings/industry-advisory-panel-4-june-2019/.

51 Article 10 ECT.

52 Article 13 ECT.

53 Article 12 ECT.

an investment must be "associated with an Economic Activity".[54] In order to constitute an Economic Activity, the investment must involve energy materials and products, such as coal, natural gas, petroleum and petroleum products, and electrical energy.[55] The investment must also "concern" the exploration, extraction, refining, production, storage, land transport, transmission, distribution, trade, marketing, sale of Energy Materials and Products, or the distribution of heat to multiple premises.[56] The Annex further includes several illustrations of Economic Activity, such as the construction and operation of power generation facilities such as wind farms and the removal and disposal of wastes from power stations. As the ECT was signed in 1994 when the energy sector was primarily focused on fossil fuels, it does not expressly protect investments in low-carbon technologies and energies, such as liquefied natural gas (LNG), distributed energy resources, demand-side response, carbon, capture and storage, and hydrogen. It is therefore unclear whether investments concerning these energy sources, carriers and CO_2 pipelines and storage are protected under the ECT. That being said, such investments may eventually be covered under the ECT as the Energy Charter Conference has emphasised the promotion of low-carbon investments in its ongoing modernisation efforts.[57]

Similar to the EU, the ECT also emphasises the central role of energy efficiency. The Protocol on Energy Efficiency and Related Environmental Aspects (PEERA) was the first set of international policy principles designed to promote energy efficiency. Pursuant to Article 1(2) of the PEERA, its objectives are "(a) the promotion of energy efficiency policies consistent with sustainable development; (b) the creation of framework conditions which induce producers and consumers to use energy as economically, efficiently and environmentally soundly as possible, particularly through the organisation of efficient energy markets and a fuller

[54]Article 1(6) ECT. Also note that the Treaty investment protection regime also covers certain investments designated by Contracting Parties that are notified to the Secretariat.

[55]And not those listed in Annex NI which covers among others, fuel wood and wood charcoal.

[56]Article 1(5) ECT.

[57]Decision of the Energy Charter Conference, 'Road Map for the Modernisation of the Energy Charter Process of 24th November 2010 Concerning Road Map for the Modernisation of the Energy Charter Process' (24 November 2010) https://www.energycharter.org/fileadmin/DocumentsMedia/CCDECS/CCDEC201010.pdf.

reflection of environmental costs and benefits; and (c) the fostering of cooperation in the field of energy efficiency". PEERA focuses mainly on policies that work well in specific country situations, how to better enforce these policies, and how to create market conditions to scale up investments in energy efficiency. It is also notable that the ECT, like the EU, promotes energy research and innovation, and that the ECT obliges Contracting Parties to eliminate existing obstacles and abstain from creating new obstacles to the transfer of technology concerning Energy Materials and Products with a view to promote access to and transfer of energy technology.[58] Accordingly, this section has shown that the ECT can play a role in accelerating the energy transition. This is because the ECT facilitates the protection of low-carbon energy investments through its dispute settlement provisions, soft law instruments, knowledge-sharing, and capacity-building programmes. The ECT also contains a dedicated set of international policy principles to promote energy efficiency. Further, although it does not expressly protect investments in low-carbon technologies and energies that are relevant for the energy transition, its recognition of environmental protection and the direction of the ongoing modernisation efforts leave the door open for the potential protection of such investments under the ECT.

8.2.3 Commentary on South East Europe

Countries like Albania, Bosnia and Herzegovina, the Former Yugoslav Republic of Macedonia, Kosovo, Montenegro, and Serbia have "practically non-existent" installed capacity of wind and solar power technologies despite their substantial potential.[59] In June 2019, Director Janez Kopač of the Energy Community Secretariat pointed out that direct subsidies for the coal sector of Energy Community Contracting Parties amounted to EUR1.2 billion during 2015–2017 and hidden subsidies

[58] ECT, Article 8.

[59] Energy policy in South East Europe in Competitiveness in South East Europe: A Policy Outlook 2018, 476 https://www.oecd-ilibrary.org/docserver/9789264298576-17-en.pdf?expires=1573830909&id=id&accname=guest&checksum=4D28ADD636FA75AFE8932D0B4287F354.

amounted to EUR1.9 billion on an annual basis, which was significantly higher than the subsidies directed towards renewables.[60] Director Kopač identified coal as a serious obstacle between Contracting Parties and closer European integration. In furtherance of the energy transition, South East European countries would, therefore, benefit from assistance in phasing out fossil-based generating capacities before the end of their economic life and replacing them with modern, sustainable and clean energy sources to achieve the goals set out in international agreements. As highlighted in Sect. 8.2.1, the EU is strongly committed to the energy transition and has already taken concrete steps to implement its vision of a resilient Energy Union with a forward-looking climate change policy. Given the importance of the convergence of the energy policies in South East European countries with those of the EU for both regions, the investment promotion and protection mechanisms of the ECT are significant in their potential to encourage renewable energy investments in South East European countries. Soft law instruments under the ECT can also assist in capacity-building programmes in South East Europe. Further, the research and innovation provisions in the ECT are important for the energy transition in light of the continuous technological advances regarding low-carbon energies and for improving access to clean, smart, and affordable energy. They also support the export and transfer of technology from developed countries to emerging economies in need of such technology. Accordingly, the ECT can play a role in accelerating the energy transition in South East Europe to support its convergence with the objectives and legal frameworks of the Energy Union.

[60] Energy Community, Barbora Poyner, 'Director Unveils Final Results of Study on Coal Subsidies and Their Impacts at EU Sustainable Energy Week' (20 June 2019) https://www.energy-community.org/news/Energy-Community-News/2019/06/20.html.

8.3 Energy Security

8.3.1 The Energy Security Concerns of the EU and the Energy Union Strategy

Despite the urgent need to transition to low-carbon energy systems to meet the climate goals of the Paris Agreement, the production and consumption of fossil fuels continue to have an upward trend.[61] As the EU's energy needs are mostly met by net imports (imports minus exports),[62] access to imported energy and energy security will, therefore, remain a significant regional concern in the short and mid-term so long as fossil fuels constitute the dominant source of energy powering the European economy. This concern is reflected in the dimension of security, solidarity and trust in the Energy Union strategy. This dimension aims to reduce the vulnerability of Member States to external shocks, interruptions and disruptions through regional cooperation and integration as well as the diversification of Europe's sources of energy.[63] Indeed, the Energy Union strategy was in part developed as a response to the energy transit crises between Russia and Ukraine where disputes relating to the price of gas and transit fees led Gazprom, the Russian gas exporter, to cease exports to Ukraine, which resulted in interruptions of transit flows in several EU Member States such as Romania, Slovakia, Austria, and Poland.[64] In particular, during the 2009 Russian-Ukraine transit crisis, gas flows transmitted through Slovakia were

[61] BP Statistical Review of World Energy 2019 68th edition https://www.bp.com/content/dam/bp/business-sites/en/global/corporate/pdfs/energy-economics/statistical-review/bp-stats-review-2019-full-report.pdf.

[62] Eurostat, '2.3 From Where Do We Import Energy and How Dependent Are We?' https://ec.europa.eu/eurostat/cache/infographs/energy/bloc-2c.html.

[63] The stress test conducted in the summer of 2014 showed that a severe disruption of gas supplies from the east (i.e. Russia) would still have a major impact throughout the EU. For more information see: Communication from the Commission to the European Parliament and the Council on the short term resilience of the European gas system Preparedness for a possible disruption of supplies from the East during the fall and winter of 2014/2015 Brussels, 16.10.2014, COM(2014) 654 final.

[64] European Parliament, Directorate-General for External Policies Policy Department, 'A Cold Winter to Come? The EU Seeks Alternatives to Russian Gas' http://www.europarl.europa.eu/RegData/etudes/STUD/2014/536413/EXPO_STU(2014)536413_EN.pdf.

fully suspended, dramatically affecting Western Hungary, Slovenia, and Croatia. Gas flows transmitted through Eastern Hungary were reduced by 80%, with heavy implications for Serbia and Bosnia and Herzegovina.[65] The EU has therefore recognised that the foundation for energy security lies in internal cohesion through a functioning internal market and increased cooperation, interconnection and solidarity at regional and European levels.[66] Such cooperation includes increasing the EU's indigenous energy production, strengthening emergency mechanisms, moderating energy demand, diversifying external supplies, and establishing necessary energy infrastructure.[67] A key pathway to implementing the security dimension of the Energy Union is the mobilisation of investment capital from public and private sectors, which is required for the renovation of buildings, and district cooling and heating, among others.[68] Such investment is also critical for EU energy production (particularly from sustainable sources), the infrastructure required for the diversification of external suppliers, and back-up infrastructure to strengthen emergency mechanisms. Some of these policy areas have already been operationalised. In 2019, the Commission reported that all Member States but one have access to more than one independent source of gas and by 2022, it is anticipated that 23 Member States will have access to global LNG markets.[69] It is also anticipated that by 2020, the European market will be able to access certain quantities of Azerbaijani natural gas through the Southern Gas Corridor.[70] This is critical for South East European countries, given their historical dependence on Russia for gas supplies.[71]

---

[65] Emil Asenov, 'Development of the Regional Natural Gas Market in Southeast Europe'.

[66] Communication from the Commission to the European Parliament and the Council, European Energy Security Strategy, COM/2014/0330 https://eur-lex.europa.eu/legal-content/EN/ALL/?uri=CELEX:52014DC0330&qid=1407855611566 (hereinafter 'the Communication').

[67] Ibid.

[68] Ibid.

[69] European Commission, Fourth Report, p. 14 https://ec.europa.eu/commission/sites/beta-political/files/fourth-report-state-of-energy-union-april2019_en_0.pdf.

[70] The Communication.

[71] OECD (n 15).

8.3.2 The Role of the ECT in Facilitating the Secure Energy Flows

Internationally, the ECT is the only dedicated multilateral agreement that directly addresses critical issues associated with energy transit, transport facilities, market access, and tariffs. Its investment protection regime also plays a critical role in attracting energy infrastructure investments. The ECT therefore has the potential to complement the Energy Union through its provisions on investment protection and transit facilitation. This section will focus on the obligation to provide uninterrupted flows of energy once established within the ECT constituency. Transit is covered under the ECT where a State through whose territory Energy Materials and Products transit ('transit State') and either the State of origin or the State of destination is a Contracting Party to the ECT. It also includes situations where the State of origin is also the State of destination. It applies to the carriage of Energy Materials and Products through Energy Transport Facilities, which are defined as "high-pressure gas transmission pipelines, high-voltage electricity transmission grids and lines, crude oil transmission pipelines, coal slurry pipelines, oil product pipelines, and other fixed facilities specifically for handling Energy Materials and Products".[72] The ECT aims to address concerns surrounding the reliability and security of energy flows. It obliges contracting States to "facilitate transit… consistent with the principle of freedom of transit", in a non-discriminatory manner and without the imposition of unreasonable delays, restrictions or charges.[73] The Contracting States are also to "encourage relevant entities to co-operate" in modernising, developing, operating, and facilitating the connection of Energy Transport Facilities as well as in "measures to mitigate the effects of interruptions" in energy supply.[74] The ECT also goes a step further to ensure that additional transit capacity can be established. Where the carriage of

[72]Energy Charter Secretariat, the Energy Charter Treaty, Trade Amendment and Related Documents https://www.italaw.com/sites/default/files/laws/italaw11107.pdf.

[73]Article 7(1) ECT.

[74]Article 7(2) ECT.

Energy Materials and Products through existing infrastructure is unrealistic on commercial terms, the contracting parties have to abstain from obstructing the establishment of new capacity subject to the national legislation[75] of the transit state and the security and efficiency of the energy systems of the parties.[76] The transit provisions of the ECT are implemented via three different tailor-made dispute settlement mechanisms. Contracting Parties can access Article 27 for disputes concerning the application and interpretation of the ECT as well as Article 7(7), which contains a conciliation procedure devised to deal with any matter arising from "existing transit". In the event of a dispute concerning transit, the transit regime of the ECT prohibits the interruption or reduction of existing transit flows[77] until the special "dispute resolution mechanism" foreseen in the same article is exhausted, or the transit contract in question provides otherwise.[78] Additionally, foreign investors can access Article 26 for resolving transit disputes provided that it also constitutes an investment dispute falling under Part 3 of the ECT on Investment Promotion and Protection. In that respect, the term "investment" is defined broadly and involves legal and contractual rights and permits or licences.[79] Furthermore, the voluntary Early Warning Mechanism of the Energy Charter foresees the establishment of an Energy Security Contact Group in the case of an emergency.[80] The objective of the Energy Charter Early Warning Mechanism is to prevent and overcome emergency situations in the energy sector related to the Transit and supply of energy materials and products. The methodology of the Energy Charter Early Warning Mechanism includes exchange of information and response to requests for information, consultations, verification and

[75]Such legislation shall be consistent with principles of freedom of transit and non-discrimination *see* the Understanding reached by the Treaty Contracting Parties with respect to Art 7(4).

[76]Article 7 (5) ECT.

[77]Article 7(6) ECT.

[78]Notably, the contracting State must ensure any entity under its jurisdiction also complies with this obligation.

[79]Article 1(6) ECT.

[80]Decision of the Energy Charter Conference, 'Model Energy Charter Early Warning Mechanism' (26 November 2014) available at https://www.energycharter.org/fileadmin/Documents Media/CCDECS/CCDEC201414.pdf.

monitoring, risk evaluation, and recommendations for action in view of an emergency situation or the threat of an emergency situation.[81] Accordingly, the transit provisions under the ECT provide assurance towards secure flows of energy sources among the ECT constituency. Further, while the transit provisions do not extend to marine transport, these principles would remain of value for future hydrogen transport options via dedicated or converted hydrogen pipelines.

8.3.3 Commentary on South East Europe

The EU and South East European countries are mutually important to each other in ensuring security of supply. It has even been observed that the European Commission's description of the Energy Community evokes the energy security dimension of the Energy Union.[82] Specifically, the Energy Community is described as being about "investments, economic development, security of energy supply and social stability; but—more than this—the Energy Community is also about solidarity, mutual trust and peace".[83] For the EU, South East European countries form part of the "Southern Corridor" which bypasses Russia, thus reducing the EU's dependence on Russian gas imports.[84] Further, the Energy Community serves as a platform for EU *acquis* to be exported to South East Europe to push for regulatory consistency and infrastructure development in the fields of electricity, oil, and gas.[85] This will potentially expand the regulatory space of the EU and absorb South East

[81]For more information, see Urban Rusnák and Iryna De Meyer, the Energy Charter Early Warning Mechanism: 2014 Russia-Ukraine-EU Transit Issues. *Oil, Gas, Energy Journal* (2015) 13 (6).

[82]Rafael Leal-Arcas, 'The European Energy Union', p. 28.

[83]Report from the Commission to the European Parliament and the Council, "Under Article 7 of Decision 2006/500/EC (Energy Community Treaty)", at p. 2, COM(2011) 105 final (10 March 2011) mentioned in The European Energy Union: The quest for secure, affordable and sustainable energy by Rafael Leal-Arcas (Claeys & Casteels Publishing, 2016), p. 28.

[84]Dimitar Bechev, '"Southeast Europe" in Securing the Energy Union: Five Pillars and Five Regions' https://www.iss.europa.eu/sites/default/files/EUISSFiles/Report%2032.pdf.

[85]Ibid.

Europe into the EU's internal energy market (although the implementation of this has been slow and fragmented).[86] South East European countries can also benefit from the Energy Union strategy, particularly the EU's aim to diversify its energy sources and to establish a fully integrated internal energy market. Cooperation and interconnection with the EU are crucial for the energy security of South East European countries to meet future energy demands that develop in tandem with economic growth. The ECT's transit and investment protection provisions, therefore, provide further means for South East European countries to enhance energy security and integrate further into the Energy Union. Further, it has been recognised that a move to sustainable sources such as renewables would further energy security as it would reduce the use of imported fossil fuels.[87] Hence, the investment protection provisions under the ECT play a role in placing low-carbon energies on par with conventional energies as well as supporting nascent low-carbon energies that are not yet cost-competitive and require further technological advancements. The ECT also contributes to energy security through the promotion of energy efficiency, which has been recognised as "a more creative and innovative conception of energy security" that focuses on energy demand as opposed to focusing purely on energy supply.[88] In other words, with greater efficiency, less energy is consumed and hence, less needs to be imported.

8.4 Conclusion

This chapter presented the relevance and potential of the ECT for supporting the convergence of South East European countries with the Energy Union strategy in terms of accelerating the energy transition and strengthening energy security. Regarding the energy transition, the EU has shown its strong commitment towards reducing GHG emissions and

[86]Ibid.

[87]Communication from the Commission to the European Parliament and the Council, European Energy Security Strategy, COM/2014/0330 https://eur-lex.europa.eu/legal-content/EN/ALL/?uri=CELEX:52014DC0330&qid=1407855611566.

[88]Rafael Leal-Arcas, p. 90.

increasing renewables in its energy mix. These actions have been pursued as part of its Energy Union strategy and have already been implemented through legislation under the CEP and in other steps, such as regulating the disclosure obligations on the carbon footprints of investments to facilitate investment in "green" products. Further, the commitment of the EU to the energy transition is evidenced in its reports and plans that track and disclose its progress towards achieving the energy transition, as part of its realisation of its Energy Union strategy. South East Europe however, has limited installed capacity of wind and solar power technologies despite substantial potential for it and has also prioritised coal over renewables in its energy policy. In that respect, the ECT can support the practicalities of converging South East Europe's energy strategy with the Energy Union strategy as its investment protection provisions can promote renewable energy investments in South East Europe. Further support is offered through ECT provisions on energy efficiency and research and innovation as well as the soft law instruments, knowledge-sharing, and capacity-building programmes under the broader Energy Charter process. The acceleration of the energy transition would also reduce the demand for imported energy, thus strengthening energy security, which was one of the factors that motivated the Energy Union strategy. In particular, both South East Europe and the EU are heavily dependent on Russian energy and have shared vulnerabilities regarding secure flows of energy. This was exemplified in the Russia-Ukraine gas crises, which resulted in disruptions to energy flows in EU Member States and South East European countries. Among others, the Energy Union strategy has also identified the need for investments to improve energy security. Accordingly, the ECT can play a part in improving energy security in South East Europe as it promotes investment and secure transit. In particular, Contracting Parties are obliged to facilitate transit, which includes the possibility of establishing additional transit capacity. This, therefore, plays a key role in ensuring continuous energy flows. Additionally, through the ECT's investment protection provisions, investments in the infrastructure necessary for continuous energy supplies are promoted. The promotion of energy efficiency under the

ECT also supports energy security as the operationalisation of these areas can reduce dependence on imported fossil fuels.

References

International Energy Agency (IEA). 'Climate Change' (undated) available at https://www.iea.org/topics/climatechange/.

Paris Agreement (adopted 5 October 2016, entered into force 4 November 2016) UNTS 54113.

United Nations General Assembly, 'Transforming Our World the 2030 Agenda for Sustainable Development' (21 October 2015) A/RES/70/1.

United Nations, 'Ensure Access to Affordable, Reliable, Sustainable and Modern Energy' (undated) available at https://www.un.org/sustainabledevelopment/energy/.

World Energy Council (WEC), 'World Energy Trilemma Index 2019' (2019) available at https://www.worldenergy.org/assets/downloads/WETrilemma_2019_Full_Report_v4_pages.pdf.

IEA, 'Energy Security' (undated) available at https://www.iea.org/topics/energy-security.

Jean-Claude Juncker, 'A New Start for Europe: My Agenda for Jobs, Growth, Fairness and Democratic Change. Political Guidelines for the Next European Commission' (15 July 2014) available at https://www.eesc.europa.eu/resources/docs/jean-claude-juncker—political-guidelines.pdf.

Donald Tusk, 'Roadmap Towards an Energy Union for Europe' (submitted to the European Commission on 10 April 2014).

European Commission, 'Communication from the Commission to the European Parliament, the Council, the European Economic and Social Committee, the Committee of the Regions and the European Investment Bank, "A Framework Strategy for a Resilient Energy Union with a Forward-Looking Climate Change Policy"' (25 February 2015) available at http://ec.europa.eu/transparency/regdoc/rep/1/2015/EN/1-2015-80-EN-F1-1.PDF (hereinafter 'Energy Union strategy').

European Commission, 'Communication from the Commission, "State of the Energy Union 2015"' (18 November 2015) available at https://eur-lex.europa.eu/legal-content/EN/TXT/?qid=1449767367230&uri=CELEX:52015DC0572.

Energy Union strategy (n 11).

European Commission, 'Energy Union and Climate' available at https://ec.europa.eu/commission/priorities/energy-union-and-climate_en.

International Renewable Energy Agency, 'Europe' (undated) available at https://www.irena.org/europe.

OECD, 'Competitiveness in South East Europe: A Policy Outlook 2018' (24 April 2018) available at https://read.oecd-ilibrary.org/development/competitiveness-in-south-east-europe_9789264298576-en#page470.

European Commission, 'Energy Community' (undated) available at https://ec.europa.eu/energy/en/topics/international-cooperation/international-organisations/energy-community.

Institute of Energy for SE Europe, 'South East Europe Energy Outlook 2016/17' (May 2017).

The Energy Charter Treaty and Related Documents, 17 December 1994, UNTS 2080, 100 34 I.L.M. 360 (ECT).

International Energy Charter (IEC), 'The Energy Charter Treaty' (updated on 18 February 2019) available at https://www.energycharter.org/process/energy-charter-treaty-1994/energy-charter-treaty/.

'European Energy Charter' available at https://eur-lex.europa.eu/legal-content/EN/TXT/?uri=LEGISSUM:l27028.

IEC, 'The European Energy Charter' (2 June 2015) available at https://www.energycharter.org/process/european-energy-charter-1991/.

Decision of the Energy Charter Conference, 'Road Map for the Modernisation of the Energy Charter Process' (24 November 2010) available at https://www.energycharter.org/fileadmin/DocumentsMedia/CCDECS/CCDEC201010.pdf.

IEC, 'Conflict Resolution Centre' (updated on 16 April 2019) available at https://www.energycharter.org/what-we-do/dispute-settlement/conflict-resolution-centre/.

Decision of the Energy Charter Conference, 'Adoption by Correspondence—Model Instrument on Management of Investment Disputes' (23 December 2018) available at https://www.energycharter.org/fileadmin/DocumentsMedia/CCDECS/2018/CCDEC201826_-_INV_Adoption_by_correspondence_-_Model_Instrument_on_Management_of_Investment_Disputes.

Decision of the Energy Charter Conference, 'Modernisation of the Energy Charter Treaty' (28 November 2017) available at https://www.energycharter.org/fileadmin/DocumentsMedia/CCDECS/2017/CCDEC201723.pdf.

Decision of the Energy Charter Conference, 'Modernisation of the Energy Charter, Treaty: Mandate, Procedural Issues and Timeline for Negotiations'

(6 November 2019) available at https://www.energycharter.org/fileadmin/DocumentsMedia/CCDECS/2019/CCDEC201910.pdf.

European Commission, 'Clean Energy for All Europeans' (26 July 2019) available at https://op.europa.eu/en/publication-detail/-/publication/b4e46873-7528-11e9-9f05-01aa75ed71a1/language-en?WT.mc_id=Searchresult&WT.ria_c=null&WT.ria_f=3608&WT.ria_ev=search.

Directive 2009/28/EC of the European Parliament and of the Council of 23 April 2009 on the promotion of the use of energy from renewable sources and amending and subsequently repealing Directives 2001/77/EC and 2003/30/EC, OJ L 140.

Directive 2012/27/EU of the European Parliament and of the Council of 25 October 2012 on energy efficiency, amending Directives 2009/125/EC and 2010/30/EU and repealing Directives 2004/8/EC and 2006/32/EC, OJ L 315.

European Commission, 'Communication from the Commission to the European Parliament, the European Council, the Council, the European Economic and Social Committee and the Committee of the Regions: The European Green Deal' COM(2019) 640 final (11 December 2019).

https://ec.europa.eu/commission/priorities/energy-union-and-climate_en.

Regulation (EU) 2018/1999 of the European Parliament and of the Council of 11 December 2018 on the Governance of the Energy Union and Climate Action, amending Regulations (EC) No 663/2009 and (EC) No 715/2009 of the European Parliament and of the Council, Directives 94/22/EC, 98/70/EC, 2009/31/EC, 2009/73/EC, 2010/31/EU, 2012/27/EU and 2013/30/EU of the European Parliament and of the Council, Council Directives 2009/119/EC and (EU) 2015/652 and repealing Regulation (EU) No 525/2013 of the European Parliament and of the Council, PE/55/2018/REV/1, OJ L 328.

Fourth Report on the State of the Energy Union (Fourth Report), p. 6 https://ec.europa.eu/commission/sites/beta-political/files/fourth-report-state-of-energy-union-april2019_en_0.pdf.

Who We Are/Industry Advisory Panel/Meetings/Industry Advisory Panel (4 June 2019) https://www.energycharter.org/who-we-are/industry-advisory-panel/meetings/industry-advisory-panel-4-june-2019/.

Article 10 ECT.

Article 13 ECT.

Article 12 ECT.

Article 1(6) ECT. Also note that the Treaty investment protection regime also covers certain investments designated by Contracting Parties that are notified to the Secretariat.

Article 1(5) ECT.

Decision of the Energy Charter Conference, 'Road Map for the Modernisation of the Energy Charter Process of 24th November 2010 Concerning Road Map for the Modernisation of the Energy Charter Process' (24 November 2010) https://www.energycharter.org/fileadmin/Documents Media/CCDECS/CCDEC201010.pdf.

ECT, Article 8.

Energy Policy in South East Europe in Competitiveness in South East Europe: A Policy Outlook 2018, 476 https://www.oecd-ilibrary.org/docserver/9789264298576-17-en.pdf?expires=1573830909&id=id&accname=guest&checksum=4D28ADD636FA75AFE8932D0B4287F354.

Energy Community, Barbora Poyner, 'Director Unveils Final Results of Study on Coal Subsidies and Their Impacts at EU Sustainable Energy Week' (20 June 2019) https://www.energy-community.org/news/Energy-Community-News/2019/06/20.html.

BP Statistical Review of World Energy 2019 68th edition https://www.bp.com/content/dam/bp/business-sites/en/global/corporate/pdfs/energy-economics/statistical-review/bp-stats-review-2019-full-report.pdf.

Eurostat, '2.3 From Where Do We Import Energy and How Dependent Are We?' https://ec.europa.eu/eurostat/cache/infographs/energy/bloc-2c.html.

Communication from the Commission to the European Parliament and the Council on the short term resilience of the European gas system Preparedness for a possible disruption of supplies from the East during the fall and winter of 2014/2015 Brussels, 16.10.2014, COM(2014) 654 final.

European Parliament, Directorate-General for External Policies Policy Department, 'A Cold Winter to Come? The EU Seeks Alternatives to Russian Gas' http://www.europarl.europa.eu/RegData/etudes/STUD/2014/536413/EXPO_STU(2014)536413_EN.pdf.

Emil Asenov, 'Development of the Regional Natural Gas Market in Southeast Europe'.

Communication from the Commission to the European Parliament and the Council, European Energy Security Strategy, COM/2014/0330 https://eur-lex.europa.eu/legal-content/EN/ALL/?uri=CELEX:52014DC0330&qid=1407855611566 (hereinafter 'the Communication').

European Commission, Fourth Report, p. 14 https://ec.europa.eu/commission/sites/beta-political/files/fourth-report-state-of-energy-union-april2019_en_0.pdf.

Energy Charter Secretariat, the Energy Charter Treaty, Trade Amendment and Related Documents https://www.italaw.com/sites/default/files/laws/italaw11107.pdf.

Article 7(1) ECT.

Article 7(2) ECT.

Article 7(5) ECT.

Article 7(6) ECT.

Article 1(6) ECT.

Decision of the Energy Charter Conference, 'Model Energy Charter Early Warning Mechanism' (26 November 2014) available at: https://www.energycharter.org/fileadmin/DocumentsMedia/CCDECS/CCDEC201414.pdf.

Urban Rusnák and Iryna De Meyer, The Energy Charter Early Warning Mechanism: 2014 Russia-Ukraine-EU Transit Issues. *Oil, Gas, Energy Journal* (2015) 13 (6).

Rafael Leal-Arcas, 'The European Energy Union', p. 28.

Report from the Commission to the European Parliament and the Council, "Under Article 7 of Decision 2006/500/EC (Energy Community Treaty)", at p. 2, COM(2011) 105 final (10 March 2011) mentioned in The European Energy Union: The quest for secure, affordable and sustainable energy by Rafael Leal-Arcas (Claeys & Casteels Publishing, 2016) p. 28.

Dimitar Bechev, '"Southeast Europe" in Securing the Energy Union: Five Pillars and Five Regions' https://www.iss.europa.eu/sites/default/files/EUISSFiles/Report%2032.pdf.

Part III

Transitional Aspects of the Energy Union

9

Drivers and Enablers of LNG Bunkering

Alexandros Lagakos

9.1 Introduction

In the upcoming years, the use of Liquefied Natural Gas (LNG) in bunkering is expected to attract an ever-increasing market share, driven by the ever more stringent environmental standards for marine fuels as well as by the ever more urgent need to curtail greenhouse gas (GHG) emissions. At the same time, the European Union has identified LNG as a crucial tool for the implementation of its Energy Union. It is explicitly stated in the EU strategy communicated by the European Commission to the European Parliament that LNG will not only serve the security of supply and competitiveness, but it will also support the EU in the achievement of its sustainability objectives. More specifically, the

A. Lagakos (✉)
Blue Grid Gas & Power, Athens, Greece

Greek Energy Forum, Athens, Greece

M. Mathioulakis (ed.), *Aspects of the Energy Union*,
Energy, Climate and the Environment,
https://doi.org/10.1007/978-3-030-55981-6_9

Commission expects that LNG will be increasingly used as an alternative fuel to diesel in the transport sector, both shipping as well as heavy-duty vehicles. In this context, the 2020 standards set by the International Maritime Organization (IMO) take the lead, setting a 0.5% cap on the sulphur content of bunker fuels used in open seas (vs a current 3.5%). It is worth noting that within port limits in most European ports, and within Emission Control Areas (ECA) in the Baltic and the United States, this cap has already been set as low as 0.1%. This new global limit imposed by the IMO will incur even higher operational costs for global shipping. LNG is one of the most attractive options amongst currently feasible solutions, allowing not only for a more benign environmental footprint but also for lower fuel costs. According to the latest data by the Society for Gas as a Marine Fuel (SGMF), by the end of 2019, a total of 395 LNG-fuelled vessels are either on the water or are on order, and 79 terminals on a global scale currently offer LNG as a marine fuel.[1] The penetration of LNG as a marine fuel is not expected to be uniform amongst shipowners. This occurs because the commercial feasibility of this fuel option depends on the individual commercial and operational mode of each ship. For example, a ferry operating within a specific region and always returning to the same home port every day may easily switch to LNG. However, an oil tanker which is chartered on a spot-voyage basis calling random ports within a broader region may not be able to procure LNG volumes so easily. Nevertheless, LNG still faces challenges such as the plausible complexity of a retrofit scheme or even the incremental cost of a newbuilding, the methane slip risk or the non-uniform regulatory framework across the globe. However, technological breakthroughs, combined with constant market developments, ensure the timely resolution of such issues. In the context of the global public discussion about Climate Change, shipping has been admittedly put in the spotlight. It's ranking in the stack order of climate-driven priorities undertaken by governments remains, however, subjective. Shipping is one of the most carbon-intensive sectors, where its share on global CO_2 emissions is four times higher than its share on

[1]"Members Directory", Society for Gas as a Marine Fuel. Accessed March 7, 2020. www.sgmf.info.

global energy consumption. At the same time, shipping offers minimal energy consumption per unit of freight, hence accounting only for 14% of GHG emissions attributable to transport in Europe. Nevertheless, there is already a wide global consensus in favour of the transition of shipping—in parallel with all other forms of transport—towards alternative fuels. Many options are in discussion (i.e. Methanol, Ammonia, Hydrogen, Batteries, Biofuels), and some of them will eventually become feasible and tangible options for shipping customers. The level of technological and commercial maturity of each propulsion method substantially varies. It seems that LNG is currently the most competitive and easier-to-adopt marine fuel, monetising the valuable experience and knowledge of the LNG shipping industry for almost half a century as well as the existing and foreseeable abundance of LNG around the world. It is a fuel that has already covered more than "half the way" towards being widely available to shipping customers, benefiting from an existing large-scale LNG infrastructure (carriers, terminals, storage tanks) across the globe. It is a marine fuel that is not just greener than oil-derived contestants but currently admirably cheaper, too. Several elements function as prime drivers for shipping customers to transition to LNG, addressing developments of environmental, economic and business nature.

9.2 Drivers for a Transition to LNG in Shipping

9.2.1 Environmental Performance

LNG features the main environmental advantage of practically eliminating particulate matters, a primary pollutant for the urban environment, hence making it an attractive proposition for passenger ships that frequently call on city ports, which typically reside at, or in great proximity to, city centres. At the same time, it emits almost 90% less SO_x and 80% less NO_x emissions vs oil-derived bunkers. Regarding CO_2, the reduction is in the range of 25% vs "business-as-usual". However, CO_2 reduction is much higher in the case of BioLNG, reaching even 100% elimination in the case of liquefying 100% renewable gas. BioLNG can be derived either via biomethane or synthetic methane. An example of the former should be the production of biomethane from animal or agricultural waste, which is already implemented as a solution through various projects in Europe. An example for the latter is a methanation plant, where synthetic methane is produced by combining CO_2 with H_2 derived via electrolysis utilising renewable electricity. This concept has already materialised into a real project by MAN and Audi, operating a methanation Power-to-Gas plant in Germany since 2013. A controversial issue regarding the environmental performance of LNG is the risk of methane slip. Although short-lived, methane is a highly impactful GHG. Nevertheless, the risk of methane slip should be seen more as an operational risk rather than a given downside. The appropriate engine configuration and prudent health and safety measures in bunkering operations could almost practically eliminate this concern, or at least substantially outweigh its impact, especially when compared to the undeniable environmental advantages of switching to LNG as a marine fuel.

9.2.2 Regulation

The aforementioned environmental records of LNG have to be assessed in the context of international environmental regulation. The ultimate milestone for the countries that adhere to the Paris Convention is the year 2050. However, this may look too distant to be able to guide the shipping industry by its own. This is the reason for the existence and continuous emergence of in-between milestones and standards the industry needs to comply with. At the beginning of 2020, IMO 2020 came in force, obliging ships across the globe to comply with a maximum 0.5% limit for SO_x emissions. This has practically left fuel-oil (between 1 and 3.5% sulphur-rich) out of competition unless a ship has a scrubber system installed onboard (a retrofitted system which washes SO_x emissions away from exhaust gases). A considerable number of vessels have already decided to implement this solution, primarily choosing to instal "open-loop" scrubbers. The latter systems sequester SO_x and dispose it into the sea and hence constitute a controversial "solution". This is the reason why, currently, an ever-increasing number of ports/countries have banned the discharge of wash water from open-loop scrubbers within their port limits or national waters. In the author's personal opinion, when places like Germany, Norway, China, Singapore or California have openly denounced the use of "open-loop" scrubbers, there is high likelihood that this will eventually become an international trend, which might evolve into more stringent measures against such retrofits in a just a few years from now. For the time being, the mainstream reaction of the market has been the utilisation of a new type of "compliant" fuel featuring 0.5% Sulphur content. The compliant fuel started being available in various locations already.

There have been some quality-relevant claims by customers, but most likely these are just signs of "childhood diseases" of a new type of fuel. Nevertheless, such a fuel remains a non-compliant fuel for certain parts of the world, such as the Emissions Control Areas (ECA) extending from the Isle of Man to the Baltic Sea or along the US coastline within a range of 200 nautical miles. These areas currently require a maximum sulphur content of 0.1%, which is only realistically achievable via the use of marine gasoil and LNG at the moment. In the near future, it is expected that more ECA zones will be issued in the world and at the same time, the currently existing ECAs will extend their environmental requirements to other pollutants such as NO_X. Such an example is the Mediterranean ECA, possibly implementable as early as in 2024. In December 2019 in Naples, the 22 Member States of the Barcelona Convention (a convention amongst the Mediterranean States missioned to protect the environment and ecosystem in their region) ratified on a Ministerial level a detailed roadmap towards the implementation of the Mediterranean ECA Zone. When this gets implemented, the same environmental requirements that already apply to the North-western and Northern part of Europe will prevail. Furthermore, in the same month, the new President of the European Commission, Mrs Ursula Von der Leyen, presented the EU Green Deal, a strategic initiative to effectively guide Europe towards a carbon-neutral future. In the mission statement, the Green Deal mentions the Commission's decision to propose the induction of shipping under the Emissions Trading Scheme (ETS), to incentivise a lower carbon footprint for the sector. It is unclear when this will happen, but when it does, it should enable a competitive advantage for non-oil derived fuels. Timing will be critical to decide which alternative fuels will have commercially matured by then, in order to pick the winner of such an ambitious policy target.

9.2.3 Price Competitiveness

Currently, price represents an undeniable competitive advantage for LNG. As a commodity denominated in energy units, LNG is substantially cheaper than oil-derived fuels across the globe. Subject to the local infrastructure availability, to the maturity of LNG bunkering at a specific location, and the price of the displaceable oil fuel, LNG as a marine fuel can currently offer substantial cost savings on an average annual basis.

9.2.4 Corporate Social Responsibility

For some specific type of ships, switching to cleaner fuels such as LNG features additional implicit value. Cruise ships are an indicative example, where the drive for less emissions and more action against Climate Change seems to be a prevailing customer requirement. This is why cruise ships have already embraced LNG as the fuel of choice, with 24 out of 124 vessels on the orderbook being LNG-fuelled. Another example is Volkswagen, which has been the first car company deploying two LNG-fuelled car carriers connecting Europe and North America.

9.3 The Enablers

9.3.1 Switching to LNG

There are two fundamental ways for a shipowner to start utilising LNG as a marine fuel; the installation of an LNG bunkering station and tank as a retrofit on an existing vessel or a newbuild vessel designed to run on LNG. The former is not necessarily an available option for all ships because it depends on the current configuration and type of the vessel's

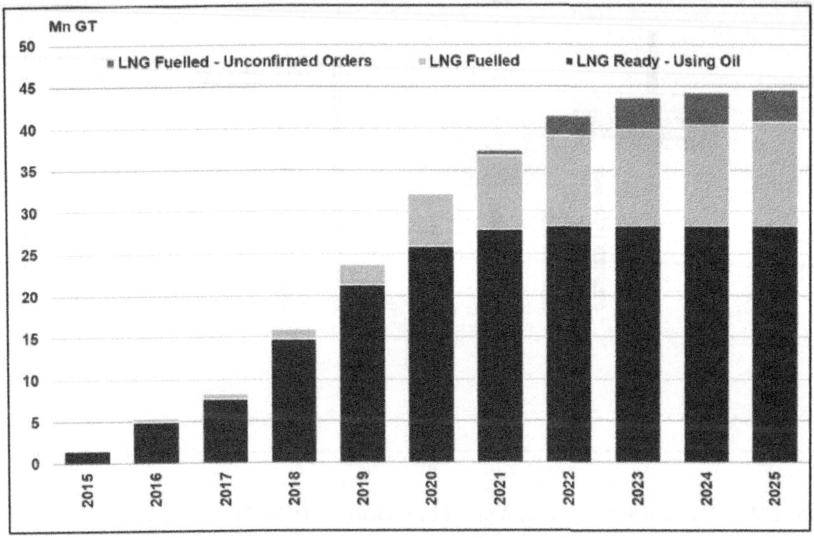

Fig. 9.1 Cumulative LNG fuelled fleet (*Source* Splash247.com)

engines. However, it is worth mentioning a recent study published by Maritime Strategies International, which touches upon the issue of "LNG-ready" vessels placed on order, seen here in Fig. 9.1.[2] These ships typically have space allotted for LNG fuel tanks along with additional supporting materials used for low temperatures, engines capable of being converted to dual fuel operation and space designated for pipe routing and ventilation. This practically means that with the right incentive, those vessels could be able to switch to LNG at a relatively short lead-time. This enables a substantial demand foundation potential for LNG bunkering within the ongoing decade.

Regarding newbuildings, it is worth highlighting a couple of useful remarks. The first one is the relatively broad penetration of LNG across various shipping segments. This becomes quite evident in an online

[2]"Volume Uptake of Gas as Fuel Awaits the Conversion of LNG-Ready Tonnage", Splash247.com. November 22, 2019. https://splash247.com/volume-uptake-of-gas-as-fuel-awaits-the-conversion-of-lng-ready-tonnage/.

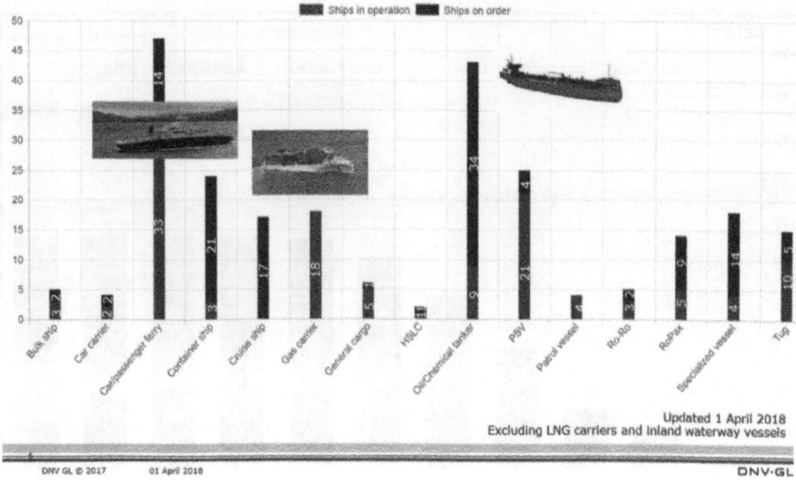

Fig. 9.2 LNG uptake by vessel segment (*Source* DNV.GL Group)

presentation by DNV-GL,[3] that, as seen here on Fig. 9.2, categorises LNG-fuelled ships in operation or on order across 15 different segments, covering almost the full spectrum of mainstream shipping applications.

The second remark has to do with the operational profile of each vessel. Until LNG is available as a marine fuel in the majority of global bunkering locations, its utilisation will typically stay amongst vessels with adequate predictability regarding their operational profiles. This predictability should allow a supplier to secure the necessary LNG quantities and assets to be able to serve such a customer within the region at a modest frequency. A typical example could be a ferry, which features well-defined itineraries and frequent bunkering at one or two specific ports. Therefore, it is a feasible case for such a vessel to switch to LNG and to achieve security of supply, provided that the regionally available LNG and its supply chain enable the right economics that justify conversion to LNG—either via a retrofit solution or as a newbuilding. The latter should explain why the largest uplift of LNG so far has been with

[3]"LNG Regulatory Update: Best Fuel of the Future, Conference & Study Tour", Ole Vidan Nilsen, DNVGL. April 10, 2018. http://www.golng.eu/files/Main/20180417/2.%20Ole%20Vidar%20Nilsen%20-%20DNV%20GL.pdf.

ferries and cruise vessels. Although it might look as counter-intuitive, the numbers reported for tankers and chemical carriers also comply to the rule mentioned above, since these cases also primarily refer to particular vessels that are destined to operate within specific geographical regions with available LNG, such as the North Sea and the Baltic Sea.

9.3.2 Methods of Bunkering

There are currently three main modes of performing LNG bunkering that include Truck-to-Ship (TTS), Ship-to-Ship (STS) and Terminal (Port)-to-Ship (PTS). Amongst these methods, a categorisation of typical LNG bunkering per different generic ship type relates, as can be seen in a report by the European Maritime Safety Agency (EMSA) in Fig. 9.3,

Vessel Type (Receiving vessel)	Bunker Quantity	Rate	Duration	Hoses or arm diameter (pol)	Adequate Bunkering Mode
Service vessels, tugboats, patrol boats and fishing boats	50 m^3	60 m^3/h	45 min	2x2" or 1x3"	TTS
Small Ro-Ro and Ro-Pax vessels	400 m^3	400 m^3/h	1 hr	2x4" or 1x6"	TTS/ STS
Large Ro-Ro and Ro-Pax vessels	800 m^3	400 m^3/h	2 hr	2x4" or 1x6"	STS
Small cargo, container and freight vessels	2,000 – 3,000 m^3	1,000 m^3/h	2 to 3 hr	2x8" or 1x12"	STS
Large freight vessels	4,000 m^3	1,000 m^3/h	4 hr	2x8" or 1x12"	STS
Large tankers, bulk carriers and container ships	10,000 m^3	2,500 m^3/h	4 hr	2x10"	STS/ PTS
Very large container ships and oil tankers	20,000 m^3	3,000 m^3/h	7 hr	2x12"	STS/ PTS

Fig. 9.3 Typical LNG bunkering per different generic ship type (*Source* European Maritime Safety Agency [EMSA])

to a number of technical parameters including bunker quantity, Rate, Duration and Hoses on Arm diameter.[4]

9.3.2.1 Truck-to-Ship (TTS)

A truck tugging either an LNG semi-trailer or an ISO container is connected to the, already berthed and moored receiving vessel, using a flexible hose. The discharge rate of the fuel typically ranges between 40 and 60 m^3 per hour and the typical payload of a truck is in the range of 50 m^3. In most cases, the bunkering operation may not interrupt the operations of the receiving vessel, hence enabling the so-called "simultaneous operations" (SIMOPS). The latter is a vital prerequisite for most marine fuel consumers, and LNG suppliers need to comply with the time restrictions of their customers. A distinct advantage of TTS is the operational flexibility it offers, given that it is performed using fully mobile assets. It is also worth highlighting the limited infrastructure requirements that TTS entails. Finally, this is a mode which also allows for the simultaneous discharge of up to 6 trucks, hence increasing the magnitude of the achievable hourly discharge rate. The latter may become feasible by using a type of discharge multiplexer, which connects directly to the receiving vessel. To its disadvantage, TTS becomes unworkable for higher bunkering volume requirements. To serve larger LNG stems, the supplier will either need to schedule a long and sequential discharge of a considerable amount of trucks or reserve a large area at the quayside to allow for multi-truck operations. The former would require a high number of available trucks and lead to complex operations, the latter—if feasible—would most likely hinder SIMOPS.

9.3.2.2 Ship-to-Ship (STS)

LNG is delivered to the receiving vessel by the LNG Bunkering Vessel (LNGBV). The LNG transfer takes place either when the receiving

---

[4]"Guidance on LNG Bunkering to Port Authorities and Administrations", European Maritime Safety Agency, 51. January 31, 2018. https://www.parismou.org/sites/default/files/EMSA%20Guidance%20on%20LNG%20Bunkering.pdf.

vessel is at the berth with the LNGBV being moored alongside, or at anchorage. The hose connecting the LNGBV and the receiving vessel is handled by the former, usually using a specialised crane. The advantages of STS include the operational flexibility, the larger delivery capacity and the higher discharge rate (500–1000 m^3/hour). The latter particularly favours receiving vessels with short port turnaround times. Generally, STS does not interfere with cargo/passenger handling operations, hence also complying with the SIMOPS requirement. The major disadvantage has to do with the fact that it entails a sizeable initial investment by the supplier, given that there are not currently many LNGBVs available in the world today (22 LNGBVs by the end of 2019). Another one relates to potential port limitations that may apply and which may hinder the feasibility or competitiveness of a particular supply chain.

9.3.2.3 Terminal (Port)-to-Ship (PTS)

In PTS, LNG is either delivered directly from an onshore storage tank or a dedicated LNG terminal. The primary advantage of this method is the achievement of very high discharge rates (1000–2000 m^3/hour). However, some evident disadvantages constitute this LNG bunkering method as the least common so far. The most obvious one is the fact that the respective storage tank or Terminal may not be accessible at the berth of the receiving vessel. This means that the receiving vessel will most likely need to make an extra shift at the berth of the storage tank or the Terminal, hence adding expenses and time as well as cancelling the SIMOPS requirement. A second one is a risk inherent in smaller onshore storage tanks, where the availability of adequate LNG quantities is not granted at the exact time the receiving vessel arrives to be bunkered.

9.4 Conclusions

As already highlighted, LNG bunkering and its continuing expansion are being fuelled by a mix of environmental, business and social drivers. However, public discussion on the future of bunkering is dominated by a wide variety of contradicting opinions, which inevitably place the future market share of LNG in a dubious position. Although nobody questions the prospective dominance of alternative fuels (anything else than petroleum products), it is still hard to find consensus on the eventual winners and the indicative timings of prevailing for each fuel. Nevertheless, there is a wide consensus on the fact that LNG is currently the most competitive and proven feasible solution within the alternative fuels' era. The significance of LNG as a marine fuel does not rely on its prospective market share in the years to come. LNG has already proven that it is a fuel that carries the capacity to revolutionise bunkering. Primarily, it is the only marine fuel that is priced in energy units rather than volume or mass units. This makes it a very attractive product, allowing the customers to pay exactly for what they eventually need, energy. Furthermore, LNG is a fuel that breaks the dependency of the shipping industry from refineries, a shrinking rather than an expanding segment of the hydrocarbons industry. Instead, it is a fuel reliant on a broad geographical dispersion of geological reserves and existing export terminals. Finally, LNG may prove to be an enabler for climate-neutral shipping by 2050. As already explained at the beginning of this chapter, LNG has the potential to become the medium for the utilisation of renewable gases in the shipping industry, allowing for a smooth transition for shipowners, which will eventually continue to use the same ships, employ the same bunkering assets and rely on the same bunkering methods. It needs to be stressed that this last point is unique, as it represents one of the few times where the "bill" for fighting Climate Change would not be sent to the shipowners but to the fuel providers onshore.

References

DNV.GL Group. "LNG Regulatory Update". April 10, 2018. http://www.golng.eu/files/Main/20180417/2.%20Ole%20Vidar%20Nilsen%20-%20DNV%20GL.pdf.

European Commission. "Communication on an EU Strategy for Liquefied Natural Gas and Gas Storage". February 16, 2016. https://eur-lex.europa.eu/legal-content/EN/TXT/PDF/?uri=CELEX:52016DC0049&from=EN.

European Maritime Safety Agency. "Guidance on LNG Bunkering to Port Authorities and Administrations". January 31, 2018. https://www.parismou.org/sites/default/files/EMSA%20Guidance%20on%20LNG%20Bunkering.pdf.

Society for Gas as a Marine Fuel. "Members Directory". www.sgmf.info.

Splash247.com. "Volume Uptake of Gas as Fuel Awaits the Conversion of LNG-Ready Tonnage". November 22, 2019. https://splash247.com/volume-uptake-of-gas-as-fuel-awaits-the-conversion-of-lng-ready-tonnage/.

10

The EU Green Deal and the Impact on the Future of Gas and Gas Infrastructure in the European Union

Constantine Levoyannis

10.1 Introduction

The build-up to the European elections of 2019 was unlike any other before it, marking a defining moment in which the topic of climate change entered the political discussion as a mainstream priority alongside security, migration and the economy. The mass mobilisation of civil society and "taking to the streets" by Europe's youth and future generations throughout the European Union (EU) sent shockwaves across the political establishment. It contributed towards putting the topic of climate change at the heart of the EU's policy agenda. Only this time it is not just for the next five years of the incumbent European Commission's mandated term, but for good; for the long-haul. With 93% of the EU population seeing climate change as a serious problem and 72% agreeing

This chapter presents strictly the author's personal views and opinions and does not in any way represent or is linked with views and opinions of his affiliated organisations.

C. Levoyannis (✉)
Greek Energy Forum, Brussels, Belgium

M. Mathioulakis (ed.), *Aspects of the Energy Union*,
Energy, Climate and the Environment,
https://doi.org/10.1007/978-3-030-55981-6_10

that reducing fossil fuel imports can increase energy security and benefit the EU economically, it is clear that a business-as-usual scenario is no longer an option. A new, more dynamic and forward-looking energy and climate strategy is needed and with it, a need to evolve from the traditional meaning of energy security. We are witnessing a paradigm shift in the EU's energy trilemma of security, sustainability and affordability, with sustainability replacing security at the pinnacle. Furthermore, in order for the EU to achieve its long-term climate and energy objectives, a more holistic cross-cutting policy framework is needed to encourage so-called sectoral integration, in other words, the linking of gas, electricity and heating sectors with industry, transport and agriculture. This breaking down of silo mentality can stimulate a multiplier effect when it comes to decarbonisation efforts. The EU needs a policy framework that can give rise to the development of new value chains, the creation of quality jobs in Europe and the promotion of European innovation champions. All these notions are becoming ever more relevant in today's complex and continuously evolving global landscape where the norms of global multilateralism and free trade continue to be put to question. A tectonic shift in policymaking does not mean that the essential elements and hard work undertaken to implement the EU's *Energy Union,* the dogma that characterised the past five years, should be swept aside. The promotion of clean energy, energy efficiency, decarbonisation, security and integration of markets remain stable as pillars of EU climate and energy policy. Only this time, energy and climate policy needs to be firmly integrated within a broader all-encompassing strategy. A shift from the Energy Union to an updated Energy Union 2.0 is insufficient to respond to the challenges of today and tomorrow. What is needed is a centrepiece growth strategy: enter the EU Green Deal. As the President of the European Commission Ursula von der Leyen stated when adopting the European Green Deal in November 2019, the Green deal is, "…about changing behaviours—reconciling the economy with our planet. It is about changing the way we produce, transport and consume".[1] In addition, she emphasised that "the old growth model

[1] "Press remarks by President von der Leyen on the occasion of the adoption of the European Green Deal Communication," European Commission, Accessed December 1, 2019, https://ec.europa.eu/commission/presscorner/detail/en/speech_19_6749.

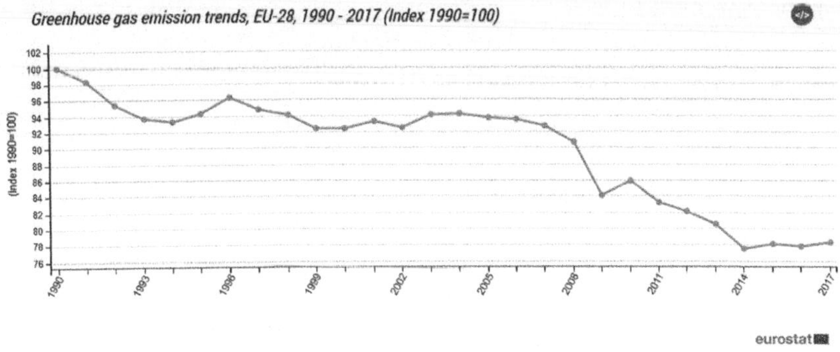

Fig. 10.1 Greenhouse gas emission trends, EU-28, 1990–2017 (*Note* Greenhouse gas emissions [including international aviation, indirect CO_2 and excluding LULUCF]. *Source* European Environment Agency [online datat code: env_air_gge] and Eurostat)

that is based on fossil-fuels and pollution is out of date, and out of touch with our planet".[2] This was a very clear message, full of ambition and intent, with a clear aim towards Europe's political establishment and Europe's citizens and industry; not only in Europe but globally. For some, the EU Green Deal has been characterised as the EU's "man on the moon moment".[3] For many, it marks a significant turning point in EU policymaking and the beginning of a new journey.

10.2 The Current EU Energy and Climate Framework and Future Initiatives

As seen in Fig. 10.1, greenhouse gas emissions have been declining in the EU since 1990. The EU has been taking pride in positioning itself as the global climate leader and pioneer in terms of climate legislation and efforts aimed at reducing greenhouse gas emissions.[4]

---

[2] Ibid.

[3] Ibid.

[4] "Greenhouse gas emission trends", Eurostat, Accessed January 3, 2020, https://ec.europa.eu/eurostat/statistics-explained/index.php/Greenhouse_gas_emission_statistics.

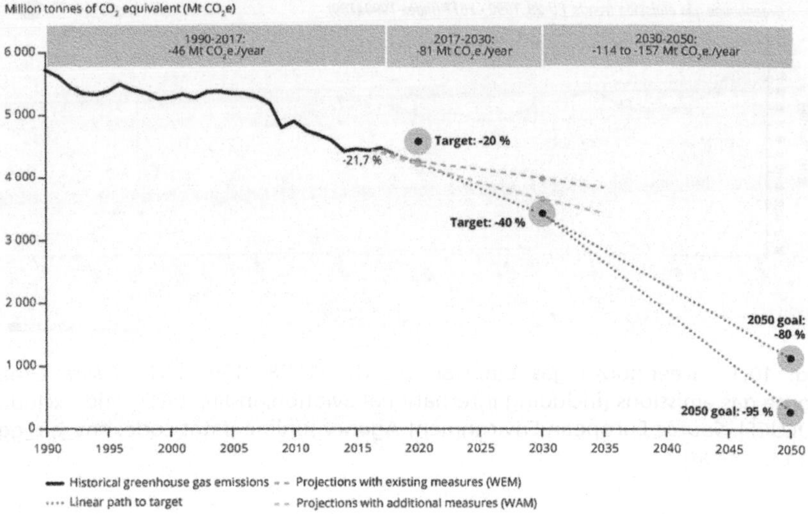

Fig. 10.2 Greenhouse gas emission trend projections (*Source* European Environment Agency)

Indeed, to meet the Paris objectives agreed at COP21 in 2015 and restrict global warming, the EU has already set ambitious climate and energy targets for 2030. Existing EU legislation entails a policy framework for 2030 in which there are three targets to achieve: a target for 32% renewables, 32.5% energy efficiency and 40% greenhouse gas emission reductions. However, as seen in Fig. 10.2, the current pathway and legal framework for 2030 can only bring us close to as much as 60% GHG emissions reductions in 2050.[5]

As part of the EU Green Deal, the European Commission has proposed to review the EU's greenhouse gas reduction target between 50 and 55% in 2030. A Communication and the accompanying impact assessment on this are expected as late as September 2020 together with a Communication on Smart Sector Integration. These non-legislative initiatives will be followed by legislative initiatives in 2021 that will focus

[5]"What is the progress in Europe towards international commitments regarding greenhouse gas emissions?" European Environment Agency, Accessed January 3, 2020, https://www.eea.europa.eu/data-and-maps/indicators/greenhouse-gas-emission-trends-6/assessment-3.

inter alia on the revision of the Renewable Energy Directive (RED2) and the decarbonisation of gas. The revision of a Directive such as RED2, only months after its entry into force, could be interpreted as a radical move. However, it accurately reflects the ambitious nature of the EU Green Deal and the gargantuan task ahead. Turning to mid-century, the target of reducing EU greenhouse gas emissions between 80 and 95% by 2050 compared to 1990 levels is no longer the EU's headline target. Following the conclusion of the meeting of EU Heads of States in December 2019, the European Council officially endorsed the European Commission's objective to achieve climate neutrality in 2050[6] and become the first climate-neutral continent in the world. This objective is set to be enshrined in an EU Climate Law, presented in March 2020.

10.3 *Quid* Energy Transition?

Without a doubt, the EU's new strategy changes the way one looks at the energy system and raises question marks over how future investment decisions should be taken. Within the context of the EU's sustainable finance agenda, now more than ever, there is increasing pressure on the shareholders of energy giants and infrastructure companies to ensure that investments are sustainable in the long term and that they are avoiding greenwashing, lock-in effects, and the risk of stranded assets. In particular, this new strategy presents challenges to the gas industry, posing the question: what role can gas and gas infrastructure play in the future climate-neutral economy? The ambition to reach climate neutrality will lead to a plethora of challenges across the gas value chain, from its production to transportation, all the way through the consumption and end-uses. As ever, challenges are not without opportunity. The gas infrastructure business, in particular, needs to carefully consider future innovative business models and its role in sectoral integration, forging stronger links to sectors such as agriculture and transport besides its

[6]One Member State (Poland), at this stage cannot commit to implement this objective as far as it is concerned, and the European Council will come back to this in June 2020, https://www.consilium.europa.eu/en/press/press-releases/2019/12/12/european-council-conclusions-12-december-2019/.

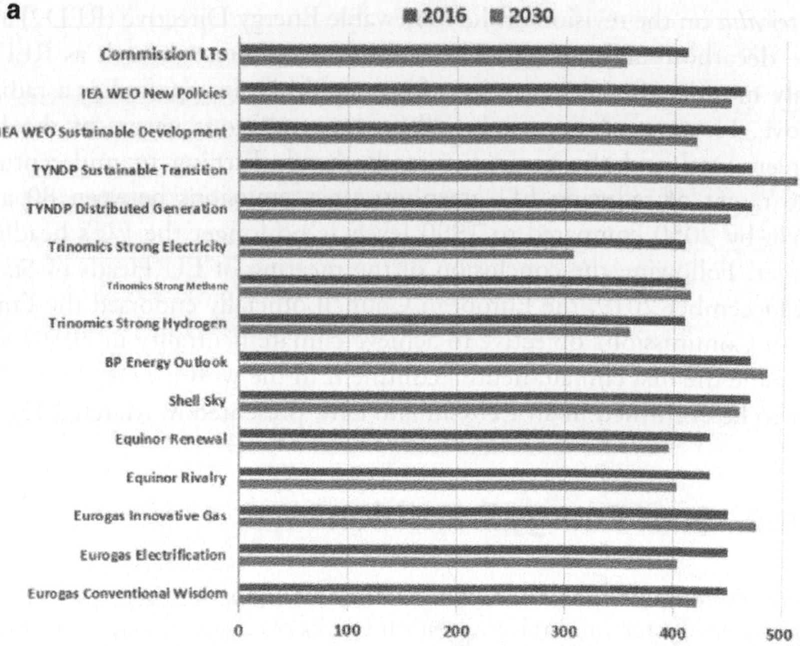

Fig. 10.3a Projected gas demand in 2030 in billion cubic metres (*Source* CEPS)

more traditional focus on markets such as power, heating and industry. It is worth reminding that today's energy mix in Europe is based 80% on molecules and 20% on electrons, meaning that the current energy system is heavily reliant on the security of supply, flexibility and affordability rendered possible by investments made into the development of European gas infrastructure and our interconnected internal market for gas. Nevertheless, as demonstrated in Figs. 10.3a and 10.3b, the golden age for gas as referred to by the International Energy Agency[7] (IEA) may

[7]"Golden Rules for a Golden Age for Gas", International Energy Agency, Accessed January 2, 2020, https://www.iea.org/reports/golden-rules-for-a-golden-age-of-gas.

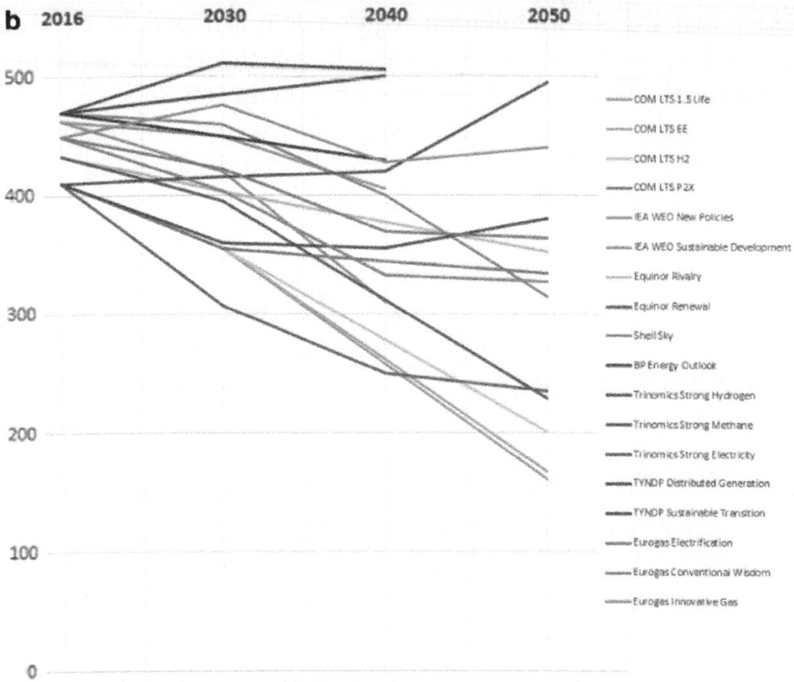

Fig. 10.3b Projected gas demand in 2040–2050 in billion cubic metres (*Source* CEPS)

have reached its zenith in Europe. Among others, the European Commission[8] and IEA projections[9] indicate that gas demand in Europe is set to decline rather drastically post 2030.

Several drivers are contributing to this, such as decreasing costs in renewables and energy efficiency as well as the EU's energy and climate targets themselves. Achieving climate neutrality, however, does not mean 100% electrification. While some policymakers entertained this vision for several years, the electricity sector has most recently conveyed the

[8] "2050 long-term strategy", European Commission, Accessed January 4, 2020, https://ec.europa.eu/clima/policies/strategies/2050_en.

[9] "World Energy Outlook", International Energy Agency, Accessed November 12, 2020, https://www.iea.org/reports/world-energy-outlook-2019.

message that EU decarbonisation requires as much as 60% electrification,[10] taking us to an energy system in which 60% is made up of electrons and 40% from molecules. The energy transition is a challenging process, and gas will have to go through its own transition on this journey. The gas business is no stranger to transitions as it has already been through two revolutions from coal gas and town gas to natural gas. For European gas infrastructure operators to continue playing an essential role in Europe, the third gas revolution is inevitable. This third revolution is already underway as gas infrastructure operators explore ways in which they can contribute to this new reality, bringing value to society and the European economy while futureproofing their assets.

10.4 The Future Role of Gas Infrastructure and the Role of Renewable Gases

For decades, the EU supported, politically and financially, the development of EU gas infrastructure (pipelines, LNG facilities, storage facilities) across the Union to develop an integrated, interconnected and competitive internal market for gas made up of diversified sources. The continued use of fossil fuels, however, has become politically intractable, with the European Parliament declaring a climate emergency in January 2020 and calling for an end to fossil fuel subsidies.[11] With gas infrastructure operators facing declining demand and stringent targets, an evolution in the narrative regarding the role of infrastructure operators is required. For example, the coal-to-gas switching narrative is floundering despite the fact that in 2019, this contributed to saving *circa* forty megatons of carbon dioxide emissions.[12] The gas industry needs to

[10] "Decarbonization pathways, European economy, EU electrification and decarbonization scenario modelling. Synthesis of key findings", Eurelectric, Accessed November 11, 2019, https://www.eurelectric.org/news/decarbonisation-pathways-electrification-part/.

[11] "European Parliament Resolution on the climate and environment emergency", European Parliament, Accessed December 12, 2019, https://www.europarl.europa.eu/doceo/document/TA-9-2019-0078_EN.html.

[12] "Global CO_2 emissions in 2019", International Energy Agency, Accessed March 12, 2020, https://www.iea.org/articles/global-co2-emissions-in-2019.

demonstrate that it can be innovative through R&D and pilot projects. Moreover, given the constraints of the 2030 targets around the corner, these pilots need to turn into real, sizeable, bankable and commercial projects sooner rather than later. The contention that gas infrastructure is, simply put, a remnant of the past age of fossil fuels and therefore, mass decommissioning is justified, is overly simplistic and would prove economically unwise. This infrastructure has been primarily paid by EU taxpayers to create value, bolster competition leading to affordable prices and, ultimately, to serve society. The energy transition will cost trillions of Euros to achieve, and policymakers have emphasised the importance of delivering a "just transition" in an affordable and sustainable manner. It has to be a just and inclusive transition; otherwise, it won't happen at all.[13] Utilising existing gas infrastructure, repurposing and redefining its role, can decrease the cost of the energy transition, thus avoiding the need for exponential investment costs in new electricity infrastructure, in some cases. Therefore, Europe's existing gas infrastructure should be viewed as a strategic asset and EU policy objectives would be better served by gearing towards the promotion and incentivising the decarbonisation or 'greening' of the commodity that is being transported and/or stored in this infrastructure. Gas infrastructure can serve to accommodate the uptake and development of renewable gases. Several studies have been undertaken at European and national level to explore the potential for renewable gases[14] with some studies highlighting the possibility to almost exclusively replace natural gas with renewable gases by 2050. In the case of France, the French Environment & Energy Management Agency (ADEME) explored the potential for a 100% renewable gas mix in 2050,[15] with the primary goal of the study being to analyse the conditions and economic feasibility of a gas system based entirely on renewable gas by 2050. As demonstrated in Fig. 10.4, the study found

[13]"Press remarks by President von der Leyen on the occasion of the adoption of the European Green Deal Communication", European Commission, Accessed December 1, 2020, https://ec.europa.eu/commission/presscorner/detail/sl/speech_19_6408.

[14]Mihnea Catuti, Christian Egenhofer & Milan Elkerbout, "The future of gas in Europe: Review of recent studies on the future of gas", *Centre for European Policy Studies (CEPS)*, August 2019, p. 15.

[15]"100% renewable gas mix in 2050? Study summary", ADEME, Accessed October 18, 2019, https://www.ademe.fr/a-100-renewable-gas-mix-in-2050.

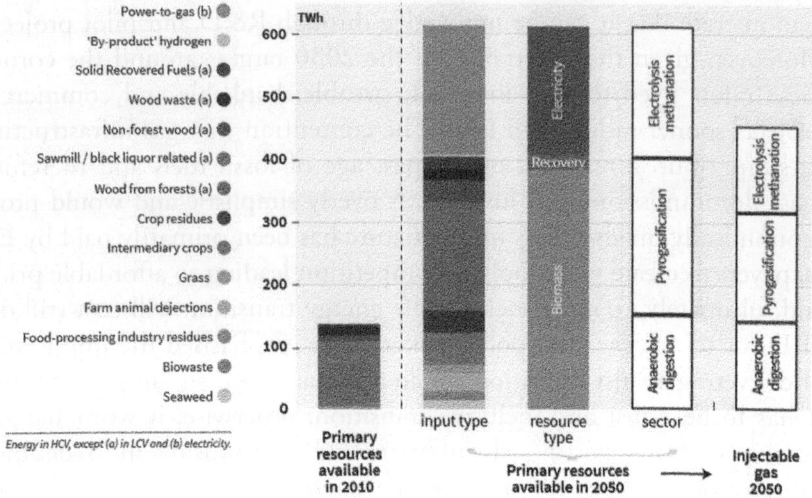

Fig. 10.4 Resource availability and potential production (*Source* ADEME)

that a 100% renewable gas mix in France is possible, based on a scenario where annual gas demand has decreased to 300 TWh in 2050, down from today's figure of 460 TWh. The study finds that there is a vital role for technologies that include anaerobic digestion, thermal gasification and power-to-gas.[16]

However, as seen in Fig. 10.5 regarding the production projections for renewable gas and hydrogen, the outcomes and differences related to production potential vary from one study to another due to the different assumptions, starting points as well as inputs and feedstocks used to produce renewable gases. Some studies have also exclusively focused on biomethane, some on hydrogen and others on both.

Given the potential for renewable gases and the versatile nature of gas as a commodity that can serve energy demand across different sectors of the economy, it is important to examine further the role that gas infrastructure can play in the years and decades to come. A study at EU level performed by Navigant on behalf of gas TSOs, predominantly from North to West Europe and European Biogas Producers found that "using

[16]Ibid., p. 5.

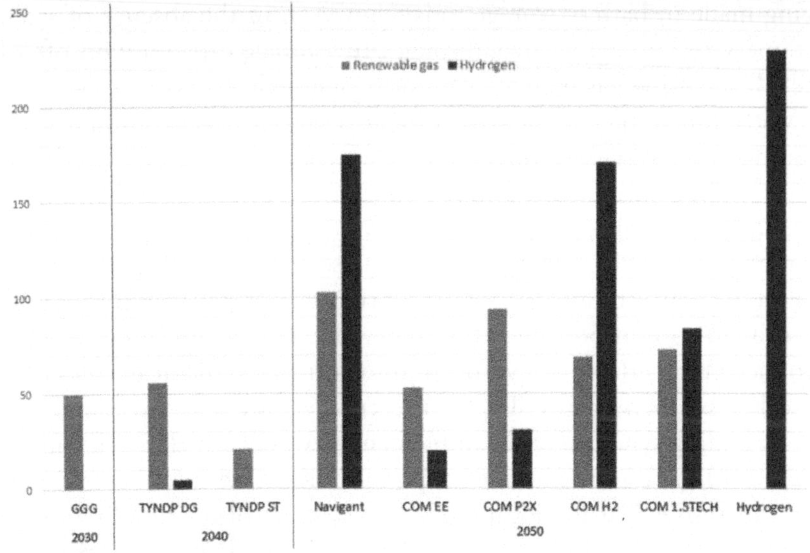

Fig. 10.5 Production projections for renewable gas and hydrogen (bcm) (*Source* CEPS)

around 2900TWh or approximately 270 billion cubic metres (natural gas equivalent) of green hydrogen and renewable methane through existing infrastructure across the EU, in smart combination with renewable electricity could save society €217 billion annually by 2050 compared to an energy system using a minimal amount of gas".[17] Indeed, the use of existing gas infrastructure to transport and store renewable gases in the EU offers an attractive value proposition, not only in economic terms ensuring a just transition, but also when it comes to one of the key energy topics of the future: energy storage. Today, there is 1200 TWh of gas storage in Europe, enough to cover average EU energy demand for three months. In electricity, we have 1.5 TWh of storage, enough for five to ten hours of average EU demand. Substantial investments are

[17]"Gas for Climate. A path to 2050: Our vision", Navigant, Accessed January 10, 2020, https://www.gasforclimate2050.eu/our-vision.

being made in battery storage today, however, in the absence of a revolutionary technological breakthrough, the storage capability on offer by the EU's existing gas network cannot be ignored.

With much of the new renewable electricity generation capacity coming from variable sources such as solar and wind and production centres often disconnected from areas and sectors with high energy demand, it is vital to have cost-effective storage and transport solutions. The gas system not only offers value as seasonal storage parameter; an insurance policy to aid the system during winter peaks, but it can also be utilised to transport and store renewable energy, for example locally produced biomethane or hydrogen from curtailed or dedicated electricity or other synthetic gases. The future energy system is more and more about finding synergies and complementarity between electricity and gas and coordination among energy players themselves and actors, which includes citizens, at the local level. The element of decentralisation and locally produced energy, in particular, is a topic that merits particular attention following the entry into force of the Renewable Energy Directive and relevant electricity market legislation which provide incentives for *prosumers* and local actors who wish to become active stakeholders in energy production. Identifying and developing these synergies between electricity and gas and integrating the element of decentralisation are vital factors for achieving a social license to operate and obtain an energy transition that is sustainable, socially just and affordable.

10.5 The Rise of Bioenergy, Circular Economy and Decentralisation

In 2018, the International Energy Agency referred to modern bioenergy as the "*overlooked giant of renewables*".[18] Bioenergy and renewable gas are projected to grow steadily in the coming years, with some Member States highlighting the role of biogas and biomethane prominently in their national energy and climate plans (NECPs) for 2030. Typically, there

---

[18] "Renewables 2018 analysis and forecasts to 2023 executive summary", International Energy Agency, p. 3.

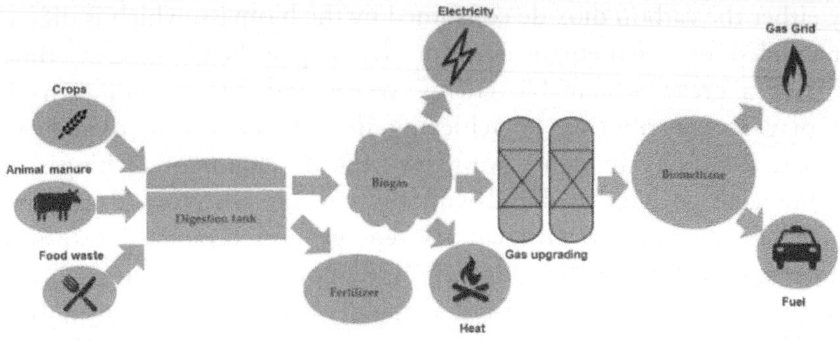

Fig. 10.6 Schematic overview of the anaerobic digestion process (*Source* Ecofys)

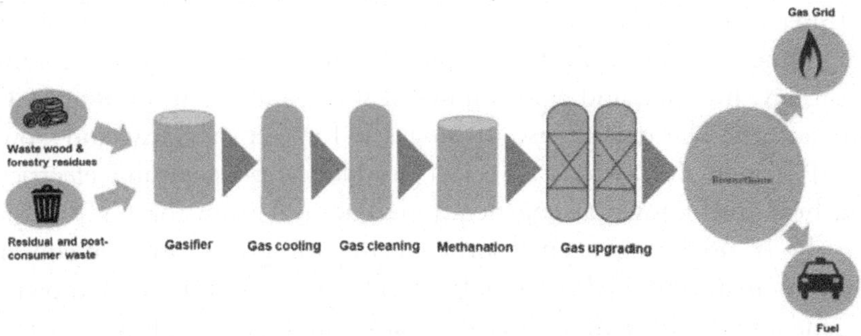

Fig. 10.7 Schematic overview of the thermal gasification process (*Source* Ecofys)

are two ways in which biogas can be produced: anaerobic digestion and thermal gasification. Anaerobic digestion, presented here in Fig. 10.6, is a mature technology[19] while thermal gasification, presented in Fig. 10.7, is still in the pilot phase.[20] Producing biomethane is made possible through upgrading from biogas via a purification process involving CO_2 removal.

It is important to note that the use of biogas, made from organic waste streams, does not add to the carbon dioxide load in the atmosphere; carbon dioxide produced during combustion of biogas is offset

[19]Ecofys, "Gas for Climate. How gas can help to achieve the Paris Agreement target in an affordable way", February 15, 2018, p. 13, available at: https://www.gasforclimate2050.eu/files/files/Ecofys_Gas_for_Climate_Feb2018.pdf, accessed February 15, 2020.

[20]Ibid., p. 14.

by either the carbon dioxide consumed by the biomass, which is digested or avoided fugitive methane emissions from open slurry storage.[21] Biogas is thus a green sustainable energy vector and has a significant role to play in contributing to achieving the EU long-term objective of climate neutrality in 2050. Moreover, the development of biogas and biomethane go hand in hand with the circular economy, thus constituting a low hanging fruit in the energy and ecological transition. Bioenergy contributes not only to greenhouse gas reductions in the energy sector and its associated demand sectors but also to sustainable food production as well as emissions reductions in the agricultural industry—one of the largest emitting sectors in the EU. Indeed, biogas producers handle a lot of livestock manure that is turned into digestate, a fertiliser which can increase the yield of a crop.[22] Furthermore, replacing one tonne of artificial fertiliser replaced with digestate saves one tonne of oil, one hundred and eight tonnes of water and seven tonnes of CO_2 emissions.[23] The default use of biogas is for Central Heat and Power (CHP) production, which is in fact production of renewable electricity and heat, also known as cogeneration. The heat from the CHP engine can also be used to drive an absorption chiller to give a source of cooling, resulting in a combined cooling, heat and power (CCHP), also known as trigeneration.[24] In terms of industrial demand, bioenergy and waste meet one-quarter of energy demand in the EU cement industry in accordance with robust waste management policies.[25] At the other end of the chain, upgrading to biomethane is developing rapidly; there are now close to six hundred biomethane-upgrading facilities worldwide, of which over five hundred are in Europe. Renewable gas for both heat and transport is the current growth sector in many countries. In the buildings sector, while

[21]"The role of anaerobic digestion and biogas in the circular economy," IEA Bioenergy Task 37, 2018, p. 8.

[22]Nutrients from food waste are also utilised to this end.

[23]"How anaerobic digestion benefits everyone", Anaerobic Digestion and Bio Resources Association, Accessed December 16, 2020, http://adbioresources.org/about-ad/how-ad-benefits-everyone.

[24]"The role of anaerobic digestion and biogas in the circular economy," IEA Bioenergy Task 37, 2018, p. 9.

[25]"Renewables 2018 analysis and forecasts to 2023 executive summary", International Energy Agency, p. 5.

there is a strong tendency towards the installation and use of heat pumps instead of gas boilers, heat pumps struggle to perform efficiently enough to heat homes when temperatures drop below four to five degrees celsius. It, therefore, makes sense to consider hybrid heat pumps—heat pumps plus existing gas boiler that acts when needed—as a solution in existing buildings. In the transport sector, electrification of the light-duty vehicles segment is a top priority for the EU. Lack of technological advancement, however, leaves room for renewable gas penetration into the heavy-duty sector.

The renewable energy directive prescribes a 14% transport sub-target for advanced biofuels. Within this broader target, there is a dedicated target of reaching at least 1% advanced biofuels and biogas by 2025 and 3.5% by 2030. These targets could be reviewed when the Renewable Energy Directive is reviewed in 2021. Much will also depend on the results of the review clause of the EU regulation on CO_2 standards for light-duty vehicles which is now expected to take place in 2021 instead of 2023. Should the review find that the EU is not on track to meet the 2025 target of 15% CO_2 reductions, there could be more opportunity for biomethane to play a role in this segment. The development of bioenergy is Green Deal compliant. It contributes to the EU's vision for a climate-neutral economy in 2050 but should also be viewed within the context of the circular economy and the EU's new *Farm to Fork Strategy*. It can further contribute to achieving the EU's targets for waste reduction outlined in the *Waste Framework Directive*. The potential to create new revenue streams for farmers, while also helping them reduce waste, is non-negligible. It can also help boost employment prospects, with one study finding that by 2050, 100,000–150,000 direct jobs in agriculture and forestry could be created to provide feedstock to biomethane production facilities, with another 200,000–300,000 direct industrial jobs created, related to the development and operations of digesters, thermal gasification plants and electrolysers in the hydrogen business.[26] Lastly, a favourable and sustainable framework for the development of decentralised bioenergy in Europe will also contribute towards reducing reliance on imported fossil fuels in favour of a more environmentally

[26]"Job creation by scaling up renewable gas in Europe," Navigant, November 2019, p. 2.

friendly, local solution. It will also support the promotion and protection of EU industry and manufacturing. In particular, Europe is a leader in the development of anaerobic digesters and should continue to be in the lead on this technology. The World Bank estimates a nine per cent increase in world population by 2030, which leads to a seventy per cent increase in waste generation. This suggests that anaerobic digesters could well become an export technology for Europe.

10.6 The Return of Hydrogen

The prospect of hydrogen is not new. In 1874, Jules Verne wrote in his book the Mysterious Land, "I believe water will one day be employed as fuel, that hydrogen and oxygen which constitute it, used singly or together, will furnish an inexhaustible source of heat and light, of an intensity of which coal is not capable". One hundred forty-six years later, the prospects for hydrogen have never been better and political support has never been more promising. Unlike bioenergy, it is not prone to the prejudices created in the past decade relating to the conflict between food crops and energy crops for bioenergy production. As a clean energy carrier, hydrogen is an energy carrier, a fuel and a feedstock that can also be used across major energy demand sectors. Hydrogen can be produced from both natural gas and renewable electricity and is at the crossroads of numerous crucial technologies related to the energy transition. Hydrogen can be produced through water electrolysis with low carbon or renewable power, and Steam Methane Reforming (or auto thermal reforming and/or methane cracking) with CCS/CCU. Alternative ways to produce hydrogen using renewable gases or waste include Biogas/biomethane reforming and biomass gasification/pyrolysis. Electrolysis, in particular, is set to play an important role in the future energy system. While Europe currently claims to hold a robust presence and role as a leader in the electrolysis market (integrators, component providers, OEMs), currently, the technology remains relatively expensive. High capital costs necessary for this technology, require broader markets and additional development to achieve industrial scale-up, necessary to bring costs down. The absence of an adopted policy and regulatory framework

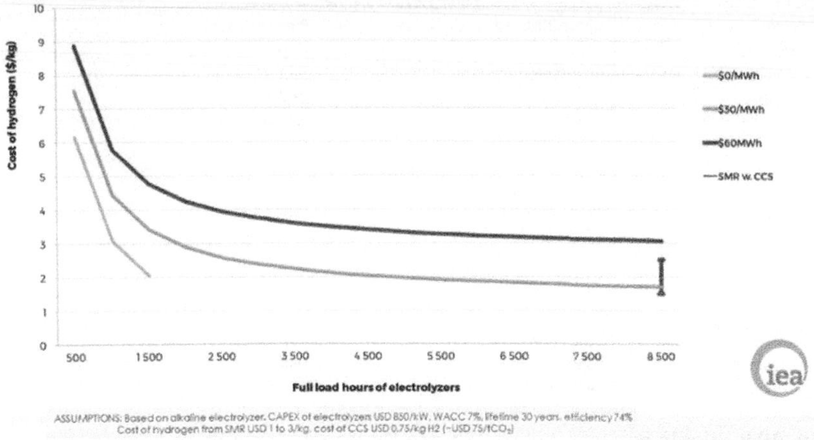

Fig. 10.8 Cost of hydrogen from electrolysis for different electricity costs and load factors ($/kg)—Energy Transitions Commission (2018) (*Source* Energy Transitions Commission)

for electrolysis is constraining market development.[27] However, as shown in Fig. 10.8, costs are expected to decrease dramatically due to several factors including the uptake of power-to-gas technology, the increase in the share of renewable energy in the EU and the associated decrease in the cost of renewable electricity. Combining these developments with the availability of existing gas infrastructure and geological storage for the transportation and storing of hydrogen, Europe can further increase its global leadership in hydrogen.

Indeed, power-to-gas is a crucial enabler for linking the electricity and gas sectors. It consists of converting renewable electricity into gas to provide a storage and flexibility solution; flexibility being the cornerstone of tomorrow's power system. Wind and solar facilities produce electricity, but not always when consumers need it. Power-to-gas can use this surplus electricity to produce hydrogen. Europe's gas infrastructure can play a pivotal role in bringing the hydrogen economy closer to reality as it can facilitate the upscaling of renewable energy sources;

[27]"Hydrogen from renewable power: Technology outlook for the energy transition," International Renewable Energy Agency, 2018 Abu Dhabi.

Cable versus pipeline		
Name	Cable (BritNed)	Pipeline (BBL)
Capacity	1GW	20GW
Construction costs	EUR 600M	EUR 500M
Distance	260 km	230km
Volume (year)	8TWh	120TWh

Cable versus pipeline		
Name	Pipeline (IUK)	Cable (NEMO)
Capacity	FWD (import BE): 27 153 MWh/h REV (export BE): 33 476 MWh/h	1000 MW
Construction costs	EUR 713M	EUR 690M
Distance	235 km	140km
Volume (year)	240TWh	8TWh

Fig. 10.9 Cable versus pipeline (*Source* Hydrogen Europe)

provide a cost-effective option for transport, long-term seasonal storage, grid balancing and security of supply while at the same time contributing to the decarbonisation of gas infrastructure and gas end-uses. The integration of hydrogen into the gas grid will facilitate the penetration of hydrogen across the economy, thanks to the interconnected nature of EU's internal gas market, given access to varying geographies and scales—from clusters, to the local and regional level as well as across borders. The development and uptake of hydrogen can be further accelerated through the identification of suitable locations of large demand centres with energy needs stretching across various sectors. Industrial clusters or ports, for example, could be prime takers of hydrogen in the future. Already, several projects are being examined in this regard.[28] Under this framework, the transportation and storing of hydrogen can begin as a more local phenomenon which could extend to transportation over larger distances, under certain conditions. The advantage of using the gas system compared to electricity lies in its ability to transport larger volumes of energy over long distances. A comparison between a power cable and a natural gas interconnector, presented here in Fig. 10.9, can offer a useful insight into the fundamental differences between the two. Figure 10.9 compares the BBL natural gas interconnector and

[28]See Hydrogen for Climate Action Projects, https://www.hydrogen4climateaction.eu/projects.

the BritNed high-voltage direct-current submarine power cable. Both infrastructures needed approximately the same investment capital for the construction phase, while both connect the Netherlands to the UK. However, while BritNed has a capacity of 1 GW, the BBL pipeline has a total capacity of more than 20 GW. The comparative advantage of the gas infrastructure is thus evident.[29]

Gas infrastructure can spur the potential development of a hydrogen economy in Europe through a phased approach. At first, this can begin through the blending of hydrogen with natural gas in existing pipelines. Several European Transmission System Operators have been studying the technical and economic considerations of hydrogen injection into the natural gas grid. As presented in Fig. 10.10, most literature findings converge on the fact that one can blend up to 20% of hydrogen with natural gas in existing infrastructure, with associated adaptation costs increasing the higher the blend.[30]

However, as seen in Fig. 10.11 that follows, in a mix above 20%, it makes more sense to either retrofit or convert old gas pipelines for hydrogen use or invest in new dedicated hydrogen pipelines.[31]

10.7 Conclusion and Future Challenges

The development of renewable gases across the European Union can contribute to achieving climate neutrality by 2050 and creating a new *raison d'etre* for European gas infrastructure operators; one that is more sustainable and connected to the EU Green Deal. Aspiring for a hybrid energy system that links electricity and gas and makes the best use of their competitive advantages makes sense from both an environmental and energy efficiency standpoint but also an economic point of view.

[29] "Hydrogen Europe vision on the role of hydrogen and gas infrastructure on the road toward a climate neutral economy; A contribution to the transition of the gas market", Hydrogen Europe, April 2019, pp. 7–8.

[30] "Technical and economic conditions for injecting hydrogen into natural gas networks," French gas operators, 2019, p. 22, https://www.grtgaz.com/fileadmin/plaquettes/en/2019/Technical-economic-conditions-for-injecting-hydrogen-into-natural-gas-networks-report2019.pdf.

[31] Ibid., p. 23.

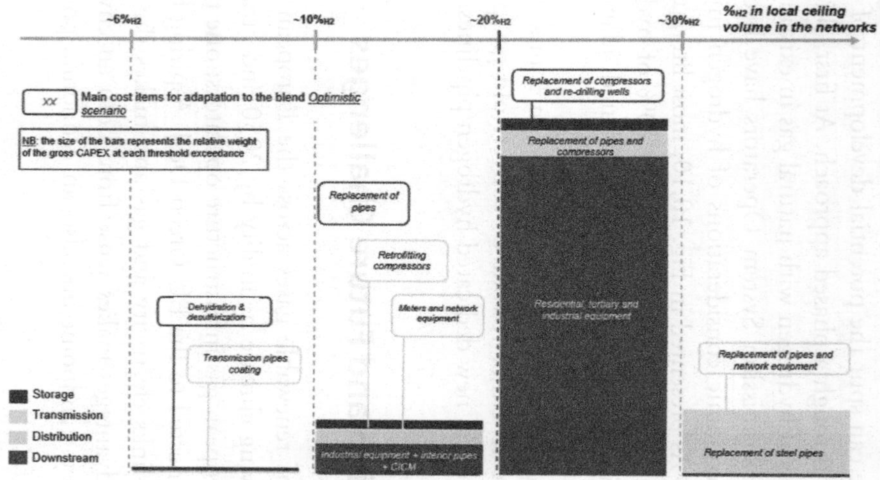

Fig. 10.10 Percentage of H_2 in local ceiling volume in the networks (*Note* The operating costs of H_2/CH_4 separation upstream of NGV stations or groundwater aquifiers take effect from 1% H_2. *Source* French gas operators)

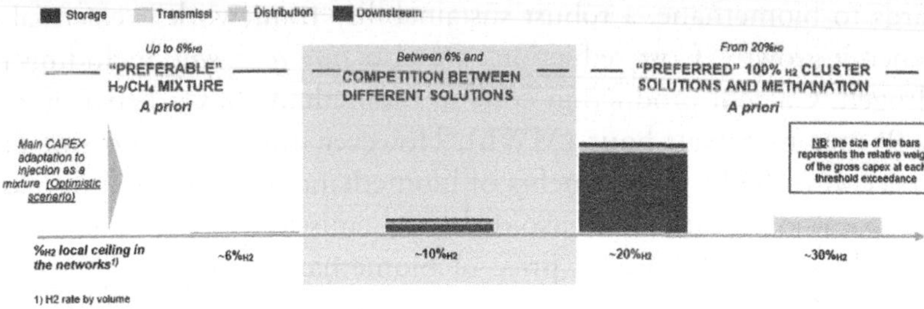

Fig. 10.11 Higher than 20% of H_2 in local ceiling volume in the networks (*Source* French gas operators)

Inevitably, making this a reality comes with several challenges. In general terms, the concept of "two-speed Europe" is a notion that also extends to the EU Green Deal. While the EU Member States have committed to the destination (climate neutrality in 2050), the journey of each Member State will be different, with some countries having to endure more hardship than others. To this extent, the EU has developed tools and mechanisms to assist more vulnerable countries through proposals that include a Regulation for a *Just Transition Fund Mechanism* and a *Sustainable Investment Plan.* From a market perspective, the advent of renewable gases poses specific difficult questions regarding the functioning of the internal gas market. Traditionally, the development of the internal gas market has revolved around natural gas. Investments and market developments over the past decades have culminated to the creation of a homogeneous market for natural gas. While biomethane can easily be integrated into the existing market, the integration of hydrogen creates complexities that will need to be addressed in future legislation. Whether a separate market for hydrogen will be proposed in 2021 as part of a gas decarbonisation package is still unknown at this stage. The existing Gas Regulation and Gas Directive may need to be reviewed together with the *gas quality Directive.* Speculation aside, policymakers should need to be diligent in facilitating solutions that are interoperable while supporting well-functioning markets internally and cross-border. Regarding scalability, at present, the numbers for renewable gases remain relatively modest, and policy measures are needed to boost the uptake of new sustainable gases to substitute fossil fuels. With regards to biomethane, a robust sustainability framework is essential to bioenergy growth. Cost reductions are also *sine qua non*, including for hydrogen. Current production costs of biomethane hover between €80 and 95 per megawatt-hour (MWh). However, these costs do not take into account the broader benefits of biomethane, such as rural development, local job creation and avoided CO_2 emissions. If all those factors are put together, the overall price of biomethane could fall to around €35–55/MWh, which is close to the current average wholesale price of electricity and not far from the natural gas prices, which currently are at historic lows. Favourable feed-in tariffs can contribute to the uptake of biogas/biomethane, however other solutions are needed as well, for example blending obligations in heating and for industry, *akin* to the obligations prescribed for the transport sector. More generally, the EU needs to legislate on clear definitions for what is considered a renewable gas. This is as a starting point for any other discussions, for example on proposing targets for renewable gases which can help give signals to investors and end-users. The EU also needs to carefully review policies related to guarantees of origin as they relate to renewable gases.

At present, guarantees of origin are not linked to sustainability criteria, meaning that there is no guarantee that the energy being delivered is green. Moreover, any discussion on the future of gas cannot take place without evoking and addressing the issue of methane emissions. More data is required to understand the issue better. The EU is currently developing a *methane strategy* which is set to be unveiled during 2020 or, at the latest, in 2021. Monitoring and reporting obligations across the gas industry are likely to be among the proposals outlined in this strategy. The coming year is also crucial, specifically, with regards to the future of energy infrastructure investments as we await the evaluation and review of the *Regulation on Trans-European Networks for Energy* (TEN-E). There has been much controversy associated with the adoption of lists of projects of common European interest; a list that is governed by the TEN-E regulation. The new TEN-E needs to be aligned with the Green Deal and the EU's long-term climate and energy objectives. The decarbonisation value of a project as well as its contribution to sectoral integration ought to be considered as new criteria. Finally, while the EU Green Deal sets out a cross-cutting approach, the fact remains that legislative proposals will continue to be proposed separately and dealt with by different committees in the European Parliament. Stakeholders also need to take into consideration the trade and industrial competitiveness components of the EU Green Deal. Europe has to compete on a global level and cannot afford to fall behind in the technological race as it did with photovoltaic panels. The adherence of policymakers to the holistic nature of the EU Green Deal and the need for consistency and coherence across policy areas is a prerequisite for success.

References

ADEME, "100% renewable gas mix in 2050?" Study summary, Accessed October 18, 2019, https://www.ademe.fr/a-100-renewable-gas-mix-in-2050.

Anaerobic Digestion and Bio Resources Association, "How anaerobic digestion benefits everyone", Accessed December 16, 2020, http://adbioresources.org/about-ad/how-ad-benefits-everyone.

CEPS, "The future of gas in Europe: Review of recent studies on the future of gas", August 2019, https://www.ceps.eu/wp-content/uploads/2019/08/RR2019-03_Future-of-gas-in-Europe.pdf.

Ecofys, "Gas for Climate. How gas can help to achieve the Paris Agreement target in an affordable way", February 15, 2018, https://www.gasforclimate2050.eu/files/files/Ecofys_Gas_for_Climate_Feb2018.pdf.

Energy Transitions Commission, "Electricity, hydrogen & hydrogen-based fuels in a zero-carbon economy", June 2018, p. 13, Accessed February 16, 2020, http://energy-transitions.org/sites/default/files/ETC-Consultation-Paper-Electricity-and-hydrogen-based-fuels-website.pdf.

Eurelectric, "Decarbonization pathways, European economy, EU electrification and decarbonization scenario modelling. Synthesis of key findings", Accessed November 11, 2019, https://www.eurelectric.org/news/decarbonisation-pathways-electrification-part/.

European Commission, "Press remarks by President von der Leyen on the occasion of the adoption of the European Green Deal Communication", Accessed December 1, 2019, https://ec.europa.eu/commission/presscorner/detail/en/speech_19_6749.

European Commission, "2050 long-term strategy", Accessed January 4, 2020, https://ec.europa.eu/clima/policies/strategies/2050_en.

European Environment Agency, "What is the progress in Europe towards international commitments regarding greenhouse gas emissions?", Accessed January 3, 2020, https://www.eea.europa.eu/data-and-maps/indicators/greenhouse-gas-emission-trends-6/assessment-3.

European Parliament, "European Parliament Resolution on the climate and environment emergency", Accessed December 12, 2019, https://www.europarl.europa.eu/doceo/document/TA-9-2019-0078_EN.html.

Eurostat, "Greenhouse gas emission trends", Accessed January 3, 2020, https://ec.europa.eu/eurostat/statistics-explained/index.php/Greenhouse_gas_emission_statistics.

French gas operators, "Technical and economic conditions for injecting hydrogen into natural gas networks", June 2019, https://www.grtgaz.com/fileadmin/plaquettes/en/2019/Technical-economic-conditions-for-injecting-hydrogen-into-natural-gas-networks-report2019.pdf.

Hydrogen Europe, "Hydrogen Europe vision the role of hydrogen and gas infrastructure on the road toward a climate-neutral economy; A contribution to the transition of the gas market", April 2019, https://hydrogeneurope.eu/sites/default/files/2019_Hydrogen%20Europe%20Vision%20on%20the%20role%20of%20Hydrogen%20and%20Gas%20Infrastructure.pdf.

International Energy Agency, "Global CO_2 emissions in 2019", Accessed March 12, 2020, https://www.iea.org/articles/global-co2-emissions-in-2019.

International Energy Agency, "Golden Rules for a Golden Age for Gas", Accessed January 2, 2020, https://www.iea.org/reports/golden-rules-for-a-golden-age-of-gas.

International Energy Agency, "Renewables 2018 analysis and forecasts to 2023 executive summary".

International Energy Agency, "The role of anaerobic digestion and biogas in the circular economy", Bioenergy Task 37, 2018.

International Energy Agency, "World Energy Outlook", Accessed November 12, 2020, https://www.iea.org/reports/world-energy-outlook-2019.

Mihnea Catuti, Christian Egenhofer & Milan Elkerbout, "The Future of Gas in Europe: Review of recent studies on the future of gas", *Centre for European Policy Studies (CEPS)*, August 2019, p. 15.

Navigant, "Gas for Climate. A path to 2050: Our vision", Accessed January 10, 2020, https://www.gasforclimate2050.eu/our-vision.

Navigant, "Job creation by scaling up renewable gas in Europe", November 2019.

11

Energy in Transition: Investing in Green Energy and Circular Economy to Enhance Energy Security

Spyros Kiartzis, Katerina Tsita, and Evangelia Paschalidou

11.1 Introduction

The energy transition is already underway, driven by rising demand, constrained supplies and concerns about greenhouse gas (GHG) emissions and their link to climate change.[1] The predominant task of

[1] Eric Spiegel, Neil McArthur and Rob Norton, *Energy shift game-changing options for fueling the future* (McGraw Hill, 2009), pp: 62–164, ISBN: 978-0-0715083-4-6.

S. Kiartzis (✉)
Hellenic Petroleum S.A., Maroussi, Greece
e-mail: skiartzis@helpe.gr

K. Tsita
Department of Economics, Aristotle University of Thessaloniki, Thessaloniki, Greece

E. Paschalidou
International Institute for Industrial Environmental Economics (IIIEE), Lund University, Lund, Sweden

M. Mathioulakis (ed.), *Aspects of the Energy Union*,
Energy, Climate and the Environment,
https://doi.org/10.1007/978-3-030-55981-6_11

policymakers, business leaders and the many other stakeholders involved in energy is to find ways to accelerate the process for delivering the vast amounts of low-carbon energy that will be needed in the coming decades to support economic growth. In this context, redefining energy security can be done by upgrading from the level of national self-interest to a global scale that reflects the increasing internationalization and interdependence of energy policies. EU Energy Union strategy aims to provide secure, sustainable, competitive and affordable energy at Union's consumers.[2] Europe's strategic long-term vision for a prosperous, modern, competitive and climate neutral economy has enhanced its energy security. This is achieved through interconnecting national markets, further diversifying energy sources, deploying indigenous renewables, implementing energy efficiency measures and promoting an enabling environment for investment.[3] All future scenarios in which climate and energy security are enhanced, envisage increased use of renewable energy sources. They are free from constraints on fuel supply, produce no or few GHG emissions and are geographically widely available; technological advances are expected to bring down costs. However, achieving this level of emissions reduction will require rapid and widespread diffusion of renewables technologies and the introduction of new energy systems and infrastructures that are more suited to the more dispersed and intermittent generation associated with renewable energy.[4] A significant milestone in the attempt to transform our energy systems is the ambitious agreement reached by the United Nations climate change conference (COP21) in Paris.[5] Fluctuations in energy products and raw materials prices, climate change concerns and

---

[2]COM/2015/080, "A framework strategy for a resilient Energy Union with a forward-looking climate change policy." 2015. Accessed February 22, 2020. https://eur-lex.europa.eu/legal-content/EN/TXT/?uri=COM:2015:80:FIN.

[3]COM (2019) 175 Final, "Fourth report on the State of the Energy Union." 2019. Accessed February 22, 2020. https://ec.europa.eu/commission/sites/beta-political/files/fourth-report-state-of-energy-union-april2019_en_0.pdf.

[4]Antony Froggatt and Michaela A. Levi, "Climate and energy security policies and measures: Synergies and conflicts." *Foreign Affairs* 85, no. 6 (November 2009): 1129–1141.

[5]Neste, "Taking action on climate change Neste outlook for changing markets." Accessed February 22, 2020. https://www.neste.com/sites/default/files/attachments/taking_action_on_climate_change_version2.pdf.

political problems in providing reliable access to oil reserves lead to new fuels in the transport sector. Alternatives should be economically and environmentally viable and socially accepted. Unconventional fossil carbon sources produce less expensive fuels when compared to renewable energy and even more compatible with the existing transport infrastructure but would lead to increased carbon emissions.[6] Biofuels are a promising solution for the transport sector unbundling from conventional fuels. A variety of technology developments and alternative fuels can only answer the problems of growing transportation fuel use and GHG emissions. Coordinated efforts are required to develop more efficient vehicles and benign fuels and to find more sustainable ways to satisfy transportation demands. Apart from regulatory instruments, a key factor is the role that incentive-based policies such as feebates, taxes, pay-as-you-drive insurance and scrappage incentives can play.[7] These policies contribute to achieving (a) Rush development and deployment of appropriate technologies and generate market pull for those technologies through synergies and (b) Be transparent and appear fair to all stakeholders, especially those bearing the highest costs of the necessary transitions. To this direction, fleet electrification has been proposed as a strategy for reducing air pollutant emissions and improving air quality in urban areas. Several new propulsion systems such as plug-in hybrid electric vehicles (PHEVs), range extenders as well as electric vehicles (EVs) have emerged and entered the market or are ready to enter the market in the near future.[8] However, cost-disadvantages of newly emerging propulsion systems, as well as their limited driving range, have to be overcome

---

[6]Kahn Suzana Ribeiro, Shigeki Kobayashi, Michel Beuthe, Jorge Gasca, David Greene, David S. Lee, Yasunori Muromachi, Peter J. Newton, Steven Plotkin, Daniel Sperling, Ron Wit and Peter J. Zhou, "Transport and its infrastructure." In *Climate change 2007: Mitigation. Contribution of Working Group III to the Fourth assessment report of the Intergovernmental Panel on Climate Change* [B. Metz, O.R. Davidson, P.R. Bosch, R. Dave, L.A. Meyer (eds)]. Cambridge University Press, Cambridge, UK and New York, NY, 2007: 325–327.

[7]Anup Bandivadekar, Kristian Bodek, Lynette Cheah, Christopher Evans, Tiffany Groode, John Heywood, Emmanuel Kasseris, Matthew Kromer and Malcom Weiss. "On the road in 2035: Reducing transportation's petroleum consumption and GHG emissions." *Report LFEE 2008-05RP* MIT, 2008.

[8]Evanthia A. Nanaki, Spyros Kiartzis and George A. Xydis, "Are only demand-based policy incentives enough to deploy electromobility?" *Policy Studies* (2020). https://doi.org/10.1080/01442872.2020.1718072.

to achieve a shift in the transportation sector. In this context, finding effective policy measurements for the promotion of the use of EVs is of great significance for a sustainable transportation system. Transport electrification alone will be unable to cope with the decarbonization challenge. The significant contribution of heavy vehicles, marine and air transportation into CO_2 emissions cannot be reduced without enhanced biofuels introduction into the fuel mix. Replacing diesel and gasoline with low-carbon fuels is the proper sustainable solution. Since the objective is low-carbon fuels, the next-generation biofuels should be added to the existing crop-based ones. Although next-generation biofuels industry is still in its infancy, strong EU policy support via blending obligation targets is expected to provide a long-term perspective and enlarge the market.[9] Next-generation biofuels are mainly derived from lignocellulosic biomass from agricultural and forest residues, wastes and used oils. Next-generation biofuels do not compete with the food chain since they are produced from cheap, abundant and locally available raw material while contributing to the management and utilization of waste.[10] However, there are two significant risks, i.e. technology maturity and feedstock security of supply that can be addressed through Forward contracts. Technology maturity and the corresponding increased cost issues can be resolved by constructing commercial-scale plants to reduce capital expenditure.[11] The commercialization of new technologies related to energy and transportation infrastructure is a time-consuming process.

[9]Katerina G. Tsita, Spyros J. Kiartzis, Nikolaos K. Ntavos and Petros A. Pilavachi, "Next generation biofuels derived from thermal and chemical conversion of the Greek transport sector." *Thermal Science and Engineering Progress* (2019). https://doi.org/10.1016/j.tsep.2.09.100387.

[10]Stylianos Stefanidis, Petros Pilavachi, Konstantinos Kalogiannis and Aggelos Lappas (in Greek), "Pyrolysis of agricultural and forest residues for the production of fuels and chemical products." *Workshop of the Energy Committee of the Academy of Athens—Management and Energy Exploitation of Residues in Greece*, 147–162 (2015). Athens, Greece.

[11]Simone Alfano, Federico Berruti, Nicolas Denis and Alberto Santagostino, "The future of second-generation biomass." 2016. Accessed February 22, 2020. https://www.mckinsey.com/~/media/McKinsey/Business%20Functions/Sustainability/Our%20Insights/The%20future%20of%20second%20generation%20biomass/The-future-of-second-generation-biomass.ashx.

Investment decisions made today is a part of a long-term plan of collaboration between a broad range of stakeholders.[12] Many companies are already well advanced in the employment of low-emission strategies, while motivated by consumer choices and existing regulations enhancing their corporate social responsibility agenda.[13] These strategies include self-generation of energy, using on-site renewables and other low-carbon technologies, electrification of vehicle fleets and water and waste management strategies. Incorporating climate change into our thinking in business, or in government, is not a one-time change that we're adapting to but it's moving out of stability into permanent instability.[14] Therefore, the duration of any decision becomes more complicated. This chapter begins with the introduction of the ways in which energy security and climate change overlap through renewable energy contribution. The analysis continues with a presentation of the new era in transport that relies on advanced biofuels and electrification and continues with a discussion on green energy investments and circular economy's contribution. The chapter concludes with a brief review of the options available for investing in a low-carbon energy economy, also highlighting the emerging energy security dilemmas.

11.2 Energy Security at Crossroads

IEA defines energy security as "the uninterrupted availability of energy sources at an affordable price". Energy security can be seen from two perspectives: either from a long-term one, concerning primarily the well-timed investments in supplying energy accordingly to the economic

---

[12]OGCI, "Catalyst for change collaborating to realize the energy transition." 2017. Accessed February 22, 2020. https://oilandgasclimateinitiative.com/wp-content/uploads/2017/10/OGCI-2017-Report.pdf.

[13]The Economist Intelligence Unit. "Ready for the change—Pathways to a low-emissions future." 2019. Accessed February 22, 2020. https://futurepresent.economist.com/ready-for-change-pathways-to-a-low-emissions-future/.

[14]Spencer Glendon and Simon London, "Redefining the power industry." *McKinsey Quarterly*, 2019. Accessed February 22, 2020. https://www.mckinsey.com/~/media/McKinsey/Business%20Functions/Sustainability/Our%20Insights/Redefining%20the%20power%20industry/Redefining-the-power-industry-vF.ashx.

developments and the environmental needs, in a sustainable way, or; from a short-term one, meaning the resiliency of the energy system to sudden supply-demand fluctuations.[15] Furthermore, due to physical unavailability of energy or uncompetitive and rigid prices, the lack of energy security can be associated with their negative social and economic impacts. The aforementioned risk of physical unavailability is only restricted to extreme events when referring to the international oil market, while supply security concerns are primarily linked to economic losses as a result of extreme price spikes. In electricity and natural gas markets, though, where the transmission systems must be kept continuously in balance, the concern for physical unavailability of supply is more widespread. For example, this is related to cases where there are capacity constraints or where the supply-demand balance cannot be restored from the price mechanism of the markets. Moreover, in the World Economic Forum's Global Risks Report,[16] environmental risks are still prevailing over the results, since they are considered for three of the top five risks by likelihood and four by impact. The highest concern risk was extreme weather, along with an increased worry about possible environmental policy failure. The results of climate inaction are becoming increasingly apparent. Climate change has driven substantial change in the world's infrastructure needs since there is now more awareness of the risks it poses and greater consensus on the need for collective policy responses. The low-carbon transition will shape the profile of infrastructure investment in multiple ways. In the energy sector, investment in renewables is likely to accelerate; and in the transport sector infrastructure will need to be adapted to manage increasing shares of EVs, as well as huge projected increases in road, air and sea traffic. Moreover, sensor-based technologies are likely to be widely deployed across all kinds of networks and grids, increasing demand for the digital infrastructure on which they rely. The rapid rollout of sustainable infrastructure is likely to lead to continuing financial innovation as more investors move into this market. Digital

---

[15]OECD/IEA, "Energy supply security: Emergency response of IEA Countries 2014." 2014. Accessed February 22, 2020. https://webstore.iea.org/energy-supply-security-the-emergency-response-of-iea-countries-2014.

[16]World Economic Forum, "The global risks report 2019—14th edition." 2019. Accessed February 22, 2020. http://www3.weforum.org/docs/WEF_Global_Risks_Report_2019.pdf.

transformation of energy networks requires long-term investments in technology and infrastructure. Globally, the need to pursue a new vision of energy security—one based on availability, reliability, affordability and sustainability—poses enormous challenges.[17] The above four pylons are directly affected by fostering network digitalization. The more digital energy networks become, the more fragile they turn to cyber-attacks. Enhancing cybersecurity in the energy sector is a critical priority. The electricity grid is the largest physically interconnected machine in the world. It consists of thousands of power plants, substations, kilometres of high-voltage transmission lines and millions of kilometres of low-voltage distribution lines. Cyber threats are initiated from sophisticated actors that are prioritizing as a top security concern. A vulnerability assessment framework is required to systematically evaluate the vulnerabilities of SCADA (Supervisory Control and Data Acquisition) systems at all levels: system, scenarios and access points. As grids get smarter, cybersecurity protections are imperative and must not interfere with energy delivery functions. In smart grids, advanced metering infrastructure enables faster outage restoration while distribution automation can also trigger grid self-healing functions. Finally, energy storage can provide contingency reserves for grid stability.

The reality of renewable energy's contribution to energy and climate security is more complicated. Although from a fuel perspective, renewable energy enhances the security of supply, significant concerns are raised on the import dependence for critical technologies and materials needed for renewable power and related storage.[18] Large amounts of renewable resources, including water and non-renewable materials such as metals, steel and concrete, are required to capture energy using the above technologies. Largescale biofuel farming can result in ecological damage due to ecosystem alteration and loss of habitat, leaching of fertilizers and pesticides, and energy consumption for crop planting, harvesting and processing. Thus, methodologies to reduce energy consumption by improving energy conversion efficiencies and

[17]Carlos Pasqual and Jonathan Elkind (eds), *Energy security—Economics, politics, strategies and implications* (Brookings Institution Press, 2010), pp: 120–130, ISBN: 978-0-8157-6919-4.

[18]Antony Froggatt and Michaela A. Levi, "Climate and energy security policies and measures: Synergies and conflicts." *Foreign Affairs* 85, no. 6 (November 2009): 1129–1141.

lifestyle changes should be an integral part of policies to limit GHG emissions.[19] Bioenergy has a key role in energy security issues, and related policy goals are enhanced through technology advancements and level economic playing fields, for crop production, conversion and end-use. Sustainable bioenergy provides flexibility to address multiple energy needs —power, fuels and heat— with locally available, nationally adaptable solutions that adjust to local resource availability, seasonal needs and diversity priorities.[20] However, bioenergy has risks associated with weather extremes, economic competitiveness, and crop-related disease or pest infestation. As international trade expands, bioenergy issues will play an increasingly larger role in the geopolitical dialogue, including the complexities across multiple energy segments and the interconnectivity with other geopolitical issues including food, water, trade, human rights and conflict.

11.3 A New Era in Transport

In order to be sustainable, transport must cut its energy consumption and its dependence on oil. Among renewable sources of energy, biofuels present their own special set of additional complications. They have the potential to displace significant quantities of gasoline, diesel and jet fuel, hence reducing dependence on oil. First-generation biofuels are obtained by the simple transformation of agricultural commodities grown for human consumption. These fuels are already available, but their production may be in competition with the food industry in terms of arable land. Second-generation biofuels are obtained by a more complex transformation of non-food biomass (straw, wood, etc.). They limit competition with food production by using agricultural waste products. Third-generation biofuels use algae as a feedstock. They offer the advantage of relieving the pressure on arable land and water resources.

---

[19] Alexander P. Mathews, "Renewable energy technologies: Panacea for world energy security and climate change?" *Procedia Computer Science* 32 (2014): 731–737.

[20] Glaucia M. Souza, Reynaldo L. Victoria, Carlos A. Joly, and Luciano M. Verdade (eds), *Bioenergy & sustainability bridging the gaps* (São Paulo: Scientific Committee on Problems of the Environment [SCOPE], 2015), pp: 61–64, ISBN: 978-2-9545557-0-6.

These fuels are currently at a research and early industrialization stage. One major positive point in favour of biofuels is their ability to create more jobs than the oil industry of other oil alternatives. It is estimated that the same manpower will be needed to convert biomass into fuel as is currently employed in oil refineries. Furthermore, biomass production should generate more jobs than drilling operations. In addition, the creation of such jobs will be even more substantial due to artisanal agricultural methods.[21] Second-generation biofuels, however, have yet to be proven on a substantial scale.

If investments over the next decade can improve production processes and bring down costs, biofuels will have the potential to make a significant contribution to dealing simultaneously with energy security and climate change. Cellulosic ethanol for commercial use is still in its early days. The industry has been focusing its powers in developing scalable and economically viable cellulosic ethanol for nearly a decade. That only happened on 2013, and, besides the fact that numerous new cellulosic plants have been announced, most of them are being held behind either due to financing reasons or are still in the early planning stages.[22] Today there is an increase in Hydrotreated Vegetable Oil (HVO) production from waste and residue fat fractions derived from food, fish and slaughterhouse industries, as well as from non-food grade vegetable oil fractions.[23] During the last 5 years, HVO production has been developing rapidly. Some oil majors, such as ENI or TOTAL or dedicated renewable fuel producers, e.g. UPM or Preem, are investing in HVO production units. Although Neste is still dominating the world HVO scene, new companies investigate the possibilities to invest in Green Diesel production. Despite the higher capital expenditures and

[21] Remi Rousso and Virginie Boutueil, "Towards low-carbon mobility: Tackling road transport emissions." *Les Cahiers de la Chaire Economie du Climat*, September 2011. Assessed February 25, 2020. https://www.chaireeconomieduclimat.org/wp-content/uploads/2015/06/11-09-16-Cahier-ID-n11-Russo-Boutueil.pdf.

[22] Rohan Nath and Kevin Adams, "The U.S. biofuel race—Hype or reality?" *BCG Perspectives 10/13, The Boston Consulting Group Inc.*, October 28, 2013. Accessed February 22, 2020. https://www.bcg.com/fr-ca/publications/2013/energy-sustainability-us-biofuel-race-hype-or-reality.aspx.

[23] Neste Oil, "NExBTL® Renewable diesel Singapore plant; Tallow pathway description." 2018. Accessed February 22, 2020. http://www.arb.ca.gov/fuels/lcfs/2a2b/apps/neste-aus-rpt-031513.pdf.

implementation difficulty compared to regular FAME production from esterification or HVO co-processing, all the companies decide to invest in pure HVO production plants instead of trying co-processing.[24] Pure HVO production gives more possibilities since it can be used on its own as drop-in or jet fuel, and can be blended with off-spec diesel to improve its parameters. Current transportation fuels are based on petroleum and whose importation has political and societal ramifications. Hydrogen is viewed by many as the ultimate "end-game" fuel. A transition from petroleum to dimethyl ether (DME) to hydrogen may be more cost-effective than a step change to hydrogen. DME can be introduced and exploited with existing technologies, and enable the eventual implementation of advanced technologies, such as fuel cells.[25] Because DME is produced from natural gas, coal or biomass, DME can address energy security, energy conservation and environmental concerns. As fuel processors and fuel cells are introduced to the public, DME can be further exploited as a non-toxic, non-corrosive, environmentally benign hydrogen carrier produced from domestic resources. Hydrogen can be considered as alternative energy to substitute fossil fuels in the future due to its clean characteristics.[26]

Due to its environmental merits, the share of biomass-derived hydrogen in the automotive fuel market will grow fast.[27] If biomass gasification to hydrogen is combined with carbon sequestration, biomass could even serve as a measure for returning CO_2 from the atmosphere into the earth.[28] Hydrogen fuel cell vehicles driving range, and the time needed for refuelling are more comparable to conventional cars than

---

[24]Greenea, "New players join the HVO game." Accessed February 22, 2020. https://www.greenea.com/wp-content/uploads/2017/02/HVO-new-article-2017-1.pdf.

[25]Troy A. Semelsberger, Rodney L. Borup and Howard L. Greene, "Dimethyl ether (DME) as an alternative fuel." *Journal of Power Sources* 156, no. 2 (2006): 497–511.

[26]Ya-Chieh Li, Yung-Feng Liu, Chen-Yeon Chu, Pao-Long Chang, Chiung-Wen Hsu, Ping-Jey Lin and Shu-Yii Wu, "Techno-economic evaluation of biohydrogen production from wastewater and agricultural waste." *International Journal of Hydrogen Energy* 37, no. 20 (2012): 15704–15710.

[27]Havva Balat and Elif Kirtay, "Hydrogen from biomass—Present scenario and future prospects." *International Journal of Hydrogen Energy* 35 (2010): 7416–7426.

[28]Dominic Rutz and Rainer Janssen, *Biofuel technology handbook* (Munich: WIP Renewable Energies, 2007). http://www6.zetatalk.com/docs/Biogas/Biofuel_Technology_handbook_1vs_WIP_2007.pdf.

those of EVs. The main issues with fuel cell vehicles and hydrogen are associated with technical drawbacks of all presently available options for hydrogen storage onboard vehicles and the high costs of fuel cells.[29] There is a concern that electrification of the transportation sector will provide a secure low-carbon solution to oil dependence, making biofuels unnecessary. Currently, electrification is generally limited to light-duty vehicles and sometimes medium-duty vehicles; heavy vehicles, boats and aircraft will not be able to rely on electricity. EVs only account for 1% of new passenger car sales today, but their market share is set to grow rapidly. As battery costs continue to decline, EVs could account for up to 10% of the world passenger car fleet by 2030, displacing some 1.5 mbpd of fossil fuel demand.[30] The deployment of EVs in urban areas is not only beneficial regarding the reduction in CO_2 and GHG emissions, but also as they can work as energy storing device able to absorb renewable energy, they can contribute to solving the problem of intermittency and non-dispatch ability of this power.[31] EVs can utilize their storage capacity as a vehicle to grid function in order to balance the electricity demand of various locations in the network or to ease the integration of intermittent renewable energy sources to the grid.[32]

[29] RETRANS, "Opportunities for the use of renewable energy in road transport, policy makers report." 2010. Accessed February 22, 2020. https://ledsgp.org/wp-content/uploads/2015/10/opportunities-for-the-use-of-renewable-energy-in-road-transport.pdf.

[30] Neste, "Taking action on climate change Neste outlook for changing markets." Accessed February 22, 2020. https://www.neste.com/sites/default/files/attachments/taking_action_on_climate_change_version2.pdf.

[31] Sandra Bellekom, René Benders, Steef Pelgröm and Henk Moll, "Electric cars and wind energy: Two problems, one solution? A study to combine wind energy and electric cars in 2020 in The Netherlands." *Energy, Elsevier* 45, no. 1 (2012): 859–866.

[32] Evanthia A. Nanaki, Spyros Kiartzis and George A. Xydis, "Are only demand-based policy incentives enough to deploy electromobility?" *Policy Studies* (January 24, 2020). https://doi.org/10.1080/01442872.2020.1718072.

11.4 Industry of the Future: Investing in Green Energy

The energy transition is happening, but the pace of change depends on technical, business and societal factors. This transition can be translated as building economies using enough energy to deliver prosperity for society with reduced carbon emissions.[33] A low emissions economy is tough to be established. A multitude of solutions is needed, with the input of a large number of stakeholders. Investment and innovative policymaking are needed for the transition to a low emissions market to happen. Commercialization and scale-up of new technologies is a time-consuming process; especially for energy infrastructure where lead times are long.[34] It also depends on the investment and policy choices made today. Companies around the world are publicly backing the Paris Agreement. Many of these companies, exemplifying themselves by contributing actively to emissions reduction, are setting in their strategies higher targets than the required policy standards. Climate change as a macro trend can no longer be ignored in companies' long-term planning while a lower carbon footprint for services is increasingly seen as an attractive customer offering. Oil and gas producers need to be key players in helping tackle the challenge of climate change since they account for more than half of the energy that powers our economies today.[35] Climate change generates multiple environmental pressures stemming from increased demand for water and energy resources, generation of air and water pollution, waste generation, land consumption and degradation. Water and energy systems and their operating environments are intrinsically linked. An integrated approach to policy, planning, management and development of water resources and energy systems optimizes

---

[33] Cait Murphy, "Pathways and obstacles to a low-carbon economy." *McKinsey & Company*, April 2017. Accessed February 22, 2020. https://www.mckinsey.com/business-functions/sustainability-and-resource-productivity/our-insights/pathways-and-obstacles-to-a-low-carbon-economy/.

[34] OGCI, "Catalyst for change collaborating to realize the energy transition." 2017. https://oilandgasclimateinitiative.com/wp-content/uploads/2017/10/OGCI-2017-Report.pdf.

[35] OGCI. "Taking action accelerating a low emissions future." 2016. Accessed February 22, 2020. https://oilandgasclimateinitiative.com/wp-content/uploads/2017/10/2016-report-taking-action-accelerating-a-low-emissions-future.pdf.

potential benefits and innovative opportunities for national and regional economic development. Integrated regional systems planning for water and energy use can support a transition to de-carbonizing the energy supply-chain from transboundary water and energy flows providing solutions both in meeting growing energy demand and for the management of waters as regional public goods.[36] Currently, an intense debate takes place on how to close the gap between current climate policy and the long-term goal of the Paris Agreement, achieving an economy with net zero emissions. Most discussion has focused on the supply side including (a) the need to develop and deploy new industrial production processes, (b) a turn to non-fossil feedstock and fuels and (c) the utilization of carbon capture and storage to offset any remaining emissions. Less attention has been given to the demand side, introducing a circular economy that could reduce emissions through better use and reuse of the materials that already exist in the economy.[37] Several opportunities are lying on this concept, with strategies like recirculation of a greater share of materials, waste reduction in production stages, redesigning more lightweight products and structures, the extension of products lifecycle and new business models developed based on a sharing economy.

The linear economy model in which we take, manufacture, use and dispose of, is in a transition to a circular economy model where we reduce, reuse and recycle.[38] Circular economy in the energy industry reflects upon designing, processes and solutions that are aiming to a maximized efficiency in the energy production from natural resources, in the end-use of energy, in the valuation of excess energy and side streams.[39] Circular economy prevents value losses from materials flows

---

[36] Kevin Rosner and Jakob Granit, "The water-energy-security nexus: Regional approaches to a global challenge." *Friedrich Ebert Stiftung*, April 2012, Accessed February 22, 2020. https://www.gwp.org/globalassets/global/toolbox/references/the-water-energy-security-nexus_regional-approaches-to-a-global-challenge-fes-2012.pdf.

[37] Per-Anders Enkvist and Per Klevnas, "The circular economy—A powerful force for climate mitigation." *Material Economics*, 2018. Accessed February 22, 2020. https://materialeconomics.com/publications/the-circular-economy-a-powerful-force-for-climate-mitigation-1.

[38] Biofuels International, 2019. Issue 3, Vol. 13, 20–21.

[39] Deloitte, "Circular economy in the energy industry." *Environmental Research Pool*, January 2018. Accessed February 22, 2020. https://energia.fi/files/2287/Deloitte_2018_-_Circular_economy_in_the_energy_industry_-_Summary_report.pdf.

and creates competitive advantages and economic opportunities on a sustainable base. As an approach, it minimizes waste and resource use and maintains the value of materials and energy consumed in the production value chain. Economies will benefit from substantial net material savings, mitigation of volatility and supply risks, drivers for innovation and job creation, improved land productivity and soil health, and long-term resilience of the economy.[40] A successful transition towards a green and resource and energy-efficient economy will reshape labour markets. There is significant potential for creating new jobs in the production of energy from renewable sources, energy efficiency, waste and water management, air quality, restoring and preserving biodiversity, climate change adaptation and the development of green infrastructure.[41] In the petrochemicals sector, circular economy addresses the plastics waste and resource use issues. The vision of the new plastics economy is that instead of becoming waste, plastics will re-enter the economy as valuable technical or biological nutrients. The negative externalities associated with plastics would be significantly reduced by the new plastics economy.[42] Apart from the energy sector, shifts in the energy landscape will cause implications for a wide variety of companies changing the roles they play or the products they offer. Grabbing opportunities in the field of energy efficiency, electricity and decarburization is a priority for many business sectors while they can be found in a premature state in organizational or awareness mandates. Consumer preferences are shifted towards sustainable industries including attitudes on greener products; while regulators are gearing up to speed the change.[43] The future of energy will be reshaped based on two important tipping points, the further acceleration

[40]World Economic Forum, "Towards the circular economy: Accelerating the scale-up across global supply chains." Accessed February 22, 2020. http://www3.weforum.org/docs/WEF_ENV_TowardsCircularEconomy_Report_2014.pdf.

[41]European Investment Bank, "The EIB circular economy guide." Accessed February 22, 2020. https://www.eib.org/attachments/thematic/circular_economy_guide_en.pdf.

[42]Neste, "Taking action on climate change Neste outlook for changing markets." Accessed February 22, 2020. https://www.neste.com/sites/default/files/attachments/taking_action_on_climate_change_version2.pdf.

[43]Namit Sharma, Bram Smeets and Christer Tryggestad, "The decoupling of GDP and energy growth: ACEO guide." *McKinsey Quarterly*, April 2019. Accessed February 22, 2020. https://www.mckinsey.com/industries/electric-power-and-natural-gas/our-insights/the-decoupling-of-gdp-and-energy-growth-a-ceo-guide.

of the rump-up of renewables and the decline at the cost of batteries that will both result to a peak and decline in global GHG emissions, despite the continued global economic growth. Industries of the future will both be created within the current geopolitical structure and transform it. In the twenty-first century, the dominant divide is between those that have open political and economic models and those that are closed.[44]

The shift from centrally generated electricity from fossil fuels to electricity generated, on a local scale, from renewable energy sources, is placing the consumer as the centre of interest. That way demand response programmes are introduced, resulting in a transition from the traditional demand-driven system to a supply-driven one. This way, the energy consumer becomes a vital part of the energy market, benefits from new technologies and grid digitalization and reduces his expenses. In fact, we are moving far away from state monopolies to a more liberal energy model where civil society enters into the sophisticated energy regulation era. The evolution of consumers to prosumers has the potential to increase energy democratization. Climate change is part of our life, thus reducing its impacts on lives and livelihoods will demand a sustained and collective effort. Both governments and the private sector need to reschedule the way they: (a) build infrastructure, (b) use climate and weather data and (c) mobilize financial resources to offset potential risks.[45] Although the economic case for such a transformation is clear, putting it into practice requires creativity, collaboration and flexible funding. Viable instruments for substantial financial support for the transition to the low-carbon economy are the green bonds. This market needs to address several challenges and barriers that currently faces widening the scope of low-carbon investments.[46]

---

[44]Alec Ross, *The industries of the future* (Simon & Schuster, 2017), pp: 12–14, ISBN: 978-1-4711-3526-2.

[45]Alice Hill and Leonardo Martinez-Diaz, "Adapt or Perish preparing for the inescapable effects of climate change." *Foreign Affairs* 99, no. 1 (January/February 2020): 102–109.

[46]Eftichios S. Sartzetakis, "Green bonds as an instrument to finance low carbon transition." *Bank of Greece Working Paper*, March 2019. Accessed February 22, 2020. https://www.bankofgreece.gr/Publications/Paper2019258.pdf.

11.5 Conclusions

This chapter provides an insight into the future transformation of the energy markets and discusses technological, regulatory and financial challenges correlated to energy security. Redefining energy security can be achieved by upgrading from the level of national self-interest to a global level that reflects the increasing internationalization and interdependence of energy policies. Global demand growth, as projected, cannot be sustainably met from the exploitation of conventional and unconventional resources. As the implications of a global effort to drastically cut emissions can often be positive for energy security issues, a wide range of conflicts will still arise. Critical developments in the energy transition are dramatically speeding up. Many of the global trends that shape the future of energy are, in fact, driven by a multitude of local trends, which will occur at different magnitudes, speeds in specific geographies and sectors. Climate change and energy security are a priority for governments, public opinion and companies worldwide, as they integrate decarbonization dimensions into their strategy. The objective of a climate change policy is to create incentives that will drive changes in technology and consumption patterns and lead to new developments in low-carbon energy production. As a result, new energy carriers are shifted from the shale revolution towards low-carbon fuels. Regardless of the applied climate policies, timely investment into oil and gas supply remains a cornerstone of energy security for the years to come. Strong growth in the production and integration of renewable energy is expected to diversify the energy mix; where supply is as important and as vulnerable as is energy transmission and distribution.

The rapidly increasing share of renewables in electricity generation and its decreasing cost, new storage technologies and the introduction of EVs in the market, are leading towards a new system in which consumers produce, use and sell their electricity; a system that reflects the concept of distributed generation. Globally the need to pursue a new vision of energy security—one based on availability, reliability, affordability and sustainability—poses enormous challenges. The above four pylons are directly affected by fostering network digitalization that will challenge all established business models. New mobility concepts have a direct impact on fuel demand. Moving away from fossil fuels is a challenge for energy players to identify new energy sources, integrate technology and build new business models to capture value. Fleet electrification is proposed as a strategy for reducing air pollutant emissions and improving air quality in urban areas. Bioenergy is a valuable option for energy transition with positive synergies with other policy priorities, including water and food security, while improving economic growth and stability. Next-generation biofuels can play a key role in the new era in transport since they can serve all different modes (road, rail, marine and air). Renewable fuels attract capital due to shrinking investments in the oil sector, overcoming technology barriers and feedstock quest dilemmas. At the end of the day, biomass feedstock remains a commodity market with a variety of alternative processing routes to produce biofuels. The linear economy model is in a transition to a circular economy model where we reduce, reuse and recycle to enhance energy security. Creating a low-emission economy needs investment and innovative policy decisions to accelerate the creation of new low-carbon markets. The regulatory framework should support mechanisms that allow investments of projects that increase social welfare while fulfilling the energy policy targets. "Business as usual" is no longer an option for the energy sector. The energy shift is already underway, switching to fuels with lower carbon emission intensity.

References

Alec Ross, *The industries of the future* (Simon & Schuster, 2017), pp: 12–14, ISBN: 978-1-4711-3526-2.

Alexander P. Mathews, "Renewable energy technologies: Panacea for world energy security and climate change?" *Procedia Computer Science* 32 (2014): 731–737.

Alice Hill and Leonardo Martinez-Diaz, "Adapt or Perish preparing for the inescapable effects of climate change." *Foreign Affairs* 99, no. 1 (January/February 2020): 102–109.

Antony Froggatt and Michaela A. Levi, "Climate and energy security policies and measures: Synergies and conflicts." *Foreign Affairs* 85, no. 6 (November 2009): 1129–1141.
Anup Bandivadekar, Kristian Bodek, Lynette Cheah, Christopher Evans, Tiffany Groode, John Heywood, Emmanuel Kasseris, Matthew Kromer and Malcom Weiss. "On the road in 2035: Reducing transportation's petroleum consumption and GHG emissions." *Report LFEE 2008-05RP* MIT, 2008.
Biofuels International, 2019. Issue 3, Vol. 13, 20–21.
Cait Murphy, "Pathways and obstacles to a low-carbon economy." *McKinsey & Company*, April 2017. Accessed February 22, 2020. https://www.mckinsey.com/business-functions/sustainability-and-resource-productivity/our-insights/pathways-and-obstacles-to-a-low-carbon-economy/.
Carlos Pasqual and Jonathan Elkind (eds), *Energy security—Economics, politics, strategies and implications* (Brookings Institution Press, 2010), pp: 120–130, ISBN: 978-0-8157-6919-4.
COM/2015/080, "A framework strategy for a resilient Energy Union with a forward-looking climate change policy." 2015. Accessed February 22, 2020. https://eur-lex.europa.eu/legal-content/EN/TXT/?uri=COM:2015:80:FIN.
COM (2019) 175 Final, "Fourth report on the State of the Energy Union." 2019. Accessed February 22, 2020. https://ec.europa.eu/commission/sites/beta-political/files/fourth-report-state-of-energy-union-april2019_en_0.pdf.
Deloitte, "Circular economy in the energy industry." *Environmental Research Pool*, January 2018. Accessed February 22, 2020. https://energia.fi/files/2287/Deloitte_2018_-_Circular_economy_in_the_energy_industry_-_Summary_report.pdf.
Dominic Rutz and Rainer Janssen, *Biofuel technology handbook* (Munich: WIP Renewable Energies, 2007). http://www6.zetatalk.com/docs/Biogas/Biofuel_Technology_handbook_1vs_WIP_2007.pdf.
Eftichios S. Sartzetakis, "Green bonds as an instrument to finance low carbon transition." *Bank of Greece Working Paper*, March 2019. Accessed February 22, 2020. https://www.bankofgreece.gr/Publications/Paper2019258.pdf.
Eric Spiegel, Neil McArthur and Rob Norton, *Energy shift game-changing options for fueling the future* (McGraw Hill, 2009), pp: 162–164, ISBN: 978-0-0715083-4-6.
European Investment Bank, "The EIB circular economy guide." Accessed February 22, 2020. https://www.eib.org/attachments/thematic/circular_economy_guide_en.pdf.

Evanthia A. Nanaki, Spyros Kiartzis and George A. Xydis, "Are only demand-based policy incentives enough to deploy electromobility?" *Policy Studies* (January 24, 2020). https://doi.org/10.1080/01442872.2020.1718072.

Glaucia M. Souza, Reynaldo L. Victoria, Carlos A. Joly, and Luciano M. Verdade (eds), *Bioenergy & sustainability bridging the gaps* (São Paulo: Scientific Committee on Problems of the Environment [SCOPE], 2015), pp: 61–64, ISBN: 978-2-9545557-0-6.

Greenea, "New players join the HVO game." Accessed February 22, 2020. https://www.greenea.com/wp-content/uploads/2017/02/HVO-new-article-2017-1.pdf.

Havva Balat and Elif Kirtay, "Hydrogen from biomass—Present scenario and future prospects." *International Journal of Hydrogen Energy* 35 (2010): 7416–7426.

Kahn Suzana Ribeiro, Shigeki Kobayashi, Michel Beuthe, Jorge Gasca, David Greene, David S. Lee, Yasunori Muromachi, Peter J. Newton, Steven Plotkin, Daniel Sperling, Ron Wit and Peter J. Zhou, "Transport and its infrastructure." In *Climate change 2007: Mitigation. Contribution of Working Group III to the Fourth assessment report of the Intergovernmental Panel on Climate Change* [B. Metz, O.R. Davidson, P.R. Bosch, R. Dave, L.A. Meyer (eds)]. Cambridge University Press, Cambridge, UK and New York, NY, 2007: 325–327.

Katerina G. Tsita, Spyros J. Kiartzis, Nikolaos K. Ntavos and Petros A. Pilavachi, "Next generation biofuels derived from thermal and chemical conversion of the Greek transport sector." *Thermal Science and Engineering Progress* (2019). https://doi.org/10.1016/j.tsep.2.09.100387.

Kevin Rosner and Jakob Granit, "The water-energy-security nexus: Regional approaches to a global challenge." *Friedrich Ebert Stiftung*, April 2012, Accessed February 22, 2020. https://www.gwp.org/globalassets/global/toolbox/references/the-water-energy-security-nexus_regional-approaches-to-a-global-challenge-fes-2012.pdf.

Namit Sharma, Bram Smeets and Christer Tryggestad, "The decoupling of GDP and energy growth: ACEO guide." *McKinsey Quarterly*, April 2019. Accessed February 22, 2020. https://www.mckinsey.com/industries/electric-power-and-natural-gas/our-insights/the-decoupling-of-gdp-and-energy-growth-a-ceo-guide.

Neste, "Taking action on climate change Neste outlook for changing markets." Accessed February 22, 2020. https://www.neste.com/sites/default/files/attachments/taking_action_on_climate_change_version2.pdf.

Neste Oil, "NExBTL® Renewable diesel Singapore plant; Tallow pathway description." 2018. Accessed February 22, 2020. http://www.arb.ca.gov/fuels/lcfs/2a2b/apps/neste-aus-rpt-031513.pdf.

OECD/IEA, "Energy supply security: Emergency response of IEA Countries 2014." 2014. https://webstore.iea.org/energy-supply-security-the-emergency-response-of-iea-countries-2014.

OGCI, "Catalyst for change collaborating to realize the energy transition." 2017. https://oilandgasclimateinitiative.com/wp-content/uploads/2017/10/OGCI-2017-Report.pdf.

OGCI. "Taking action accelerating a low emissions future." 2016. Accessed February 22, 2020. https://oilandgasclimateinitiative.com/wp-content/uploads/2017/10/2016-report-taking-action-accelerating-a-low-emissions-future.pdf.

Per-Anders Enkvist and Per Klevnas, "The circular economy—A powerful force for climate mitigation." *Material Economics*, 2018. Accessed February 22, 2020. https://materialeconomics.com/publications/the-circular-economy-a-powerful-force-for-climate-mitigation-1.

Remi Rousso and Virginie Boutueil, "Towards low-carbon mobility: Tackling road transport emissions." *Les Cahiers de la Chaire Economie du Climat*, September 2011. Assessed February 25, 2020. https://www.chaireeconomieduclimat.org/wp-content/uploads/2015/06/11-09-16-Cahier-ID-n11-Russo-Boutueil.pdf.

RETRANS, "Opportunities for the use of renewable energy in road transport, policy makers report." 2010. Accessed February 22, 2020. https://ledsgp.org/wp-content/uploads/2015/10/opportunities-for-the-use-of-renewable-energy-in-road-transport.pdf.

Rohan Nath and Kevin Adams, "The U.S. biofuel race—Hype or reality?" *BCG Perspectives 10/13, The Boston Consulting Group Inc.*, October 28, 2013. Accessed February 22, 2020. https://www.bcg.com/fr-ca/publications/2013/energy-sustainability-us-biofuel-race-hype-or-reality.aspx.

Sandra Bellekom, René Benders, Steef Pelgröm and Henk Moll, "Electric cars and wind energy: Two problems, one solution? A study to combine wind energy and electric cars in 2020 in The Netherlands." *Energy, Elsevier* 45, no. 1 (2012): 859–866.

Simone Alfano, Federico Berruti, Nicolas Denis and Alberto Santagostino, "The future of second-generation biomass." 2016. https://www.mckinsey.com/~/media/McKinsey/Business%20Functions/Sustainability/Our%20Insights/The%20future%20of%20second%20generation%20biomass/The-future-of-second-generation-biomass.ashx.

Spencer Glendon and Simon London, "Redefining the power industry." *McKinsey Quarterly*, 2019. https://www.mckinsey.com/~/media/McKinsey/Business%20Functions/Sustainability/Our%20Insights/Redefining%20the%20power%20industry/Redefining-the-power-industry-vF.ashx.

Stylianos Stefanidis, Petros Pilavachi, Konstantinos Kalogiannis and Aggelos Lappas (in Greek), "Pyrolysis of agricultural and forest residues for the production of fuels and chemical products." *Workshop of the Energy Committee of the Academy of Athens—Management and Energy Exploitation of Residues in Greece*, 147–162 (2015). Athens, Greece.

The Economist Intelligence Unit. "Ready for the change—Pathways to a low-emissions future." 2019. https://futurepresent.economist.com/ready-for-change-pathways-to-a-low-emissions-future/.

Troy A. Semelsberger, Rodney L. Borup and Howard L. Greene, "Dimethyl ether (DME) as an alternative fuel." *Journal of Power Sources* 156, no. 2 (2006): 497–511.

World Economic Forum, "The global risks report 2019—14th edition." 2019. http://www3.weforum.org/docs/WEF_Global_Risks_Report_2019.pdf.

World Economic Forum, "Towards the circular economy: Accelerating the scale-up across global supply chains." Accessed February 22, 2020. http://www3.weforum.org/docs/WEF_ENV_TowardsCircularEconomy_Report_2014.pdf.

Ya-Chieh Li, Yung-Feng Liu, Chen-Yeon Chu, Pao-Long Chang, Chiung-Wen Hsu, Ping-Jey Lin and Shu-Yii Wu, "Techno-economic evaluation of biohydrogen production from wastewater and agricultural waste." *International Journal of Hydrogen Energy* 37, no. 20 (2012): 15704–15710.

12

Challenges and Opportunities in Renewable Energy Developments

Ivan Pineda

12.1 Introduction

The decarbonisation of the electricity sector has been one of the most transformative changes in Europe's economy. Renewable energy has played a central role in substituting carbon-intensive power generation for emissions-free electricity. Between 2008 and 2018, Europe installed more renewables than fossil fuel generation. A total of 259 GW of renewables were added to the power system, of which wind energy has been the largest single technology, with 110 GW of new installations.[1] During the same period, electricity generation from renewables almost

[1]WindEurope, "Wind energy in Europe in 2018. Trends and statistics", p. 14, available at: https://windeurope.org/wp-content/uploads/files/about-wind/statistics/WindEurope-Annual-Statistics-2018.pdf.

I. Pineda (✉)
WindEurope, Brussels, Belgium
e-mail: Ivan.Pineda@windeurope.org

M. Mathioulakis (ed.), *Aspects of the Energy Union*,
Energy, Climate and the Environment,
https://doi.org/10.1007/978-3-030-55981-6_12

doubled, from roughly 590 TWh/year to around 1072 TWh/year.[2] This reduced 21% of the CO_2 emission intensity in electricity generation, from 376 gCO_2/kWh,[3] to 267 gCO_2/kWh.[4] At the other end, Europe has removed an average of 10 GW of fossil fuel plants every year. In the decade post-2020, further decommissions will continue, particularly those power plants fired by hard coal and lignite. Six countries in the EU are now coal-free, and 14 have announced their intentions to phase-out coal by 2030 or earlier.[5] While some countries would be tempted to replace coal assets with fossil gas, the business case for building new gas-fired power plants will become very challenging.[6] These developments would not have been possible without the landmark decision of EU governments in 2007 to set climate and energy targets: 20% reduction of greenhouse gas emissions, 20% renewables in final energy demand and a 20% of energy demand reduction by the year 2020.[7] These targets were the first-of-a-kind globally and led the way for many other governments around the world to act accordingly. In 2018, ten years after, the EU agreed on a further set of policies to accelerate the energy transition in the decade post-2020. The EU's 2030 climate and energy framework increased EU-wide targets and policy objectives of at least 40% reductions in greenhouse gas emissions (from 1990 levels), at least 32% share for renewable energy in final energy demand, and at least 32.5% improvement in energy efficiency.[8] While policy set in

[2]Agora Energiewende and Sandbag, "The European Power Sector in 2019", p. 9, available at: https://sandbag.org.uk/wp-content/uploads/2020/02/Sandbag-European-Power-Sector-Review-2019.pdf.

[3]European Environmental Agency, "Total greenhouse gas emission trends and projections in Europe", last modified, 19 December 2019, available at: https://www.eea.europa.eu/data-and-maps/indicators/greenhouse-gas-emission-trends-6/assessment-2.

[4]Agora Energiewende and Sandbag, "The European power sector in 2019", available at: https://ember-climate.org/global-electricity-review-2020-data/.

[5]Agora Energiewende and Sandbag, "The European power sector in 2019", p. 23.

[6]Peter Zeniewski, "A long-term view of natural gas security in the European Union", International Energy Agency, 13 March 2019, available at: https://www.iea.org/commentaries/a-long-term-view-of-natural-gas-security-in-the-european-union.

[7]European Commission, "Communication 20 20 by 2020", 23 January 2008, available at: https://eur-lex.europa.eu/legal-content/EN/TXT/?uri=CELEX:52008DC0030.

[8]European Commission, "2030 climate & energy framework", available at: https://ec.europa.eu/clima/policies/strategies/2030_en.

motion the departure from fossil fuels for power generation, the success of renewables is an outcome of other factors. Renewables have decreased costs at a remarkable pace. Solar PV panels have decreased costs by 75% since 2010, and onshore wind energy is today the cheapest source of power generation in many parts of the world. This has made a stand-alone business case for investors. Even in times of low oil prices, the cost competitiveness of renewables is seen as a hedge against volatile fuel commodities by many private investors. In addition, governments have other compelling reasons to move away from fossil fuels. Concerns over the security of supply and fuel import bills are crucial for many countries. Not to mention the pressing challenge to improve local air quality and to alleviate water scarcity, both issues affecting many parts of the population. Renewables can solve these issues, but their large-scale deployment still faces many challenges that policy can address. This chapter describes some of the major challenges and opportunities for renewable energy expansion with a focus on wind energy and with a European perspective. It argues that the future success of renewables is based on scaling up the electrification of the economy, facilitating permitting and administrative procedures, increasing the power system flexibility and plugging the financing gap between long-term ambitions and investment needs.

12.2 Renewables' Future Success Depends on Scaling up Electrification

The EU's 32% renewable energy target to 2030 in final energy demand translates into around 55% of electricity coming from renewables. This may sound ambitious, but renewables were close to 35% of electricity generation already in 2019, including hydropower. This suggests that the scale envisioned for renewables expansion could be increased further, particularly in light of the additional ambition needed to meet the Climate Paris Agreement. This is important because electricity is only around 25% of Europe's final energy demand. Europe consumes most of its energy for heating, cooling and transport. And the vast majority of it remains fossil fuel-based. Oil products cover 94% of transport (32% of final energy demand), and 80% of heating and cooling (46% of the final energy demand) comes from fossil fuels too. Renewables make up only 18% of the supply in heating and a mere 8% in transport.[9] Therefore, integrating heating, cooling and transport sectors into a renewable electricity system will be the next big transformation in the economy.

12.2.1 Electrification

Electrification is the integration of final energy uses into the power sector. It can take place directly by substituting fossil fuels with electricity in heating, industrial processes or transport, or indirectly by converting electricity into another energy carrier, like hydrogen or hydrogen-based fuels, which then can be used by consumers. However, for electrification to contribute to climate change mitigation, it would have to come from zero-carbon energy sources. Using renewable electricity wherever is available and whenever is possible can deliver the bulk of decarbonisation needed for the economy. To achieve this, it is crucial to continue the deployment of renewable energy technologies and even accelerate it. This expansion, together with the increased number of applications and uses, will allow the continuation of technology cost reductions, which have progressed rapidly in the last years. There will be certain uses which Europe cannot electrify as fast as needed, or that would be too expensive in the short-term and could impact the competitiveness of some sectors, or that simply it is not technically possible. In these cases, known as hard-to-abate sectors, using renewables to produce other energy carriers will be necessary.

The European Commission has analysed the feasibility of different pathways to achieve net-zero emissions by 2050 as part of its communication on the European long-term strategic vision for a prosperous, modern, competitive and climate neutral economy.[10] The analysis found that to meet the Paris Agreement, Europe would need to increase the

---

[9]European Commission, "An EU Strategy on heating and cooling SWD (2016) 24 Final", p. 3, 2016, available at: https://ec.europa.eu/energy/sites/ener/files/documents/1_EN_ACT_part1_v14.pdf.

[10]European Commission, "In depth analysis in support of the Commission Communication COM (2018) 773. A Clean Planet for all. A European long-term strategic vision for a prosperous, modern, competitive and climate neutral economy", 28 November 2018,

share of electricity in energy consumption to between 50 and 60%. The European Commission estimates that renewables would have to amount to 61–85% of that electricity generation. Wind energy would be the most dominant source of generation, 51–55%, up from 15% today. The 2050 decarbonisation strategy proposed by the European Commission shows that the other half of final energy demand, which will not come from electricity, would be met by a combination of biomass, fossil gas, hydrogen and hydrogen-derived e-fuels. Combined, hydrogen and e-fuels could meet up to 24% of final energy demand in 2050. Europe has long spearheaded the global fight against climate change, but policymakers will be facing considerable challenges to lead society towards any of these futures. Among these are the acceptability to change the way energy is produced and consumed, the pace at which this change should take place, the short-term impacts on the competitiveness of specific sectors in a globalised economy, and the amount of upfront investments needed in the short-to-medium-term compared to the long-term outcomes. While such a future represents a formidable challenge, it is also a remarkable opportunity. If policymakers make a clear choice for renewables-based electrification, Europe will hold the key to a successful decarbonisation strategy while retaining its competitive edge in clean technologies. During the transition, Europe will decrease fossil fuels imports while improving air quality and reducing energy bills for citizens and businesses. All this will secure future European's living standards. Renewables-based electrification is therefore not only key to the decarbonisation of Europe's energy system, but also a strategy to secure economic growth. For example, by reducing health-related economic impacts. Air pollution costs between 3 and 9% of the EU's GDP, equivalent to €330 bn and €940 bn annually.[11] Using more renewable electricity instead of fossil fuels would reduce the exposure of the population to dangerous pollutants like SO_2, NO_x and other particles. Another example is the economic costs of energy dependence. 20%

---

available at: https://ec.europa.eu/clima/sites/clima/files/docs/pages/com_2018_733_analysis_in_support_en_0.pdf.

[11]European Commission, "Commission staff document. Executive summary of the impact assessment SWD (2013) 532 final", 18 December 2013, available at: https://eur-lex.europa.eu/legal-content/EN/TXT/PDF/?uri=CELEX:52013SC0532&from=EN.

of all EU imports are energy-related. Imported fuels cover 54% of all the energy the EU consumes. It imports 90% of its crude oil, 66% of its natural gas, 42% of its coal and 40% of its uranium and other nuclear fuels.[12] Powered by renewable electricity, the EU's economy has the potential to look very different from today. Its success will rely on making the most of Europe's local energy resources, on its competent workforce and on its ability to innovate and stay at the forefront of nascent industries. Decarbonising Europe's economy requires therefore activating the right policy levers. There are three broad categories of policies that could accelerate a renewables-based electrification, namely energy efficiency, electrification incentives and CO_2 pricing.

12.2.1.1 Energy Efficiency

Traditionally energy efficiency has meant 'using less energy', but policies should pursue a goal of 'producing the same or more while using less'. Policymakers can shape energy efficiency with two general levers: by setting obligations on energy savings or by promoting market-based instruments that incentivise users to optimise their energy consumption. Europe has combined both approaches, with the EU Institutions mainly setting the general framework of obligations,[13] and individual governments implementing market-based instruments such as auctions.[14] The EU has a non-binding 32.5% energy efficiency target by 2030. To meet this target, policymakers will use a range of measures across different sectors. In residential and commercial buildings, it aims at boosting

[12]European Commission, "Energy security strategy", p. 1, 2015, available at: https://eur-lex.europa.eu/legal-content/EN/ALL/?uri=CELEX%3A52014DC0330.

[13]Energy efficiency obligations schemes for utilities of 1.5% per year to 2030 according to the International Energy Agency (IEA), "Renewable energy for industry", 2017, available at: https://webstore.iea.org/insights-series-2017-renewable-energy-for-industry.

[14]According to the IEA, there are 12 market-based instruments for energy efficiency in the EU: AT, BG, DK, FR, IE, IT, LU, MT, PL, SL, ES, UK. Three more countries are to start shortly instruments: HT, HE, LT. The International Energy Agency (IEA), "Renewable energy for industry", 2017.

energy efficiency through the implementation of the Energy Performance of Buildings Directive (EPBD).[15] Buildings account for 40% of Europe's energy consumption, and two-thirds of Europe's buildings were built before energy performance standards were set up. Furthermore, the building renovation rate is only around 1% per year. Increasing this renovation rate has the potential to reduce between 5 and 6% of the EU's total energy consumption and to reduce CO_2 emissions by about 5%.[16] One solution is heating residential buildings with heat pumps. For every kW of electricity consumed by a heat pump, about 4 kW of energy is generated. This high coefficient of performance is expected to improve as technology progresses, while gas boilers would struggle to surpass the top end of their efficiency of 80% from the latest models. Improving energy efficiency in industry is more challenging due to the variety of energy uses, which depend on the specific industrial processes. Until now, the EU has addressed this by setting energy performance standards of manufactured products through the Eco-design Directive and energy labelling. These cover not only household appliances, but also industrial equipment, such as electric motors, water pumps, industrial refrigeration and power transformers. The Eco-design and Ecolabelling policies focus on the number of electricity machines use, rather than on the primary energy use. While the European Commission estimates that Eco-design will deliver energy savings equivalent to the annual energy consumption of Italy by 2020,[17] the transformation of Europe's energy system necessitates mining a greater fraction of the energy savings potential available in the industry. However, tackling energy efficiency in industry is very complex, especially in the production of base materials such as steel, cement and chemicals. The EU is just starting to plan policies for these sectors. In the transport sector, the policy has focused on tax reduction incentives, stricter emissions and fuel efficiency standards, as

[15]European Commission, Directive on the Energy Performance of Buildings (2018/844/EU), 26 April 2018, available at: http://data.consilium.europa.eu/doc/document/PE-4-2018-INIT/en/pdf.

[16]European Commission, "New rules for greener and smarter buildings will increase quality of life for all Europeans", 15 April 2019, available at: https://ec.europa.eu/info/news/new-rules-greener-and-smarter-buildings-will-increase-quality-life-all-europeans-2019-apr-15_en.

[17]European Commission, "Putting energy efficiency first: Consuming better, getting cleaner", 30 November 2016, available at: http://europa.eu/rapid/press-release_MEMO-16-3986_en.htm.

well as charging infrastructure deployment. Most of these policies aim at passenger vehicles rather than heavy and long-haul road transportation and shipping. Energy savings could be significant, as battery electric vehicles have 80–90% energy conversion efficiency (tank-to-wheel), which is a higher rate than internal combustion engines' 30–40% efficiency. This means EVs can drive three to four times the same distance than a combustion engine vehicle with the same amount of energy.

12.2.1.2 Electrification Incentives

Switching energy consumption from fossil fuels to electricity generated with renewables is not possible with existing market signals. Current energy prices often do not reflect true costs. This is partly because fossil fuels benefit from subsidies and tax exemptions, in particular in the country of production.[18] These subsidy regimes incentivise users to opt for fossil fuel-based heating and transport over electricity. Policymakers, therefore, hold a key lever to address market distortions through fuel and electricity pricing. For example, the ratio between electricity versus gas and oil prices has remained quite stable over time, to the detriment of electric appliances. By revising taxes and levies in retail electricity prices, which average 38% of the final price to consumers, policymakers could reverse this.[19] Especially because some taxes and levies are unrelated to the energy use. Governments use them for wider tax collection. These taxes hinder the electrification of households and industries. In contrast, tax breaks could shape consumer demand towards clean technologies. Many countries have introduced tax reductions (Germany,

---

[18]According to NGOs, the G20 nations provide four times more public financing to fossil fuels than to renewable energy. Oil Change International, Friends of the Earth, WWF and the Sierra Club, "Talk is cheap: How G20 governments are financing climate disaster", 5 July 2017, available at: http://priceofoil.org/2017/07/05/g20-financing-climate-disaster/.

[19]European Commission, "Energy prices and costs in Europe—Commission staff working document", p. 38, 2018, available at: https://ec.europa.eu/energy/sites/ener/files/documents/swd_-_v5_text_6_-_part_1_of_4.pdf.

Austria) or bonus payments/premiums for the buyers of electric vehicles (France, the UK, Norway), leading to a significant rise in sales,[20] but recent attempts towards a European approach to energy taxation have stalled. In 2011, the European Commission presented a proposal to revise the Energy Tax Directive to support the EU's wider environmental and energy goals. The proposal was withdrawn in 2015 following unsuccessful negotiations in the Council of the EU. The European Commission has re-opened discussions in 2019 and could present new policies in 2021.[21]

12.2.1.3 CO_2 Pricing

For many years, the European Emissions Trading System (ETS) has been characterised by a structural imbalance between the demand and supply of CO_2 allowances. This resulted in low carbon prices, which made the system unable to incentivise investments in clean technologies. Instead, investments in renewables have been driven by other policy interventions. In 2018 the EU Member States and the European Parliament agreed on a series of measures to address this. First, they introduced a Market Stability Reserve (MSR), which will remove some of the surplus allowances from the market. This started operating in 2019. In addition, EU lawmakers adopted a broader reform of the ETS to bring it in line with the EU's 2030 greenhouse gas emissions reduction target of 40%. From 2021 onwards, the linear reduction factor will increase from 1.74 to 2.2%. This political deal also included a considerable strengthening of the MSR. EU lawmakers agreed to double the annual surplus permit take-out rate and cancel those that sit in the MSR from 2023 onwards. This should help to restore demand and supply balance in the carbon market and push carbon prices up. These new policies combined could see prices go up to €30/tCO_2 by 2030. The EU ETS will remain

[20] 100,000 registered EVs in France in 2017 following the introduction of a 'superbonus' up to €10,000 in 2015. Vivian Zhou, "3 electric car incentives you need to know in Europe", The Current, 4 July 2017, available at: https://blog.evbox.com/electric-car-incentives.

[21] European Commission, "EU Green Deal - Revision of the Energy Taxation Directive", 2021, available at: https://ec.europa.eu/info/law/better-regulation/have-your-say/initiatives/12227-Revision-of-the-Energy-Tax-Directive

the central EU policy instrument to regulate CO_2 emissions and deliver effective transition signals. The recently adopted reforms have sent a clear message to the market that scarcity is on the way. An additional tightening of the ETS cap is essential to align the EU's CO_2 reduction trajectory with the Paris climate goals. A strengthened linear reduction factor will ensure the ETS can drive out Europe's most polluting assets more quickly bringing the EU's CO_2 reduction trajectory in line with its long-term climate goals. In addition to properly pricing CO_2 emissions, the ETS could ensure a fairer distribution of costs among energy users. Today, the ETS primarily covers electricity generation, some large industrial plants and emissions from flights between airports located in the European Economic Area. This puts electricity at a competitive disadvantage for energy use in sectors not covered by carbon pricing. For example, the electricity used for shore-side electricity (SSE) is more expensive than the tax exempted fuel commonly used in the auxiliary engines of ships.[22] Sweden asked for—and obtained—a derogation from the energy taxation directive, in order to make SSE tax-free. This put electricity usage in ships on a level playing field with other fuels.[23]

12.3 Bridging the Gap Between Scale and Implementation: Permitting and Administrative Procedures

Energy and climate change policies may set very high long-term ambitions, but policymakers need to ensure they address current and future challenges to realise their objectives. Particularly at the national level if there is a general lack of political support for renewable energy, this could translate into restrictive planning conditions such as stringent minimum distances to housing for wind energy projects. On the

[22] Ecofys, "Potential for shore side electricity in Europe", 7 January 2015, available at: https://www.scribd.com/document/366724117/Ecofys-2014-Potential-for-Shore-Side-Electricity-in-Europe.

[23] European Council Implementing Decision authorising Sweden to apply a reduced rate of electricity tax to electricity provided to vessels at berth in a port, 8 April 2011, available at: https://eur-lex.europa.eu/legal-content/EN/TXT/?uri=CELEX%3A52011PC0158.

contrary, when there is political support for developing renewables, then the permitting is facilitated through best practices and more science-based regulation. In general, obtaining permits for building infrastructure is becoming tougher all-around Europe. Especially for wind energy projects and particulary in places where installations started 20–30 years ago: France, Germany, The Netherlands, the United Kingdom and Spain. Spatial planning constraints that limit wind farm developments include strict minimum setback distances to housing, stringent noise limits, high nature safeguarding distances, radar (civil and military) safeguarding distances, wind turbine blade tip height restrictions or long lead times to obtain the necessary permits and authorities and communities consent to build projects, especially in the case of onshore wind.

As renewables scale-up, stakeholders' expectations could become challenging to meet in every single project. Unfulfilled expectations and poorly managed impacts could cause project opposition. Despite this, renewables benefit from a positive perception across most of the population in Europe; Nearly half of Europeans think that the EU should invest and develop in clean energy technologies as a priority in the next ten years.[24] Nevertheless, developers face increasing delays in obtaining permits to build projects. This delays the pace at which renewables are deployed and jeopardises the achievement of policy targets. The main impacts of permitting delays are liked to increased project costs, limited use of the latest available technology and bank financing. Regarding increased project costs, for example, a legal dispute causes the total project cost to increase due to diverging from the planned construction schedule, and modifying some contracts could be subject to penalties. Concerning the limited use of the latest available technology, the application for building a project submitted to the permitting authority includes, among other items, the proposed wind turbine model. If the proposed project is blocked for several years, the proposed turbine model becomes obsolete. Consequently, the project developer may need to file a new permit application if it wants to use a different (e.g. the latest) wind turbine model. This leads to additional costs and time. Moreover,

[24]European Commission, "Eurobarometer survey confirms public support for energy policy objectives", 11 September 2019, available at: https://ec.europa.eu/info/news/eurobarometer-survey-confirms-public-support-energy-policy-objectives-2019-sep-11_en.

in many countries, permits are issued for a determined period and may expire before even getting the consent to start building. Regarding bank financing, it has to be noted that banks are more likely to lend money for projects if the consenting risks are lower, or if there would be financial mechanisms in place, e.g. insurance to cover the risk of being denied consent, or longer consenting periods than initially envisaged. There are key indicators for planning and consenting wind energy projects that could facilitate or make more difficult their deployment. These are the setback distance, nature safeguard distances, noise limits and overall timing to obtain a permit. In some countries, restrictions imposed by civil and military aviation are critical too, but for simplicity, these are left out from this text.

12.3.1 Setback Distance

The minimum distance to housing (also known as 'setback distance') means the necessary minimum distance between the wind turbine generators and residential properties. Minimum distances can be a fixed distance or a distance relative to the turbine height (e.g. ten times the turbine tip height or rotor diameter). The rationale for setting minimum distances to housing is usually to avoid health hazards that may result from noise exposure, shadow flickering or accidents. Landscape (visual) impacts from siting wind farms close to communities are also considered in the planning phase. However, setback distances may be prone to subjective judgements and in some cases to serve political agendas despite evidence of no hazards imposed by wind energy projects to surrounding settlements or individual houses/buildings.

12.3.2 Nature Safeguard Distances

Most countries have established nature safeguarding distances to protect endangered species of birds and bats, including nesting, roosting and nursery areas in the vicinity of wind farms. Distances range between 1 and 10 km. However, legislation across countries varies significantly.

12.3.3 Noise Limits

Most European countries have national legislation in place to deal with the noise caused by all human activities, including wind turbines. Allowable environmental noise limits are usually categorised by area and timing. Maximum levels vary from 39 to 55 decibels dB(A) during the day and 5 to 10 dB(A) at night. However maximum allowable noise limits vary widely between countries. One of the reasons for the large spread is the various methodologies used to measure noise (where, when, how is the measure done). Many bases their noise limits on international bodies' recommendations, but this approach has shortcomings. For example, the World Health Organisation's (WHO) Environmental Noise Guidelines[25] set recommendations on environmental noise exposure for five relevant sources: road traffic, railway, aircraft, wind turbine and leisure. For wind turbines, it recommends noise levels to below 45 dB (A). However, this recommendation is only conditional because it recognises there is not enough scientific evidence.[26] Moreover, the guidance does not specify a methodology to measure or assess compliance. And it does not include recommendations for night noise exposure. WHO recognises that the quality of evidence of night-time exposure to wind turbine noise is too low to give a recommendation. It only specifies that authorities should implement suitable measures to reduce noise exposure from wind turbines in the population exposed to levels above the guideline values. However, it falls short of recommending any particular type of intervention over another.

[25]World Health Organisation, "Environmental noise guidelines for the European Region", 2018, available at: http://www.euro.who.int/__data/assets/pdf_file/0008/383921/noise-guidelines-eng.pdf?ua=1.

[26]According to WHO, a conditional recommendation requires a policy-making process with substantial debate and involvement of various stakeholders. There is less certainty of its efficacy owing to lower quality of evidence of a net benefit, opposing values and preferences of individuals and populations affected or the high resource implications of the recommendation, meaning there may be circumstances or settings in which it will not apply.

12.3.4 Time for Permitting

The overall permit granting process takes into account both the administrative (or planning) procedure and the grid access procedure. These two procedures could run in parallel or one after the other. The overall time for permitting starts when the first application is made to a permitting authority or grid operator and ends when all main permits and grid access is obtained. Article 16 of the revised Renewable Energy Directive states that the permit granting process shall not exceed '2 years for new projects with a maximum delay of 3 years ("2+1" rule)', and '1 year for repowered projects with a maximum delay of 1 year ("1+1" rule)'. These permit granting deadlines include neither Environmental Impact Assessment nor legal challenges.[27]

12.4 Bridging the Gap Between Scale and Implementation: Increase the Power System Flexibility

The market design legislative proposals in the 2018 'Clean Energy Package for All' from the European Commission, comprised a wide set of changes across four main areas: power system flexibility, institutional framework, playing-level field, investment environment. It introduced an emphasis on short-term trading, the role of demand response and storage. And it focused on improving regional cooperation, reinforce regulatory oversight, and the access, operation and planning of the power grids through European Network Codes. The *package*, comprising eight pieces of regulation, was meant to level playing field participation conditions for renewable market players. It addressed balancing responsibilities, dispatch and re-dispatch rules and access to ancillary services. Also, it was supposed to address resource adequacy and capacity mechanisms. This set of regulations made significant progress in integrating

[27]EU Directive on the promotion of the use of energy from renewable source (recast) EU 2018/2001, 11 December 2018, available at: https://eur-lex.europa.eu/legal-content/EN/TXT/?uri=uriserv:OJ.L_.2018.328.01.0082.01.ENG.

more renewables into the power system under the policy goal of setting an electricity market for renewables and renewables set for the market.

12.4.1 Market for Renewables and Renewables for the Market

One of the key issues for integrating renewables in a liberalised energy market is to adapt the rules for electricity trading so that these reflect the intrinsic characteristics of renewable energy generation. Wind speed and solar radiation are variable across different time frames, and their intensity needs to be forecast for electricity generation. Therefore, the rules for electricity trading needed to address this variability and uncertainty of forecast across different time frames. This is important as market rules set the way electricity prices are formed, determining the revenues for all generators. The EU-funded project MARKET4RES was one of many studies that identified the adaptations needed to facilitate renewable energy integration into the market. It recommended the following[28]:

12.4.2 Faster Markets

Large amounts of wind and other variable renewables in the power system require that the timing of markets allow for changes such as updated power demand, weather patterns variations and the consequent changes in electricity generation. Concretely, this means that Transmission System Operators (TSOs) should operate the system as close as possible to real time (gate closure time) while ensuring system security and stability assessments. This gives variable generators the opportunity to correct their deviations as their forecast improves considerably closer to the time of energy delivery.

[28] Daniel Fraile et al., "Guidelines for implementation of new market designs in Europe with high shares of RE-E penetration (post-2020)", MARKET4RES D6.2, October 2016, available at: http://market4res.eu/wp-content/uploads/Market4RES_WP6_D6-2-FINAL.pdf.

12.4.3 Larger Markets

The variability of wind power output smooths out when the generation profile of several wind farms is aggregated across large geographical areas. To take advantage of this, system operators need close cooperation to enlarge markets and connect these areas. System operators also need sophisticated methods to allocate transmission capacity closer to real time.

12.4.4 Smaller Products

Traditionally fossil fuel-fired power generation sells electricity with very large blocks of energy days, weeks, months and even years ahead of delivering it. Trading wind energy, on the other hand, is easier using small timeframe products. This is because its predictability improves at shorter timeframes. Smaller products allow for the possibility to adjust offers closer to real time. For example, in Germany, the 15-min electricity trading in the intraday market has successfully facilitated the management of renewable energy variability compared to the previous hourly based contracts. It has also helped to reduce the risks from balancing responsibilities. Wind energy forecast errors reduced by half between a day-ahead forecast and one hour-ahead forecast.

12.5 Bridging the Gap Between Scale and Implementation: Financing Renewables

Up to now, the wind industry remains a major investment opportunity, with €19 bn raised in 2019 for the construction of new wind farms.[29] But most EU countries are failing to provide investment predictability beyond 2020 despite being required to do so by the Renewable Energy

[29]WindEurope, "Wind energy in Europe in 2019. Trends and statistics", p. 23, 2020, available at: https://windeurope.org/wp-content/uploads/files/about-wind/statistics/WindEurope-Annual-Statistics-2019.pdf.

Directive and the Governance Regulation. Such uncertainty slows downs the commitment of funds available for investing in renewables and threatens the realisation of climate objectives. Another key issue for the integration of renewables in liberalised energy markets was the compatibility of state aid instruments to support the deployment of technologies. These instruments needed to become more market-based so that any distortion to the competition is minimised. Renewable energy support mechanisms moved away from feed-in tariffs, to market premiums awarded through competitive auctions. This means that as technologies mature, investors are in a better position to take on more risks that used to be covered by governments. As seen in Fig. 12.1, more than 65% of the installed capacity will be partially exposed to power markets, and more than 25% will be fully exposed. Some projects are even expected to be built without any financial government support.

Short-term spot market prices on their own may not provide a basis for investment. In the absence of policies that address structural market failures, there would be no merchant investments in any power generation asset, including conventional technologies. The bar for investment in wind energy assets is even harder to clear. They are upfront capital-intensive assets, and current hedging options are limited or not widely used. Full exposure to merchant conditions raises financing costs by deterring low-risk investors. In addition, the market value of wind energy

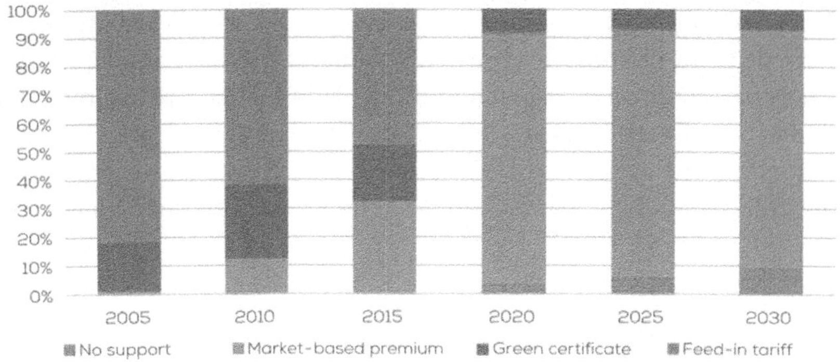

Fig. 12.1 Share of new wind capacity per type of support (*Source* WindEurope [2018])

tends to decrease as its penetration rate increases as wind farms capture lower spot market prices than other generators. This revenue attrition of wind energy projects requires to complement, their spot market revenues with a long-term revenue stabilisation mechanism.

12.5.1 Revenue Stabilisation Mechanisms

Because an Energy-only Market approach is unlikely to deliver long-term investment signals,[30] market-friendly revenue stabilisation mechanisms will still be needed by investors to secure their revenues. There are two main approaches: production-based and investment-based mechanisms.[31] The former provides operators with an operating premium for the actual output sold on the market, which can be fixed and defined ex-ante (Denmark, onshore), or sliding,[32] i.e. evolving as the difference between a strike price and a reference market price. The investment-based scheme provides compensation based on the installed capacity and the expected energy production (Spain). The former exposes more to the volume risk and the latter expose more to the price risk. Both fill the gap between the anticipated project costs, and it is expected revenues over a given period. The delivery conditions of the mechanisms will influence the risk allocation between investors and consumers. The key features to consider are: (a) whether the support remunerates all the energy produced (France, Germany) or only a certain number of full load hours (Denmark). The former minimises investors' risks while the latter incentivises demand-oriented feed-in and facilitates participation in markets such as balancing or re-dispatch (although at a higher risk for investors). (b) the benchmark against which the reference market price is calculated, either the average baseload price (the Netherlands) or the specific price received in each specific hour (e.g. Italy), and (c) the frequency that

30See WindEurope, "Creating a business case for wind after 2020", January 2017, available at: https://windeurope.org/wp-content/uploads/files/policy/position-papers/WindEurope-Long-Term-Investment-Signals.pdf.

31MIT Energy Initiative, "Revisiting support policies for RES-E adulthood", 2016. Green certificates are excluded since their use is restrained to a few countries.

32Germany (Market Premium), The Netherlands (SDE+), France ("complément de remuneration"), the UK (CfD) etc.

the premium is adjusted to: Investors' risk is minimised if the premium covers hourly price deviations, while longer periods encourage to perform better than the market outcome with sophisticated bidding strategies (month in Germany, year in Spain). To ensure greater control of policy costs, some countries are also implementing a price floor over the whole contract period, as seen in Fig. 12.2 for the Netherlands. Others, like the UK seen in Fig. 12.3, have a two-sided Contract for Difference, whereby the premium can become negative, i.e. the operator pays back any excess revenues above the strike price.

These revenue stabilisation mechanisms are awarded via competitive auctions. Since 2014, Member States have launched auctions for wind energy and other renewables as prescribed by the EU's State Aid Guidelines. However, the transition to auctions was not straightforward. Many countries struggled with auction design and made changes in view of accommodating different policy objectives. EU Governments

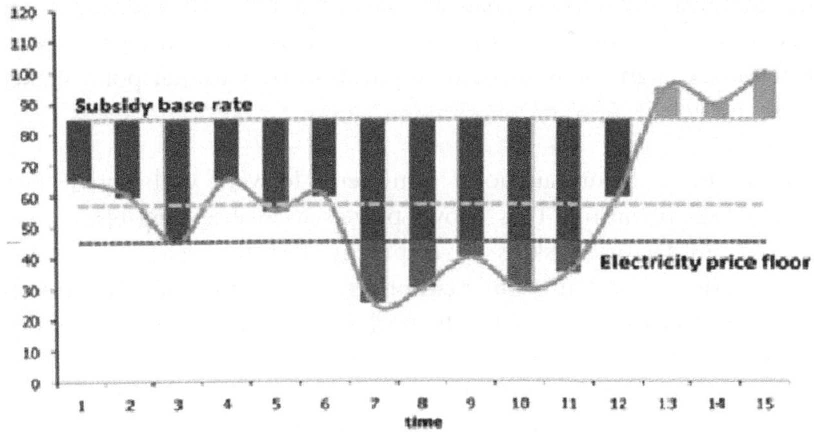

Fig. 12.2 Netherlands SDE+ scheme (*Note* The compensation (blue bars) is determined as the difference between the expected projects costs (green dotted line) and the yearly average electricity price (orange line). A price floor (red dotted line) is determined as 2/3 of the long-term average electricity price (orange dotted line). During years 7–11, the compensation is capped and does not suffice to cover expected projects costs (red bars). However, during years 13–15, the plant income is larger than expected (green bars). *Source* Adapted from Energy Research Centre of The Netherlands [2011])

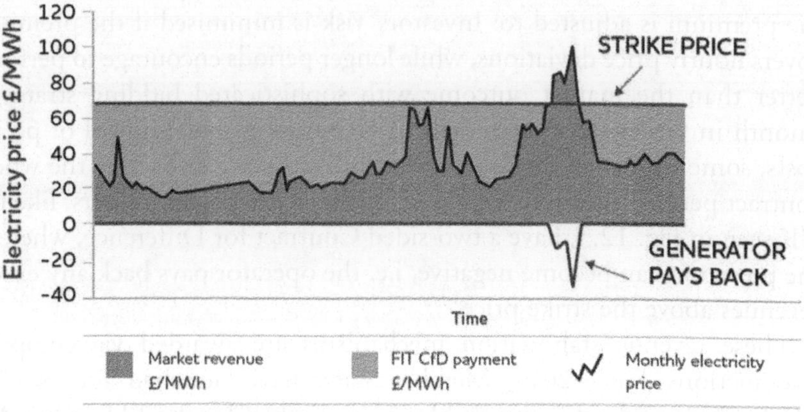

Fig. 12.3 Contract for difference (*Source* UK Government White Paper. EMR Settlement Limited [2011])

and industry are still learning, and additional practice will be necessary to make auctions fit national circumstances. In addition to the good practice tender design principles,[33] there are non-regret features of auction design for immediate application by national policymakers: These include:

- Visibility on future auctions: Long-term forward visibility is key to industrial planning. This allows players to realise long-term investments in factories, skills development, test facilities, research and innovation. These investments optimise economies of scale and efficiencies across the supply chain that allow the industry to drive costs down. A long-term schedule for public support allocation over multiple years, including the timing, capacity and budget for auctions as prescribed in the post-2020 Renewable Energy Directive.
- Transparent auction design and rules: When auction rules are complex, there is more risk for participants, who reflect this in their bids. This increases final prices. Transparency, simplicity and clarity of

[33]EWEA, "Design options for wind energy tenders", December 2015, available at: http://www.ewea.org/fileadmin/files/library/publications/position-papers/EWEA-Design-options-for-wind-energy-tenders.pdf.

rules instead go a long way in attracting bidders and cutting costs. The process towards launching an auction should include consultation and open dialogue between governments and investors on the auction design. Stakeholders should strive to secure simple and straightforward selection criteria, for example, payment arrangements and awarding mechanism. Price-only is the preferred award mechanism. It offers an objective comparison between bids as opposed to qualitative assessment between projects. Crucially for investors, is that the remuneration awarded is not subject to retroactive modifications.

- Technology baskets: Auctions could be technology-specific or technology-neutral. Regardless of their design, auctions must take into account the technical characteristic of technologies, for example, their risk profile, project lead times, size and costs, and put into comparable competition technologies. Member States should have the possibility to run technology-specific auctions to tap into a broad mix of technologies with different generation profiles that complement each other on the grid.
- Compulsory consented permits to participate in the auction: auctions are successful only if awarded projects are built. Having the consent and other relevant permits from authorities to build the project, ensures that bidders compete against each other on a level playing field and projects will be delivered if awarded. Mandatory permitting is of particular importance to onshore wind. In contrast, governments should apply flexibility on compulsory permitting for offshore wind due to longer lead times. Crucially, the simplification of administrative and permitting procedures is key to the investors' ability to participate in auctions. Authorities should strive to shorten permitting timelines and align them with auction schedules.
- Set construction deadlines: These guarantee the construction of projects to meet government objectives. Taking into account lead times ensures that there would be no mismatch between the latest technology available on the market and what is installed. The consequences of missing construction deadlines because of unforeseen events, however, should be calibrated and should not result in disproportionate penalties for the developers.

As technologies and markets evolve, future auction design should consider:

- Geographically balance of projects: Locational signals depending on proximity to demand, wind resource availability or grid infrastructure availability could balance the geographic distribution of wind energy projects in line with national circumstances.
- Repowering: Many wind energy projects will reach the end of their operational lifetime between 2020 and 2030. Future auctions should ensure that the decommissioned projects are added on top of the announced auction volumes. Moreover, auctions should allow repowered projects to compete for state aid compatible support schemes.
- Increased grid connection requirements: Current auctions do not reflect the increasing set of grid connection requirements. The lack of clarity regarding future grid connection requirements might undermine the ability of developers to realise their projects at the costs foreseen in their bids today. Reducing the lead time for projects and introducing a reasonable transition period to comply with grid codes could form part of future auction qualification criteria.
- Cross-border auctions: With increased emphasis on regional cooperation in the post-2020 European targets for renewables, cross-border auctions could increase. The voluntary opening of support schemes between countries could serve as a first step for exploring new ways for deploying renewables. However, Member States should tackle the impact of different regulatory regimes on the competitiveness of projects if they are to attract bidders.

12.5.2 Availability and Cost of Capital

Financing costs make up a considerable portion of the overall cost of a project. Therefore, it is critical for policymakers to understand the availability of capital and the implications in the cost of capital of their policies. The majority of the installed wind capacity in Europe has been financed on a non-recourse basis, typically referred to as project finance. Around 75% of Capex for onshore wind and 30% for offshore

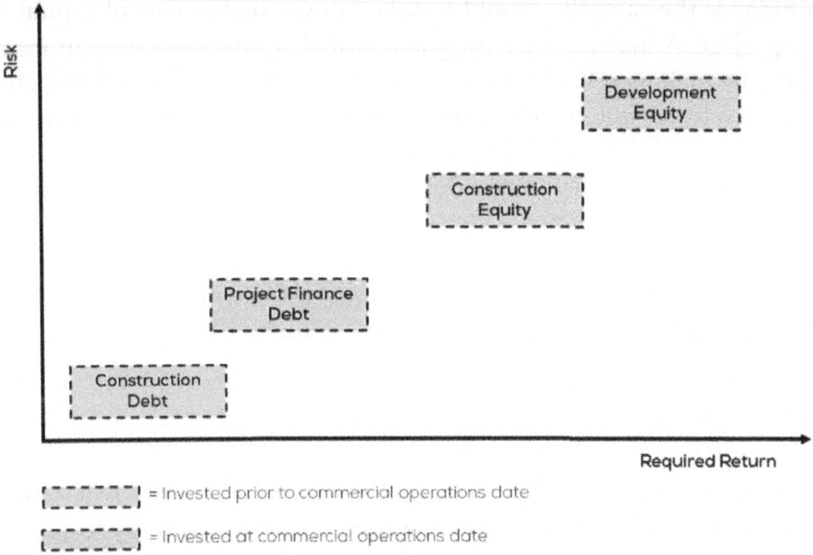

Fig. 12.4 Comparison of the relative risks and returns for typical wind energy financing sources (*Source* WindEurope, adapted from NREL [2018])

wind projects are raised with project finance.[34] Due to the non-recourse characteristic, meaning that in case of bankruptcy lenders do not have recourse to other assets but the project asset itself, is generally considered a lower-risk investment, as demonstrated here in Fig. 12.4, and therefore comes with lower-cost financing compared to equity.

A merchant environment might introduce limitations on the availability of debt. Lenders are used to some elements of merchant financing, predominantly in the certificate-based support scheme countries such as the Nordic markets or the now discontinued UK's ROC regime. However, full merchant exposure might tighten the requirements for raising debt, introduce higher cover ratios and financial covenants that a project might not be able to fulfil. Sponsor equity will then have to fill this gap. But in economic conditions when debt costs less than equity,

[34]WindEurope, "Financing and investment trends. The European wind industry in 2017", April 2018, available at: https://windeurope.org/wp-content/uploads/files/about-wind/reports/Financing-and-Investment-Trends-2017.pdf.

as today is the case, this would translate into a higher cost of capital for the project. A merchant environment will also introduce limitations on equity by making it harder for investors to exit projects post-construction and recycle capital in additional projects. While many investors are becoming more comfortable with the risks that wind and other renewables have, they remain risk-averse to full merchant exposure. Many financial investors cannot hedge merchant risk with large portfolios of assets given that their primary business is not energy production like large utilities. Financial investors would typically own smaller portfolios of assets geographically spread over a few countries.

12.5.3 PPAs

Some emerging merchant financing in the wind sector will require some form of additional revenue stabilisation complementary to support mechanisms. Corporate renewable power purchase agreements (PPAs) come with certain benefits for generators to hedge the volatility of their revenues like price visibility over a long period and a guaranteed offtake of their energy. These elements are important to lower the cost of debt financing. Lenders would typically need protection to ensure debt repayment obligations are met. Consequently, lenders tend to prefer lower revenues over a long period rather than higher but uncertain ones. Corporate renewable PPAs to date is still a niche market and not to the scale necessary to replace other policy-driven revenue stabilisation mechanisms. They limited to a handful of countries. The Nordic region followed by the UK and the Netherlands are the biggest markets for such deals. These markets share a good track record in renewable energy development, coupled electricity markets, sufficient demand for green electricity from corporates and, the legislation does not pose a barrier to signing these PPAs. Another key factor is the existing government support mechanism for renewables in the country. In countries with feed-in tariffs, the developer would not need to lock in long-term power prices with a corporate off-taker. The government feed-in tariff guarantees the price. Conversely, green certificate-based support regimes such as in the Nordic region or the (now phased-out) UK ROC scheme need a long-term power hedge to secure price certainty. This explains the significant uptake of corporate renewable PPAs in these countries. For corporate PPAs to continue growing, it is key is to ensure that future revenue stabilisation mechanisms allow for the revenue from a corporate renewable PPA to stand in conjunction with any form of government support. To ensure there are no double compensation project owners factor-in both the expected revenue of the PPA and the Guarantees of Origin (GO). Off-takers will tend to buy these two bundled products: the electricity produced as well as the GO that verifies the source of electricity. The of a government off-taker with a corporate off-taker comes with the benefit of reduced government expenses for renewable energy support.

12.6 Conclusions

Europe has been a pioneer in the deployment of renewable energy. These technologies have been known to humans for a long time, but it has taken a few crises for policymakers and investors to put them at play on a large scale. For example the 1973 oil crisis, when governments around the world realised the fragility of the supply chain of this fossil fuel, or the current climate change crisis, which is demanding urgent mitigation and adaption measures to avoid dangerous temperature increases and other meteorological phenomena. During the drafting of this book, another crisis, this time health-related due to the COVID-19 pandemic, is hitting the world. The repercussions are still unknown, but the global economy is expected to hit the worst recession since the end of the last world war as countries locked down and stopped all economic activities for months. Against this background, there are calls for the economic recovery to be based in line with policies implemented to mitigate or avoid previous crises. A 'green recovery' is touted and Europe is again pioneering it. As seen in this chapter, Europe is facing formidable challenges to meet its international commitments to mitigate climate change. It will be even harder to do so during a new economic crisis. Still, the opportunities stand even stronger and Europeans should grab them to not lose their first-mover advantage.

First, European policy should purse an renewables-based electrification. Using renewable electricity wherever it is available and whenever it is possible, can deliver the bulk of decarbonisation needed for the economy. The investments such strategy requires can lift economies from crises. So, setting only energy policies to this end will not suffice. As analysed in this chapter, economic, environmental, trade and even fiscal policies would need to align towards this strategy. I presented three levers to start with: energy efficiency in buildings and industry, electrification incentives including taxation and a stronger CO_2 pricing. Second, deploying renewables at the speed and scale needed to significantly flat the climate change curve will require streamlined administrative procedures and permitting. Even if we achieve the proper economic and energy policies, governments need to ensure that they address issues 'on the ground'. Obtaining permits for building, and improving, infrastructure is becoming tougher all-around Europe. As renewables scale-up, stakeholders' expectations could become challenging to meet in every single project. Unfulfilled expectations and poorly managed impacts can cause project opposition and policymakers react with even more stringent permitting requirements. Third, as renewables expand, the flexibility of the energy system will need to be increased. Policy should be adapted nimbly to adjust the rules for electricity trading to reflect the characteristics of renewable energy generation. Larger and faster markets and smaller energy trades are more suitable to address uncertainty and variability of weather-dependent renewables.

Lastly, enabling the conditions for the continuation of investments will be critical. Renewables are high capital-intensive assets, with low operational costs and marginal costs of electricity generation close to zero. Because an energy-only market approach is unlikely to deliver long-term investment signals, market-friendly revenue stabilisation mechanisms will still be needed by investors to secure their revenues. Together with the availability and cost of capital, revenue stabilisation is a crucial enabler for expanding renewables. As new approaches to finance renewables emerge, such as corporate PPAs, Europe will face new risks and challenges, but none of these will be greater than the dangers posed by climate change and the destruction of our economies due to recurrent crises.

References

Agora Energiewende and Sandbag, "The European power sector in 2019", 2020, available at: https://sandbag.org.uk/wp-content/uploads/2020/02/Sandbag-European-Power-Sector-Review-2019.pdf.

Agora Energiewende and Sandbagm, "The European power sector in 2019", 2020, available at: https://ember-climate.org/global-electricity-review-2020-data/.

Communication from the Commission to the European Parliament, the Council, the European Economic and Social Committee and the Committee of the Regions—20 20 by 2020—Europe's climate change opportunity, 23 January 2008, available at: https://eur-lex.europa.eu/legal-content/EN/TXT/?uri=CELEX:52008DC0030.

Ecofys, "Potential for shore side electricity in Europe", 7 January 2015, available at: https://www.scribd.com/document/366724117/Ecofys-2014-Potential-for-Shore-Side-Electricity-in-Europe.

Energy Research Netherlands, "Aanvullend onderzoek correctiebedragen SDE+-regeling", 2011, available at: https://publicaties.ecn.nl/PdfFetch.aspx?nr=ECN-E–15-070.

EU Directive on the promotion of the use of energy from renewable source (recast) EU 2018/2001, 11 December 2018, available at: https://eur-lex.europa.eu/legal-content/EN/TXT/?uri=uriserv:OJ.L_.2018.328.01.0082.01.ENG.

European Commission, "2030 climate & energy framework", https://ec.europa.eu/clima/policies/strategies/2030_en.

European Commission, "An EU Strategy on heating and cooling SWD (2016) 24 Final", 2016, available at: https://ec.europa.eu/energy/sites/ener/files/documents/1_EN_ACT_part1_v14.pdf.

European Commission, "EU Green Deal - Revision of the Energy Taxation Directive", 2021, available at: https://ec.europa.eu/info/law/better-regulation/have-your-say/initiatives/12227-Revision-of-the-Energy-Tax-Directive.

European Commission, "Commission staff document. Executive summary of the impact assessment SWD (2013) 532 final", 18 December

2013, available at: https://eur-lex.europa.eu/legal-content/EN/TXT/PDF/?uri=CELEX:52013SC0532&from=EN.
European Commission, Directive on the Energy Performance of Buildings (2018/844/EU), 26 April 2018, available at: http://data.consilium.europa.eu/doc/document/PE-4-2018-INIT/en/pdf.
European Commission, "Energy prices and costs in Europe—Commission staff working document", p. 38, 2018, available at: https://ec.europa.eu/energy/sites/ener/files/documents/swd_-_v5_text_6_-_part_1_of_4.pdf.
European Commission, "Energy security strategy", p. 1, 2015, available at: https://eur-lex.europa.eu/legal-content/EN/ALL/?uri=CELEX%3A52014DC0330.
European Commission, "Eurobarometer survey confirms public support for energy policy objectives", 11 September 2019, available at: https://ec.europa.eu/info/news/eurobarometer-survey-confirms-public-support-energy-policy-objectives-2019-sep-11_en.
European Commission, "In depth analysis in support of the Commission Communication COM (2018) 773, A Clean Planet for all. A European long-term strategic vision for a prosperous, modern, competitive and climate neutral economy", 28 November 2018, available at: https://ec.europa.eu/clima/sites/clima/files/docs/pages/com_2018_733_analysis_in_support_en_0.pdf.
European Commission, "New rules for greener and smarter buildings will increase quality of life for all Europeans", 15 April 2019, available at: https://ec.europa.eu/info/news/new-rules-greener-and-smarter-buildings-will-increase-quality-life-all-europeans-2019-apr-15_en.
European Commission, "Putting energy efficiency first: Consuming better, getting cleaner", 30 November 2016, available at: http://europa.eu/rapid/press-release_MEMO-16-3986_en.htm.
European Council Implementing Decision authorising Sweden to apply a reduced rate of electricity tax to electricity provided to vessels at berth in a port, 8 April 2011, available at: https://eur-lex.europa.eu/legal-content/EN/TXT/?uri=CELEX%3A52011PC0158.
European Environmental Agency, "Total greenhouse gas emission trends and projections in Europe", last modified, 19 December 2019, available at: https://www.eea.europa.eu/data-and-maps/indicators/greenhouse-gas-emission-trends-6/assessment-2.
EWEA, "Design options for wind energy tenders", December 2015, available at: http://www.ewea.org/fileadmin/files/library/publications/position-papers/EWEA-Design-options-for-wind-energy-tenders.pdf.

Fraile, Daniel et al., "Guidelines for implementation of new market designs in Europe with high shares of RE-E penetration (post-2020)", MARKET4RES D6.2, October 2016, available at: http://market4res.eu/wp-content/uploads/Market4RES_WP6_D6-2-FINAL.pdf.

International Energy Agency (IEA), "Renewable energy for industry", 2017, available at: https://webstore.iea.org/insights-series-2017-renewable-energy-for-industry.

MIT Energy Initiative, "Revisiting support policies for RES-E adulthood", 2016.

Oil Change International, Friends of the Earth, WWF and the Sierra Club, "Talk is cheap: How G20 governments are financing climate disaster", 5 July 2017, available at: http://priceofoil.org/2017/07/05/g20-financing-climate-disaster/.

UK Government White Paper. EMR Settlement Limited, "Contracts for difference explained", available at: https://www.emrsettlement.co.uk/about-emr/contracts-for-difference/.

WindEurope, "Creating a business case for wind after 2020", January 2017, available at: https://windeurope.org/wp-content/uploads/files/policy/position-papers/WindEurope-Long-Term-Investment-Signals.pdf.

WindEurope, "Financing and investment trends. The European wind industry in 2017", April 2018, available at: https://windeurope.org/wp-content/uploads/files/about-wind/reports/Financing-and-Investment-Trends-2017.pdf.

WindEurope, "Sustaining a cost efficient energy transition in Europe", June 2018, available at: https://windeurope.org/wp-content/uploads/files/policy/position-papers/WindEurope-Sustaining-a-cost-efficient-energy-transition-in-Europe.pdf.

WindEurope, "Wind energy in Europe in 2018. Trends and statistics", 2019, available at: https://windeurope.org/wp-content/uploads/files/about-wind/statistics/WindEurope-Annual-Statistics-2018.pdf.

WindEurope, "Wind energy in Europe in 2019. Trends and statistics", 2020, available at: https://windeurope.org/wp-content/uploads/files/about-wind/statistics/WindEurope-Annual-Statistics-2019.pdf.

World Health Organisation, "Environmental noise guidelines for the European Region", 2018, available at: http://www.euro.who.int/__data/assets/pdf_file/0008/383921/noise-guidelines-eng.pdf?ua=1.

Zeniewski, Peter, "A long-term view of natural gas security in the European Union", International Energy Agency, 13 March 2019. https://www.iea.org/commentaries/a-long-term-view-of-natural-gas-security-in-the-european-union.

Zhou, Vivia, "3 electric car incentives you need to know in Europe", The Current, 4 July 2017, available at: https://blog.evbox.com/electric-car-incentives.

13

Offshore Wind: Staying Ahead of the Curve

Marios Papalexandrou

13.1 Introduction

Wind energy accounts for 15% of the electricity consumption in the EU for 2019 with 205 GW cumulative installed capacity. The wind industry contributes €36 billion to EU's Gross Domestic Product, supports 263,000 jobs and generates €8 billion exports. The European Commission recognises the important role of wind energy towards a climate-neutral Europe by 2050. Its long-term decarbonisation strategy foresees that wind energy will have a dominant role in 2030 and beyond in driving EU decarbonisation. The European Commission has set a target for 2030 for at least 32% of final energy consumption in the EU coming from renewables as part of the Energy Union strategy. Under this vision, Member States will collectively deliver on the EU's Climate and Energy objectives. Meeting these objectives will require an ambitious

M. Papalexandrou (✉)
Aeolus Limited, Brighton, UK
e-mail: marios.papalexandrou@aeolus-energy.com

M. Mathioulakis (ed.), *Aspects of the Energy Union*,
Energy, Climate and the Environment,
https://doi.org/10.1007/978-3-030-55981-6_13

legislative framework for member states beyond 2020 and offshore wind is an important ingredient for meeting these objectives. This new source of clean, sustainable energy has been on the rise in the past three decades. Offshore wind has moved from few demonstration projects in the early 90s to a mainstream energy provider covering electricity consumption for millions of citizens. Europe is at the forefront of offshore wind developments with more than 20 GW of cumulative installations by the end of 2019, covering approximately 80% of global installations. The largest rotating machines the world has ever seen, are running offshore the European coasts in the North and the Baltic seas. However, in the past few years, offshore wind is taking big leaps in China, Taiwan, USA, Vietnam, Korea and Japan. Australia and India are two other nations that will soon join the offshore wind club. The 2020s will see the move of offshore wind from a European market to a global market. Can Europe stay at the forefront of future technological developments? What are the success stories to be shared and the challenges to be faced?

13.2 The Emergence of Onshore Wind

In order to gain a good understanding of offshore wind, it is necessary first to examine onshore wind. Offshore wind emerged in the 90s following onshore wind's development a decade earlier, in the 80s, due to an unexpected event; the oil crises of 1973 and 1979 that led to significant increases in oil prices shocked all oil-dependent developed countries and led them to reconsider and re-plan their energy strategy and security of energy supply. One of the side effects of the oil crises was moving away from oil energy production and considering alternative sources of energy. Denmark and the state of California were pioneers at that time in research and development (R&D) for wind energy. During the 80s US renewable energy R&D funding grew from less than USD1.0 million per year in the early 1970s to over USD1.4 billion by 1980. This led to 12 GW of renewable power constructed in the USA during the 1980s (IRENA 2012). Following the oil crises, legislation was in parallel developed in the USA and Denmark encouraging the development of wind farms through tax incentives and subsidies. Denmark had a long tradition using wind energy, and by the early 80s, several local manufacturers had emerged selling small turbines to local wind cooperatives. Energy legislation and subsidies scheme boosted developments in both sides of the Atlantic with Danish turbine manufacturers exporting to the USA. However, in the US, the tax incentives regime dropped following the reduction of oil prices in the 90s leading to a stall of new installations while in Denmark the political support continued by setting new targets up to 1350 MW of wind energy by 2000 (WindEurope 2020b). In parallel, many European countries from the mid to late 80s started creating a support regime for wind energy, including the UK, Germany, Spain, France and the Netherlands. To understand the technical advancements of onshore wind, it is worth following how the first wind farms looked like. The first wind farm in the USA started operation in 1980 at Crotched Mountain, consisting of 20 × 30 kW wind turbines, while in Europe the first wind farm, consisting of 5 × 20 kW turbines, started operations at the Greek island of Kythnos. By the early 90s, the size of a wind turbine grew from 20 kW to 450 kW, leading to a significant drop in costs and an increase in energy production per turbine. In parallel, the growth of commercial onshore wind in Europe during the 90s had begun moving from MW to GW scale in cumulative capacity. A decade later, average wind turbine sizes reached the MW scale.

13.3 The Birth of Offshore Wind and Its Technological Advancements

Offshore wind was born in the onshore wind's transition period when it was moving to a more mature stage. The first offshore wind farm started operations in 1991 in Denmark. The Vindeby wind farm, consisting of 11 × 450 kW turbines, operated successfully for 26 years until it was decommissioned in 2017 (ORSTED 2017). The average wind turbine size in the EU expressed in MW has increased 10-fold within the same period of Vindeby's lifetime, as presented in Fig. 13.1.

Until 2010 the offshore wind market followed the onshore wind market closely with turbine manufacturers adjusting onshore turbine models or providing upgraded models for offshore environments. After

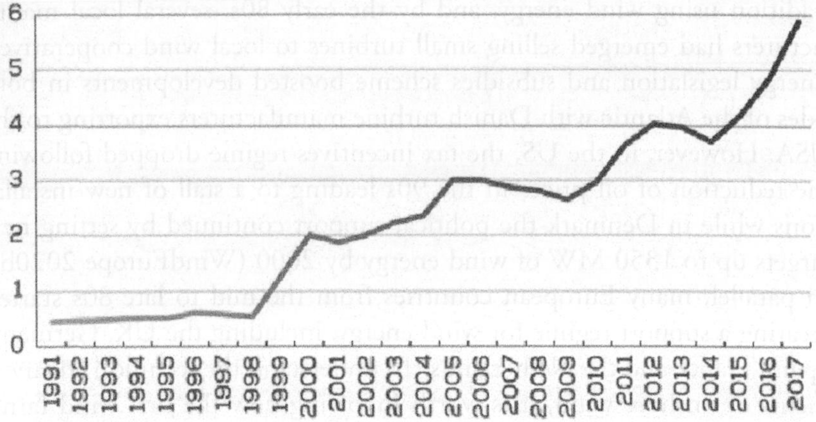

Fig. 13.1 Yearly average of newly installed EU offshore wind turbine rated capacity (MW) (*Source* WindEurope 2018)

2010, research and development of major onshore turbine manufacturers, together with market growth, created the foundations for the development of large dedicated offshore turbines which cut ties from the onshore wind turbines creating an exclusive offshore fleet. A fierce competition of turbine manufacturers commenced 10 years ago from a number of major industrial players that tried to grab a share of this emerging market. This competition continues today with a few major players MHI Vestas, Siemens Gamesa and GE competing on an international level. They are all providing commercial turbines currently exceeding 10 MW with Siemens Gamesa providing in mid-2020 the largest offshore wind turbine in the market, the SG 14-222DD, with a rotor diameter of 222 m. In 2019, the average size of an offshore wind turbine installed was 7.8 MW, which will exceed 8 MW in 2020. Turbine lifetime has also been extended the last few years from 20 years to 25+ years, lowering the offshore wind levelised cost of energy. As the turbine swept area is getting larger as depicted in Fig. 13.2, the turbine capacity factor, a measurement of turbines performance in maximum power has also increased with modern turbines having capacity factors exceeding 50% in the North Sea.

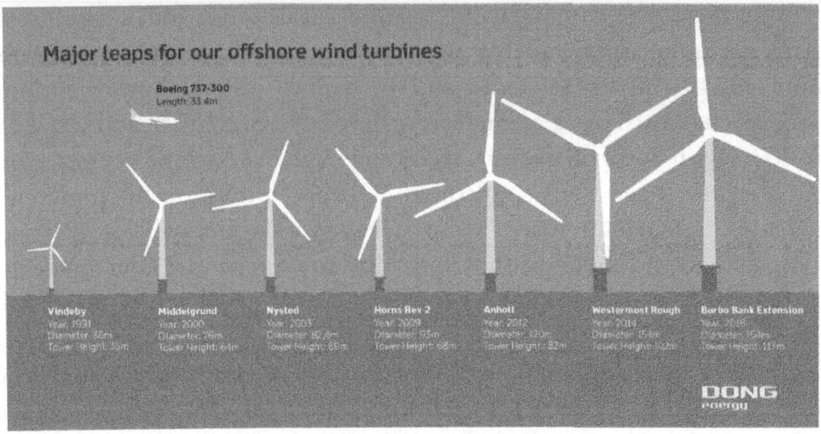

Fig. 13.2 Growth of offshore wind turbines (*Source* ORSTED 2016)

Technology advancements of wind turbine growth during this period was followed by the supply chain of the balance of plant. As turbines got larger so did wind farms which were located further offshore in deeper waters. Foundations got larger to accommodate larger and heavier turbines. Larger cables and substations were also developed to connect more powerful turbines to the onshore grid. Alternative current (AC) grid connection between the wind farms and the onshore grid has been the selection of choice for all countries apart from Germany due to the distance from shore where Direct Current (DC) connection is the preferred option as export cable lengths exceed 100 km. Installation vessels also got larger being able to accommodate more turbines and foundations per vessel and instal them faster with more powerful cranes. Cable vessels got larger as well as being able to carry more cables that will connect wind farms further offshore. In 2017 technological advancements continued with the successful installation of the first floating wind farm. Consisting of 5 × 6 MW wind turbines off the coast of Scotland, the Hywind project (EQUINOR 2020) earmarked a new era allowing turbines to be installed in water depths exceeding 100 m which unleashed the development of deepwater offshore wind farms close to coastal areas all over the world. Another 274 MW of seven floating wind projects will be connected in Europe by 2022 (WindEurope 2020a).

From the above summary, the technological trends for offshore wind farms are quite clear; moving to even larger turbines that can harness more energy, further offshore and in deeper waters. Future turbines are expected to have a nameplate capacity between 15 and 20 MW in 2030 with rotor diameters exceeding 250 m. As projects get larger and are developed further offshore, Direct Current technologies will increase their market share. More jacket foundations will be deployed compared to monopile foundations due to heavier turbines in deeper waters and floating foundations considered now novel will be well established. These technology advancements that have occurred the last few years had a huge impact on offshore wind costs. For every new 8 MW turbine which is currently installed it would require 4 × 2 MW turbines from a project that reached financial close in 2006 and 2 × 4 MW turbines from a project that reached a financial end in 2014. The amount of steel needed for the same energy produced under the same water depth conditions is massively reduced when moving to larger turbines with larger rotor diameters. Fewer turbines mean fewer foundations and cables and given installation times keep coming down, the overall project costs are reduced. Moving also to larger capacity cables optimises the energy production with lower electrical losses and fewer cables needed to be connected between the turbines and the offshore substation. A catalytic effect on cost reduction came from the size and continuation of projects being built. Thanks to the volume of continuous work occurring throughout European projects the last years, a significant cost reduction was possible, allowing innovation and industrialisation to take place. The decade of 2000 saw an average annual offshore wind installation rate below 100 MW with maximum one project constructed per year. Since 2011 it is stronger with a minimum of 1.1 GW offshore wind capacity installed per year. From 2017 this growth is even stronger with 2.5 GW minimum annual installations in the EU. The above developments have allowed the whole supply chain to move fast forward during recent years to cope with the size and volume of upcoming projects, in competitive bidding auctions. The technology advancements, the supply chain maturity, cheaper financing and the increased competition between developers and the supply chain led to a sharp price reduction in offshore wind costs, especially after 2016 when offshore wind auctions were introduced. The Cost Reduction Monitoring Framework 2016 report (ORE Catapult) showed projects reaching a final investment decision in 2015–2016 achieved a levelised cost of energy (LCOE) of £97/MWh (EUR112/MWh). This compares to projects, which completed financing in 2010/2011 with an LCOE of £142/MWh (€164/MWh) at 2011 prices (Wind Power Monthly). The global weighted average cost of electricity for plants commissioned in 2017 was approximately €120/MWh corresponding to USD140/MWh (IRENA 2018). Auction results in 2016 and 2017, for offshore wind projects that will be commissioned in 2020 and beyond, signal a step-change, with costs falling to a range between €50/MWh and €83/MWh corresponding to USD60/MWh and USD100MWh (IRENA 2018). These significant cost reductions continue with CFD auction prices announced in September 2019 in the UK falling below £40/MWh. All offshore wind auction results in 2019 ranged between €40/MWh and €50/MWh.

13.4 Leading Countries in Offshore Wind Developments

The strong growth in offshore wind mentioned above would have never happened without a strong political regime to support the delivery of this volume. Offshore wind developments as large infrastructure projects have been intimately paired with government policies to exploit offshore wind as part of EU's renewable energy targets. So which are the countries that are leading the way in offshore wind farm developments? As it has already been mentioned, Denmark was the real pioneer for the first ten years since the birth of offshore wind market and the first to install a 160 MW project, the Horns Rev, which became the largest true offshore project, located some 14–20 km offshore. Other countries that tapped into offshore wind and started installing small scale demonstration projects included the UK, Belgium, the Netherlands and Ireland. However, only the UK was consistent in delivering offshore wind farms from that time until today.

13.4.1 The UK

The UK, through leasing rounds over the last two decades, moved from a few demonstration projects of less than 100 MW each and close to shore to GW scale projects being constructed today far offshore. The UK is currently the largest offshore wind market in the world. Driven by the EU's legally binding 2020 renewable energy targets of 30% of electricity coming from renewables, the UK government has implemented the Climate Change Act, which mandates an 80% cut in carbon dioxide emissions by 2050 compared to 1990 levels. This target in combination with the declining North Sea oil and gas industry and with 1/3 of old thermal power stations coming offline after 2020 led the UK government to focus on offshore wind for future power generation (FOWIND 2014). An industrial strategy was added in 2015 as a policy priority, with an obligation for a minimum UK content on wind farm total budget, examined during the consenting process, which has led several EU manufacturing companies to establish UK subsidiaries. The seabed within the Exclusive Economic Zone of UK's continental shelf is owned and managed by the Crown Estate. Any developer needs to have secured a lease agreement with the Crown Estate to get exclusivity rights for site development. The first round of UK developments was published in 1999 by the UK government and was an open-door application process for projects limited to an area of 10 km^2 in size and with a maximum number of 30 turbines. Locations were chosen by potential developers, and 17 sites were granted a lease to exclusively develop offshore projects in April 2001, in what has become known as Round 1 of UK offshore wind development. Lessons learnt from Round 1 prompted the UK department of trade and industry to develop a strategic plan for offshore wind by selecting specific zones and pushing developments above 8 km offshore to reduce visual impact and avoid shallow feeding grounds for sea birds. The new areas were tendered to prospective developers in a competitive bid process known as Round 2 with 15 projects awarded with a combined power generating capacity of 7.2 GW. In June 2008 the Crown Estate launched the third round of site allocations by selecting 9 large zones further offshore for bidding with a combined capacity of

25 GW with successful bids announce 2 years later. A new allocation of the fourth round is currently ongoing with prospective bidders.

Once a developer has obtained a lease, a variety of permits need to be obtained together with a grid connection agreement. Once a project is constructed, generators are required to sell the transmission infrastructure through a competitive tender, to an Offshore Transmission Owner (OFTO) who receive a return on investment by charging transmission fees (FOWIND 2014). Up to April 2017, the government's financial support for offshore wind farms used to be based on renewable energy certificates (ROCs), a green certificate mechanism paid on top of the electricity price. Since 2017, support is provided by contracts for difference (CfDs). CfDs are 15-year contracts agreed at a strike price (the generation price). Under the CfD system, a fixed revenue stream of a strike price is guaranteed to the generator independent of the fluctuation of the wholesale electricity prices. However, unlike other countries, when the wholesale price is above the strike price, the wind farm will pay back the difference to maintain cost-efficiency. The UK provides near and medium-term visibility through the size of the Levy Control Framework, a fixed government budget on the support available to energy technologies through the allocation of CfDs. The UK government has allocated funding to support 10 GW of total installed capacity by 2020 and committed following a Sector Deal agreed in March 2019 to support an additional 20 GW by 2030 leading to delivery of 30 GW offshore wind installed capacity by 2030. Recently these plans were pledged to increase installed capacity to 40 GW by 2030, which should materialise when the UK government passes the relevant legislation.

13.4.2 Germany

Another success story of offshore wind is Germany. Germany has been the front runner in promotion of green energy policies for wind and solar energy for decades, however, due to its deeper waters and limited coastline, it saw its first offshore wind demonstration project, the Alpha Ventus start installation in 2009, many years after other EU countries.

Following a slow start though it has emerged to become the second-biggest market in Europe. Driven by its industrial strategy to become a world leader in clean energy together with the EU's legally binding 2020 renewable energy targets, the German government has implemented the 'Energiewende' legislation. This new energy legislation aims to transition away from fossil fuels and nuclear power after the nuclear accident in Fukushima, Japan, in 2011. The Renewable Energy Sources Act (EEG 2017) mandates an 80% renewable electricity consumption by 2050 with a significant part of it expected to be borne by offshore wind. (FOWIND 2014). The Federal Maritime and Hydrographic Agency (BSH) is the federal agency overseeing licensing for renewable energy projects in the German Exclusive Economic Zone. A thorough marine spatial plan and grid plan were developed that identified potential zones for offshore wind development where developers were able to submit applications within these zones in an open-door approach until 2017. BSH released such zones for offshore wind developments where developers, after securing exclusivity rights, submitted a consent application which was examined by the BSH. After award of a lease, the developer moved to site investigations and further site development until full consent was granted. At the same time, the Transmission System Operators (TSO) was required to develop, acquire permits and fund all offshore wind grid connection works up to an offshore connection point, with the developer responsible for connecting to this connection point. In 2017 and 2018 there was a transition period with auctions released for securing certain capacity of eligible projects.

From 2018 onwards (for projects to be commissioned after 2026) the model has changed to a more centralised approach with sites selected in advance by BSH according to the nationally coordinated marine spatial plan and offshore grid plan, with a schedule of when these sites will be auctioned. Then site investigations are conducted by BSH and site information is passed on to developers before the official auction. Winning bidders will get a subsidy contract, grid permits and the guarantee that the TSO will build and operate the transmission assets. Developers are still required to obtain full consent by BSH following the auction process. The government's financial support for offshore wind farms until 2017 was based on a feed-in tariff mechanism based on water depth and distance from shore to allocate such tariff. The tariff could be provided at a fixed price for 8 or 12 years with reduced tariff for a prolonged period. Since 2017, the support for new projects is based on a flexible market premium on top of the spot market electricity price granted through a 20-year contract. The German Offshore Wind Act has set targets of 6.5 GW by 2020 and 15 GW by 2030. The act provides visibility for future auctions and includes 500 MW per year tendered between 2021 and 2022, 700 MW between 2023 and 2025 and 840 MW annually from 2026 onwards (IEA 2017). The government has passed new legislation in May 2020 to raise Germany's 2030 offshore wind target to 20 GW by 2030.

13.4.3 The Netherlands

The Netherlands was one of the first EU countries to develop and operate large-scale commercial wind farms before 2010 with an installed capacity of 247 MW. However, despite the strong start, the Netherlands has had a stop–start history. In 2011, a change of government led to a cut in renewable energy support and stagnation in the offshore wind industry. In 2014, the market took a huge leap forward with the signing of the Dutch Energy Agreement between the government and key industry players. The financial close of the largest project to be financed by banks, the EUR2.8 billion Gemini project was another positive development (FOWIND 2014). Driven by the Energy Agreement's targets of 16% of electricity coming from renewables by 2023, the Dutch government has implemented the Energy Act, with a target to reach 4.45 GW of offshore wind capacity by 2023. Beyond 2023 a climate agreement passed in 2018 seeing additional 1 GW offshore wind installation per year between 2023 and 2030 unveiling a roadmap of 11.5 GW installed capacity by 2030. Until 2014, the Netherlands used to have an open-door approach to offshore wind development with developers securing exclusivity of sites and moving to consent by applying to the Ministry of Infrastructure and Water Management. Following extensive public consultation by the Dutch government, the Offshore Wind Energy Law came into force in 2015, where the National Water Plan allocated designated offshore wind farm zones within the Dutch EEZ. In these areas, the Netherlands Enterprise Agency (RVO.nl) acts as a one-stop-shop for offshore wind development and consultation with developers. RVO undertakes the selection of specific project sites, conducts the necessary site investigations and obtains full consent and grid permits. The Dutch transmission system operator is responsible for developing, constructing and operating the offshore transmission grid where the projects will be connected to. Then developers compete for sites through an auction process that takes place every year. Winning bidders are awarded a subsidy contract for 15 years and all necessary permits to progress with wind farm construction. RVO is executing both the offshore wind energy subsidy and permit tenders on behalf of the Ministry of Economic Affairs and Climate Policy. (RVO 2015). The government's financial support for offshore wind farms is based on a subsidy mechanism called Subsidieregeling Duurzame Energie+ (SDE+) with a guaranteed price lowered by the electricity market price (market premium). This price is capped by a minimum electricity price and is yearly corrected by an imbalance and profile factor for offshore wind projects (PWC 2017). The annual allocation of at least 700 MW of projects up to 2030 provides good visibility for developers and has allowed the Dutch TSO to standardise offshore substations reducing transmission costs.

13.5 Current Installations

Offshore wind reached a global installed capacity of 23 GW in 2018 of which 18 GW were installed in Europe, 5 GW in Asia and 30 MW in the US. A summary of recent offshore wind installations in Europe is provided in Fig. 13.3 with 3.6 GW installed in 2019. Figure 13.4 provides an overview of the global cumulative offshore wind capacity in 2018, as presented in a recent report by the Global Wind Energy Council.

The global installed capacity in 2019 reached 29 GW, with Europe and Asia installing 6.1 GW additional capacity.

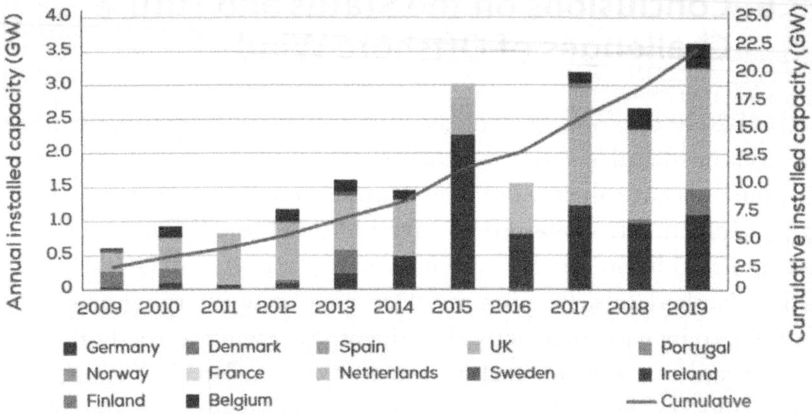

Fig. 13.3 Annual offshore wind installations (left axis) by country and cumulative capacity (right axis) (*Source* WindEurope 2020a)

MW, offshore	New installations 2017	Total installations 2017	New installations 2018	Total installations 2018
Total offshore	**4,472**	**18,658**	**4,496**	**23,140**
Europe	**3,196**	**15,630**	**2,661**	**18,278**
United Kingdom	1,715	6,651	1,312	7,963
Germany	1,253	5,411	969	6,380
Belgium	165	877	309	1,186
Denmark	0	1,268	61	1,329
Netherlands	0	1,118	0	1,118
Other Europe	63	305	0	302
Asia-Pacific	**1,276**	**2,998**	**1,835**	**4,832**
China	1,161	2,788	1,800	4,588
South Korea	3	38	35	73
Other Asia	112	172	0	171
Americas	**0**	**30**	**0**	**30**
USA	0	30	0	30

Fig. 13.4 Global cumulative offshore wind capacity in 2017 and 2018 (*Source* GWEC 2019a)

13.6 Conclusions on the Status and Future Challenges of Offshore Wind

Beyond 2020 the forecast of offshore windfarm capacity installations for the coming decades varies depending on scenarios considered. Several governments have already outlined commitments and targets for offshore wind deployment, while others have indicated aspirations to enter the sector. In addition to the current leading markets of UK, Germany, China, Netherlands, Belgium and Denmark, new emerging markets are expected to increase their activity in the offshore wind sector, both within Europe and in countries such as India, Vietnam, Japan, Taiwan, South Korea and the USA. In Europe, the UK, Germany and the Netherlands are expected to dominate the market with approximately 60 GW of installations to come online by 2030. France that has been developing wind farms for over a decade has not seen projects being realised yet, however, strong growth is expected till 2030 with at least 5 GW of new installations. Ireland has just passed a marine development plan that could pave the way for installations of at least 3.5 GW by 2030. Other Baltic states that will see offshore wind growth include Denmark with additional 2.4 GW agreed in June 2019, Estonia, Finland, Latvia, Lithuania, Poland and Sweden. Romania has also announced plans for developing offshore wind in the Black Sea. Outside Europe, US projects start building momentum with the allocation of seabed leasing rounds by the US Bureau of Ocean Energy Management (BOEM) which are expected to be constructed in the coming years. US capacity is set to grow from 30 MW today to 2 GW by 2025, and with a development pipeline of 25 GW, the total capacity could reach 10 GW by 2030 (GWEC 2019b). In Asia, China had an installed capacity of 4.6 GW by the end of 2018, making it the third-largest country in the world in offshore installed capacity. However, this is expected to change in the coming years with annual installations exceeding 5 GW per year, making China the strongest player by 2030. Taiwan is currently a hot market in offshore wind and is expected to witness up to 5.5 GW of new installations by 2025 and 10 GW by 2030. South Korea, Japan and Vietnam could also see significant growth during that period. The emergence of floating wind technology could also unlock new sites for offshore wind in deeper waters in the North part of the North Sea, the Atlantic coast and the Mediterranean Sea. France is taking a leading role in floating wind developments with the announcement of tender programmes of a 250 MW floating wind project in Brittany in 2021, followed by two of the same size in the Mediterranean in the following years. The UK, Norway and Portugal should follow a similar example moving from demonstration projects to large-scale floating projects in ongoing dedicated tenders. Floating wind installations worldwide could exceed 5 GW by 2030. Other countries that are looking into tapping into floating wind include the USA, Taiwan, Japan, South Korea and Saudi Arabia.

Although the future of offshore wind in Europe may seem bright, it is not a bed of roses. With more offshore wind coming online, new offshore wind allocated marine areas are needed to support the EU's long-term targets. However, offshore areas can get quite crowded and congested considering marine habitats and other users such as oil and gas, shipping, fisheries, aquaculture, coastal tourism, aviation and military among others. To address these future challenges, an integrated marine spatial planning in national and EU level is ongoing with cross-border and cross-industry collaboration to address stakeholder interests for allocation of dedicated offshore wind spatial areas. Such collaborations should consider efficient seabed management synergies and potential future multi-uses such as large-scale mussel farming together with offshore wind. The EU deadline for the establishment of national maritime spatial plans is set in 2021. Another big technical challenge is the ability to absorb large amounts of offshore wind and to meet the respective energy demand at any time considering the intermittency of wind energy. Denmark is the country with the largest share of wind energy in its electricity demand of 48% in 2019 (WindEurope 2020c). Denmark achieves this through strong interconnection links with its neighbouring countries. To achieve this in European level, interconnections are an important tool to deliver security of supply, market integration and the large-scale uptake of renewables. In this respect, the EU has set an electricity interconnection target of at least 10% of its installed capacity by 2020 and 15% by 2030. Projects of common interest play a pivotal role in this process, including the North Seas offshore grid with the aim to transmit offshore wind electricity between neighbouring countries. However, there are also technical limitations in the transmission capacity and with increased renewable penetration negative electricity prices that may lead to curtailment will be more common. This is where hydrogen is coming to play a role as storage fuel for offshore wind generation. Recently, since H1 2020, major oil and gas players have been looking at demonstrator projects producing hydrogen from offshore wind and the German government is expected to produce a strategy in offshore wind hydrogen later this summer (Recharge 2020). This could be a game changer in future use of offshore wind. The last challenge is access to finance and investor confidence in unprecedented challenging times for the energy sector following the coronavirus pandemic. The resilience of renewable energy during the oil price shock caused by the COVID-19 crisis shows that renewables are here to stay with more than 3 GW of offshore wind reaching financial close in the first week of June 2020 in Europe. The green recovery package set by the EU commission on 27 May 2020 sets the tone by pouring money to energy efficiency, green heating, renewable energy, hydrogen and clean cars and trains.

Based on the Global Wind Energy Council report (GWEC 2019b) 190 GW of offshore wind capacity could be installed worldwide by 2030. The majority of installations is expected in Asia (100 GW) followed by Europe (80 GW) and USA (10 GW). In this transition period, where the Chinese market is expected to become the world leader in offshore installations, Europe should ensure it keeps up with the competition for the years to come, considering also remaining resilient. How can EU countries achieve this? The answer lies in coordinated policy measures at European level. The European Commission set a European Green deal in December 2019, targeting to make Europe the first climate-neutral continent by 2050. Following the coronavirus pandemic, the Commission has set the European Green Deal as key to the EU's recovery strategy. Offshore wind will play an important part in this energy transition, which must go hand in hand with Europe's new industrial strategy to benefit its society. Under this new industrial strategy focus should be given to supporting skills and innovation, accessing cheap finance for capital-intensive investments and maximising supply chain efficiencies. With EU members setting binding renewable energy targets for 2030, offshore wind developments should be expanded in new territories including the Mediterranean Sea. An important part in this expansion is an EU maritime spatial plan. Interconnections will also play an important role in a future integrated EU electricity market via projects of common interest. Research funding support should be targeted to allow technology and supply chain providers to further reduce costs and improve technology offerings to the EU and the global market for existing and new products such as renewable hydrogen. The European Commission has also set ambitious goals for 2050: between 230 and 450 GW (WindEurope 2019). These strategies will allow Europe to stay in competitive edge beyond 2030. The offshore wind market will be a trillion-dollar industry by 2040, with hundreds of GW installed in the coming decades (IEA 2019). According to a report published by the International Renewable Energy Agency (IRENA 2016), global offshore wind capacity could reach 400 GW by 2045. As a comparison to the offshore case, the onshore wind had a global capacity of 13 GW in 1999 and took until 2016 to grow to 475 GW.

References

EEG, 2017. "Renewable Energy Sources Act". https://www.bmwi.de/Redaktion/EN/Downloads/renewable-energy-sources-act-2017.pdf%3F__blob%3DpublicationFile%26v%3D3.

EQUINOR, 2020. "How Hywind Works". https://www.equinor.com/en/what-we-do/hywind-where-the-wind-takes-us/hywind-up-close-and-personal.html.

FOWIND, 2014. "Offshore Wind Policy and Market Assessment, A Global Outlook", December 2014. http://www.gwec.net/wp-content/uploads/2015/02/FOWIND_offshore_wind_policy_and_market_assessment_15-02-02_LowRes.pdf.

GWEC, 2019a. "Global Wind Report", April 2019. https://gwec.net/wp-content/uploads/2019/04/GWEC-Global-Wind-Report-2018.pdf.

GWEC, 2019b. "Global Offshore Wind Report: Sector Has Potential to Grow to 200GW of Capacity by 2030", 26 June 2019.

IEA, 2017. "Comparative Analysis of International Offshore Wind Energy Development, International Energy Agency", March 2017. http://iea-retd.org/wp-content/uploads/2017/03/IEA-RETD-REWind-Offshore-report.pdf.
IEA, 2019. "Offshore Wind to Become a $1 Trillion Industry", 25 October 2019. https://www.iea.org/news/offshore-wind-to-become-a-1-trillion-industry.
IRENA, 2012. "30 Years of Policies for Wind Energy: Lessons from 12 Markets", 136. https://www.irena.org/DocumentDownloads/Publications/IRENA_GWEC_WindReport_Full.pdf.
IRENA, 2016. "Innovation Outlook: Offshore Wind". http://www.irena.org/DocumentDownloads/Publications/IRENA_Innovation_Outlook_Offshore_Wind_2016.pdf.
IRENA, 2018. "Renewable Power Generation Costs in 2017". http://www.irena.org/media/Files/IRENA/Agency/Publication/2018/Jan/IRENA_2017_Power_Costs_2018.pdf.
ORE Catapult. "Cost Reduction Monitoring Framework 2016". http://crmfreport.com/wp-content/uploads/2017/01/crmf-report-2016.pdf.
ORSTED, 2016. "Celebrating the Growth of Wind Turbines → Lowering the Cost of Offshore Wind. First 8 MW Turbine Installed Offshore" (@ORSTED, September 8, 2016). https://twitter.com/orsted/status/773864984027787264.
ORSTED, 2017. "World's First Offshore Wind Farm Now Dismantled." https://orsted.com/en/Media/Newsroom/News/2017/09/Worlds-first-offshore-wind-farm-now-dismantled.
PWC, 2017. "Unlocking Europe's Offshore Wind Potential", March 2017. https://blogs.pwc.de/oeffentlicher-sektor-zukunft-gestalten/files/2017/03/Offshore_Wind.pdf.
Recharge, 2020. "Berlin Mulls 'German Electrolysers' Across EU to Meet Vast Green Hydrogen Needs From Offshore Wind", accessed June 7, 2020, https://www.rechargenews.com/wind/berlin-mulls-german-electrolysers-across-eu-to-meet-vast-green-hydrogen-needs-from-offshore-wind/2-1-818553.
RVO, 2015. "Offshore Wind Energy in the Netherlands". https://www.rvo.nl/sites/default/files/2015/03/Offshore%20wind%20energy%20in%20the%20Netherlands.pdf.
Wind Power Monthly. "Offshore Wind Costs Fall 32% Since 2011", 24 January 2017. http://www.windpoweroffshore.com/article/1421825/offshore-wind-costs-fall-32-2011.

WindEurope, 2018. "Offshore Wind in Europe Key Trends and Statistics 2017", 28. https://windeurope.org/wp-content/uploads/files/about-wind/statistics/WindEurope-Annual-Offshore-Statistics-2017.pdf.

WindEurope, 2019. "Our Energy, Our Future How Offshore Wind Will Help Europe Go Carbon-Neutral". https://windeurope.org/wp-content/uploads/files/about-wind/reports/WindEurope-Our-Energy-Our-Future.pdf.

WindEurope, 2020a. "Offshore Wind in Europe Key Trends and Statistics 2019", February 2020. https://windeurope.org/about-wind/statistics/offshore/european-offshore-wind-industry-key-trends-statistics-2019/.

WindEurope, 2020b. "History of Europe's Wind Industry", accessed February 27, 2020, https://windeurope.org/about-wind/history/timeline/danish-government-sets-target-800-1350mw/.

WindEurope, 2020c. "Wind Energy in Europe in 2019 Trends and Statistics", February 2020. https://windeurope.org/wp-content/uploads/files/about-wind/statistics/WindEurope-Annual-Statistics-2019.pdf.

Part IV

Technical Aspects of the Energy Union

14

Energy Analytics: From Data Acquisition to Data-Driven Business Models

Dimitrios I. Doukas

Nomenclature

AI:	Artificial Intelligence
AMR:	Automated Meter Reading
COSEM:	Companion Specification for Energy Metering
CRM:	Customer Relationship Management
C&I:	Commercial & Industrial
DIN:	Deutsche Institute für Normung
DLMS:	Device Language Message Specification
DNO:	Distribution Network Operator
DSO:	Distribution System Operator
EBITDA:	Earnings Before Interest, Taxes, Depreciation and Amortization
ESCO:	Energy Services Company

D. I. Doukas (✉)
NET2GRID B.V., Thessaloniki, Greece
e-mail: dimitrios@net2grid.com

M. Mathioulakis (ed.), *Aspects of the Energy Union*,
Energy, Climate and the Environment,
https://doi.org/10.1007/978-3-030-55981-6_14

GDPR: General Data Protection Regulation
HAN: Home Area Network
IDC: International Data Corporation
IEC: International Electrotechnical Commission
IoT: Internet of Things
IT: Information Technology
LED: Light Emitting Diode
LoRa: Long Range
MBus: Message Bus
MeDa: Meter-Data
NILM: Non-Intrusive Load Monitoring
NIALM: Non-Intrusive Appliance Load Monitoring
PPPY: Per Person Per Year
SEP: Smart Energy Profile
SML: Smart Message Language
TIC: Télé Information Client
TSO: Transmission System Operator
VHR: Very High Resolution
WAN: Wide Area Network
Wi-SUN: Wireless Smart Utility Networks

14.1 Introduction

Digital transformation—especially since when data importance became more evident—is nowadays a trend for energy utilities and companies globally. More and more digital and "smart" hardware components replace old-generation equipment starting from the transmission level and going down to the energy end-users. Traditionally, data were acquired by a few critical points of the grid, usually at a high voltage level, whereas nowadays energy companies have access to more data from thousands of IoT endpoints within the smart grid. Most of the new components are able to connect with the internet or with each other, provide more frequent or even continuous access to data and allow third parties to build and deliver data-driven use cases to either energy companies (TSOs, DSOs, energy retailers) or the end consumers.[1] In Europe, especially, important policies and regulation were (and still are) introduced within the Energy Union framework having as a target to put the end-user at the epicentre of the new European energy system.[2] The energy landscape is changing radically, decision-makers acknowledge that and are willing to make the consumers' role more important in the future. Regulation and technology mean that, for example, smart-meters are becoming the necessary tools for the end-users to increase their impact and leverage in the future grid, either by participating in demand response schemes or by simply being able to switch energy providers– even within a day– for their own benefit.[3] In this context, more than 60% of the European utilities have already dedicated digital transformation or innovation teams in place, since they realized that Big (Energy) Data is the pillar of innovation and unlocks variable opportunities for them to become digital, divergent and decentralized. Globally and according to IDC, by 2020, 50% of the top 100 worldwide utilities will have developed a data-centric strategy leading to 15% improvement of their EBITDA.[4] It's noteworthy that utilities are expected to spend $120 billion on technology and related services to implement and manage these activities. Accurate and uninterrupted data acquisition of the required resolution is of critical importance and needs to be done in a non-intrusive and cost-efficient way,[5] but it is only one side of the equation since data richness by itself does not mean anything

---

[1]Ian Wright, Sharon Chand and Jian Wei, "Digital transformation in the energy & utilities sector", *Risk & Compliance Magazine*, October–December 2018, https://www2.deloitte.com/content/dam/Deloitte/us/Documents/process-and-operations/us-rc-reprint-digital-transformation.pdf.

[2]"Smart grids and meters", European Commission, accessed February 22, 2020, https://ec.europa.eu/energy/en/topics/markets-and-consumers/smart-grids-and-meters/overview.

[3]Thomas Pellerin-Carlin, "Putting the consumer at the centre of the European energy system", Notre Europe, Jacques Delors Institute, June 8, 2016, http://www.institutdelors.eu/wp-content/uploads/2018/01/energyconsumer-pellerincarlin-jdi-june16.pdf?pdf=ok.

[4]"IDC Energy Insights forecasts Western Europe", International Data Corporation (IDC), September 19, 2017, https://www.idc.com/getdoc.jsp?containerId=prEMEA43075617.

[5]Sinan Küfeoğlu, Gaomin Liu, Karim Anaya and Michael G. Pollitt, "Digitalisation and new business models in energy sector", Energy Policy Research Group (EPRG), The University of Cambridge, June 2019, https://www.eprg.group.cam.ac.uk/wp-content/uploads/2019/06/1920-Text.pdf.

useful though.[6] Energy companies are now in search of holistic strategies to transform Big (Energy) Data to meaningful information or in other words, turn them into valuable services and use-cases.[7] AI-based energy data analytics services are already addressing such market needs and enable several opportunities for energy companies.[8] Such opportunities include reduced cost to serve, where greater customer engagement and bills/consumption understanding lead to customer contact reduction, reduced churn, where higher customer satisfaction reduces the need to look for a new energy supplier, greater upsell/cross-sell, where greater energy consumption insight and deeper, more trusting customer relationships provide the bedrock for more successful cross- and upsell activity and customer acquisition, where energy insights are likely to become a decision criterion in searching for a new energy supplier. Although energy (big data) analytics is a relatively new area of applied research, there are already plenty of options for energy companies to pick while developing their energy insights and services proposition. It practically depends on the data they can have access to (and at what cost), if this data are suitable for their needs and the services they want to deliver and how often/fast they need their customers to be informed about their energy consumption profile. In this chapter, we will first categorize energy data depending on their volume, type and granularity, and then the alternatives for data acquisition will be presented. A discussion about energy analytics alternatives for various parts of the smart grid will follow while emphasis will be given on the residential-level sector since that's where most data originate. Energy disaggregation will be then defined as a concept as well as the foundation layer on which most residential-level energy analytics services are built. Finally, data-driven business models will be presented and categorized depending on several characteristics.

[6] Adrian Booth, Eelco de Jong and Peter Peters, "Accelerating digital transformations: A playbook for utilities", McKinsey, January 2018, https://www.mckinsey.com/business-functions/mckinsey-digital/our-insights/accelerating-digital-transformations-a-playbook-for-utilities.

[7] "Looking ahead: Future market and business models", PricewaterhouseCoopers (PWC), accessed February 22, 2020, https://www.pwc.com/gx/en/utilities/publications/assets/pwc-future-utility-business-models.pdf.

[8] Bert Lutje Berenbroek, "Platform thinking in energy", NET2GRID, accessed February 22, 2020, https://www.net2grid.com/images/n2g_future_of_energy_hlb.pdf.

14.2 Data Categorization

Energy data can be categorized based on several features and characteristics. First, depending on the data resolution (or granularity), energy data are usually categorized as low-resolution, high-resolution or very-high resolution. Low-resolution data typically refer to AMR data directly from the smart meter, for their acquisition no additional hardware is required, and the minimum granularity for them is 15-min and goes up to 1-hour, so no more than 96 measurements per day. Moving to more granular data, we can find the so-called HAN-based approaches for which additional hardware equipment is needed. The data resolution varies from one measurement per second up to one measurement every few minutes, whereas such measurements can either derive from a data-logger type of hardware that communicates with the smart meter or from a (current clamp) DIN-meter option. Lastly, VHR energy data refer to anything more granular than 1 Hz, while there are options on the market that go down even to 1 MHz. Such data can only derive from a DIN-meter (meter clamp) hardware option. Apart from the resolution, energy data are also a matter of type as well. AMR data are, most of the time, energy consumption data (E) that can be converted to active power (P) consumption data. HAN-based data depend on the data acquisition approach that was used. In case of a data-logger type of hardware, then most of the times, it is only about active power (P) consumption data. However, in the case of DIN-meter equipment, access to reactive power (Q), apparent power (S), current (I), voltage (V), power factor ($\cos\varphi$) etc. is also feasible. Similar to the latter and since all VHR energy data are DIN-meter based, access to more energy-related data attributes is achievable. Moreover, higher resolutions, in this case, allow for harmonics-based content that can be used on top of the other data streams. Last but not least, energy data can also be categorized according to their volume and cost (acquisition, storage and cost of service). Volume-wise, as a rule of thumb, AMR data refer to less than 1 MB of data PPPY, whereas 1-sec resolution HAN-based data from a 3-phase installation (active power consumption from the total as well as from the individual phases) result into approx. 1 GB of data PPPY. A proportionally higher data volume will, of course, be the result of

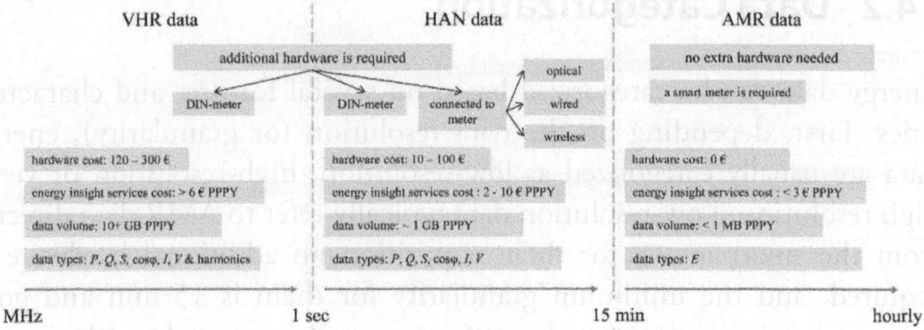

Fig. 14.1 Energy data categorization depending on their granularity

moving to even higher resolutions and that brings on the surface scalability concerns and if storing of energy data is required or not. Energy data categorization can be summarized in Fig. 14.1 that follows.

14.3 Data Acquisition

Energy data acquisition from residences can be implemented in different ways and is determined by the services an energy company wants to deliver to the end-users. It is an attempt that is heavily dependent on the metering (smart or not) hardware infrastructure that either exists in a house or needs to be installed in this context. Moreover, it is also constrained by regulation regarding the meter accessibility, e.g. in most of the countries only the DNO has access to the meter and the energy retailer, which is expected to be more interested in delivering energy insight services, cannot intervene by installing hardware that connects to the meter. The data acquisition alternatives that are illustrated in the following figures are categorized as follows.

14.3.1 Legacy Meters

Although it is already more than ten years that smart meters are actively installed in several countries, there are still countries with old types of analogue meters that cannot connect to the internet or with other equipment. Consumption monitoring for those houses is still being done manually only a few times per year, leaving the end-user with no energy insight services and with poor customer experience. In that case, two options require the installation of additional equipment.

The first option is an optical reader for legacy meters: Such readers must be physically attached to the front of the meter, and make low-resolution measurements using optical techniques (for example by reading the rotation speed of the disc of a Ferraris meter). It can be connected to an external power supply or be battery-powered since usually there is no supply in the proximity. An example can be seen in Fig. 14.2a. Legacy meters are often outside Wi-Fi radius; therefore, a point-to-point long-range communication protocol, e.g. LoRa is to be used to transmit data. Such hardware should be designed to deal with the limitations of Wi-Fi networks and be able to deliver complete datasets even when the Wi-Fi signal is poor. Low-level stack access and competence are crucial for this task. The second option is current clamp hardware: It is connected within (or close to) the house's fuse box, it gets phase-to-neutral voltage measurement(s) and current measurements with one clamp. (or more for multi-phase installations). Such a hardware component is illustrated in Fig. 14.2b. As an appliance is powered directly by the board and since the fuse box is usually within Wi-Fi range, the house's internet connection can be used to transmit data. As also

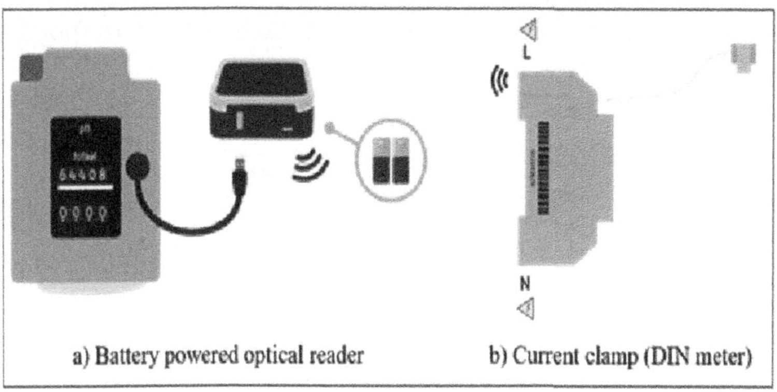

Fig. 14.2 Data acquisition alternatives for houses without a smart meter

previously mentioned, this option allows for more energy attributes to be transmitted (not only active power) and can support VHR data as well, which means that the services that can be built on top of such data richness are more sophisticated and possibly more accurate.

14.3.2 Smart Meters

Smart meters are by default able to connect (at least) to a WAN and share the so-called AMR data usually with the DNO, whereas the data resolution is 15-min, 30-min or even less frequent. That unlocks the opportunity to provide data-driven energy insight services even without any additional hardware (see Fig. 14.3a). However, as it is already proved AMR (low frequency) data are most useful to the DNO, usually for asset management and load forecasting purposes, but cannot assist energy retailers in delivering highly accurate services, e.g. consumption breakdown etc. In that case, there are a few options to get access to higher resolutions and more meaningful and content-rich data.

The first option is HAN-based data: Additional hardware components, in different form factors, can be connected to the smart meter for more granular data to be reported per house. In this case, there are more alternatives than in the case of the analogue meters. Optical readers for smart meters (see Fig. 14.3c) is the first alternative. Many digital meters are equipped with some form of optical interface. Protocols that may be supported are IEC62056-21, DLMS/COSEM, SML and simple pulse output. In the case of pulse output, each pulse corresponds to a

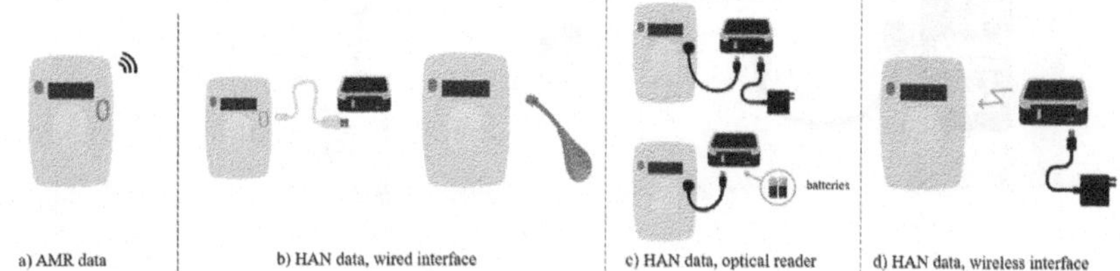

Fig. 14.3 Data acquisition alternatives for houses with a smart meter

certain amount of energy passing through the meter and can be detected by attaching a phototransistor-based reader. Other, more sophisticated protocols allow 2-way communications and provide access to a larger dataset. Although power supply and internet connections concerns might still occur, it is noteworthy that in a lot of countries the smart meter is installed in the house (not on the street side, not in the basement), which means that data acquisition is smoother. The second alternative in this category is a wired interface for smart meters (see Fig. 14.3b). This is hardware that is connected to the smart meter with a wire, in order to catch more data from the meter. Protocols that may be supported are P1, S1, ModBus, MBus, HAN connector and TIC. Depending on the smart meter, access to active, apparent and/or reactive power data is achievable. Power supply can be done through the same wire that connects the hardware to the meter, whereas there are alternatives for meters that are in our out of Wi-Fi radius. Finally, the third alternative in this category is the use of wireless interface for smart meters (see Fig. 14.3d): This is the same with the previous alternative, but in this case, instead of using a wire there is the option to communicate with a meter with a wireless communication protocol. Such alternatives in common use include cellular communications, Wi-Fi, wireless ad hoc networks over Wi-Fi, wireless mesh networks, low power long-range wireless (LoRa), ZigBee (low power, low data rate wireless), SEP, wireless M-Bus, Lemonbeat and Wi-SUN. The second option to getting access to higher resolutions and more meaningful and content-rich data is through Current clamp hardware: There is always an option, for an energy company, to overlook the smart meter as a data acquisition endpoint, either because more granular or more and different types of energy data are needed for specific services or because access to the smart meter cannot be granted by the DNO or another grid company. In that case, the option of installing hardware, like the one presented in Fig. 14.2b, in the fuse box, is the only way. The pros and cons of this option are already presented, however, for houses with smart meters this option is not often selected by energy companies since it requires more expensive hardware and is more “intrusive” by means that a technician is needed to install the hardware and is not a simple plug and play appliance like with the previous options.

14.3.3 Digital Meters

There are a few countries in which the smart meter roll-out will begin only for energy end-users that exceed a specific amount of energy consumption per year, e.g. in Germany smart meters are installed only for electricity end-users that consume more than 6000 kWh on an annual basis.[9] In that case, all consumers that consume less, so practically the vast majority of installations, do not benefit from the smart meter roll-out and digital meters are used to replace legacy ones. Such meters, e.g. the MeDa meters in Germany, are not able to transmit any consumption data and disruptive alternatives for data acquisition are necessary. In this context, there are self-install devices, manufactured at a fraction of the usual cost, that are used for the consumers to securely capture their data through a wireless interface (e.g. LoRa) and then be able to get insight into their energy usage and understand more about their appliances' efficiency. That is of critical importance because it allows energy companies to protect grid investments and help them serve with the same energy insight and data-based business models, even customers without a smart meter.

14.3.4 Super Meters

As a step forward for smart meters and energy data acquisition in the residential sector, there are already discussions about smart(er) meters, the so-called super meters that will practically integrate IoT hardware, as illustrated in Fig. 14.4, sub-components and be able to deliver not only data but actual useful information already translated into energy insight and a range of home energy management services to consumers. Super meters will allow the end-users to make more informed choices.

Energy retailers, installing and maintaining additional hardware on top of the pre-installed smart meter with no cost, will be able to deliver several benefits to energy end-users. These include improved efficiency

[9] "iONA project", Innogy, accessed February 22, 2020, https://www.iona-energy.com/iONA.

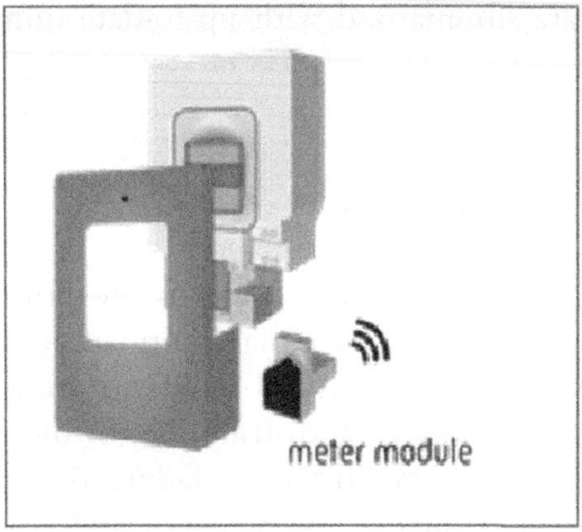

Fig. 14.4 Super meter representation

since the end-users can see in real-time how much energy their appliances consume and discover ways to reduce their energy costs, customer retention and engagement, since consumers will prefer to have a thorough understanding of their energy consumption and be recommended on ways to further reduce it instead of moving to another supplier, alerts and notifications since energy consumers can view at any time which of their appliances are ON/OFF and be alerted about critical appliances that may need repair, and demand-side management with detailed, real-time energy information can help consumers and utilities shift the timing of energy use to better address supply, reduce costs and make the grid more reliable. To sum up, things an energy company needs to consider when selecting an energy-data acquisition strategy are the following: (a) cost (DIN-meter options are usually more expensive), (b) access to power supply and internet connection (battery-powered hardware and adoption of LoRa wireless communication can be the solution for such limitations), (c) easiness-to-install (HAN equipment is usually plug-and-play, while DIN-meter options require a professional technician to install the hardware), (d) data needs (depend on the services that are to be provided to the end-users), (e) data quality issues and data interruptions (the ideal combination of technology means differs) and (f) cost of service (more data and more sophisticated services come with a greater cost). Most of these aspects are summarized with up-to-date numbers for costs in Fig. 14.1.

14.4 Energy Analytics

Energy (data) analytics as a term refers to the process of translating various kinds of energy data into actual information. Traditionally, such data were being received from critical infrastructure points of the grid, either at transmission (high) or at distribution (medium) level, to ensure secure and reliable grid operation, for failure probability modelling, to prevent risks, for predictive maintenance and cost-optimized asset management. Later, more and more data acquisition endpoints and sensors were installed even at low-voltage grid parts, enabling a more holistic monitoring of the grid by DNOs. Most of the data-collection equipment was replaced by modern versions that can connect to the internet unlocking real-time communication capabilities. The trend of moving from a few data-collection points to more and from high-voltage to low-voltage measurements continues nowadays through the smart-grid and smart-home paradigms. Now grid operators and energy retailers are given access to real-time information of various types (active and reactive power, currents and voltages, harmonics, etc.) at higher granularities from literally hundreds of thousands of endpoints within their grids. That allows them for more sophisticated energy analytics strategies on which interesting data-driven business cases can be built. Most important of these strategies is expected to be the energy disaggregation on appliance or activity level on which specific focus will be given above.

14.5 Energy Disaggregation

Energy disaggregation, also known as NILM or NIALM, is the problem of separating an aggregate energy signal into the consumption of individual appliances or activities in a consumption endpoint, most of the times for households but also for C&I consumers. It's a non-intrusive approach since no sub-measurements from specific appliances within an installation are needed, and only a single point measurement of the total consumption is used. An easy-to-understand illustration of who NILM works can be seen in Fig. 14.5. NILM scientists develop algorithms to recognize appliance types in data series in a similar way as someone would identify landmarks in the skyline of Paris.

As a concept, this technology was first introduced at the beginning of the '90s. However, non-existence of cost-efficient alternatives for electricity consumption (and ground-truth) data acquisition as well as market immaturity had, as a result, its practical use and mass deployment to be rather limited till recently. During the last decade though, there are plenty of companies competing in this field, which focus on the implementation of NILM algorithms to deliver energy-data-driven services through partnerships with energy companies (B2B approach) or directly to the electricity consumers (B2C consumers). A list of the key players in the market can be found in Fig. 14.6. Companies providing energy insight solutions and specifically NILM services can be initially split into the ones targeting the residential end-users and the ones that are dealing with non-domestic appliances. Then, for the first group, we can see that companies are categorized based on the data type and granularity they need as input to deliver energy data analytics output. As shown in Fig. 14.6, which is perfectly aligned with the information from Fig. 14.1, since we have three major groups of companies, dealing with

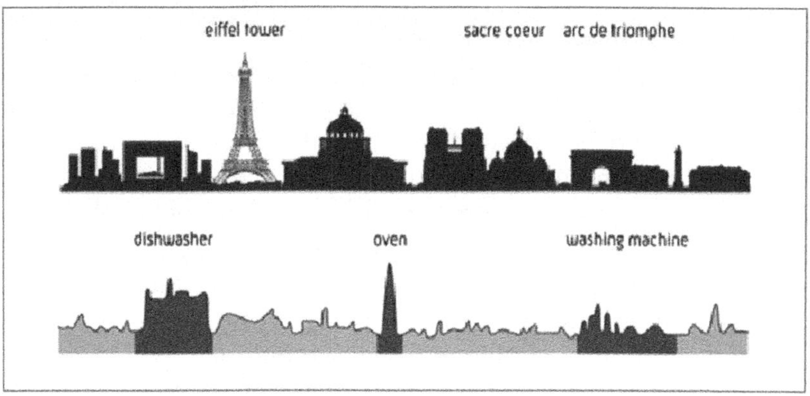

Fig. 14.5 How energy disaggregation works

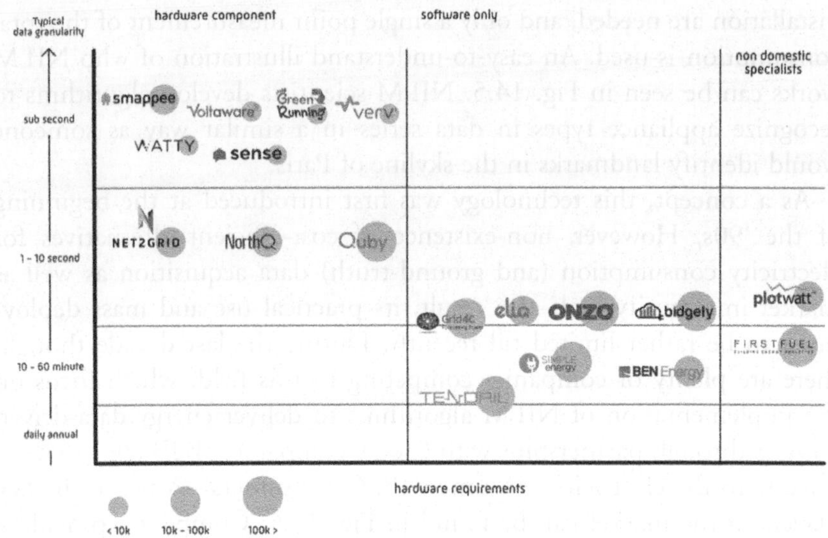

Fig. 14.6 NILM providers' categorization (*Source* Delta-EE)

AMR, HAN and VHR data, respectively. The discrimination between approaches that do (VHR and HAN) and do not (AMR) require additional hardware is also clearly depicted in the graph. Volume-wise, it is clear that companies working with AMR data (15+ min resolution) deliver services in a larger customer base so far, whereas most of the competitors in this area address the needs of 100K+ users. This is reasonable since, as explained, AMR data are faster, easier and of course, cheaper to acquire. On the other hand, since hardware for more granular energy data acquisition becomes cheaper in a very fast way, a new trend towards NILM services that are based on seconds or sub-seconds resolutions gain ground recently. The reason is that energy companies realized that with more information-rich data they could deliver more and more accurate services to their end-users with additional costs that dropped significantly (and keeps on dropping) that is usually lower than their customer retention budget.

Apart from the data-resolution-based categorization, NILM providers differ in terms of their approach on training and appliance modelling, the appliances/activities they support in their portfolios, their services response and frequency and the location their algorithms run. On the issue of training and appliance modelling, there are two main approaches: the installation-specific and the more generic. The first one refers to a strategy that results into appliances' model(s) created per installation, which means that data are collected for a specific period, typically within a few weeks, during which enough end-uses of the appliances of interest are expected to occur in order tailor-made and accurate models for them to be saved and then used for appliance identification after the training period. The second approach corresponds to the creation of pre-trained generic models per appliance type that are used for all installations. There are obvious pros/cons for both options since the first one leads to custom-made models that can deal with different appliances' types but on the other hand, comes with a necessary "waiting" period for models to be created. Pre-trained models used according to the second solution eliminate the waiting period that means better customer satisfaction, but at the same time require a huge dataset of both mains and appliances' ground truth data for the accuracy to be acceptable. So, there is a trade-off between NILM accuracy, data access and the waiting period before delivering the services. On the issue of appliances/activities supported, all NILM providers focus on the energy-intensive appliances that are responsible for the biggest part of the energy consumption, e.g. the fridge, washing machine, dishwasher, dryer, oven, etc. Since heating and transportation electrification becomes a trend with more and more electric vehicles being charged at homes and more Air/Con and heat pumps being used, such appliances emerge as high priority ones to be included in NILM portfolios. Moreover, something that is of crucial importance because of its significant contribution—usually at about 20%—in the total energy consumption, is the so-called AlwaysOn or "shadow" consumption, or in other words the energy consumption daily that derives from the minimum power consumption threshold in a household. Other appliances that can be of interest are either low-energy consumption ones that are being used really often, e.g. a microwave or a kettle, or energy-intensive ones that can be found in a few countries only, e.g. saunas in the Nordic countries.

Apart from the pre-mentioned appliances, several NILM providers also report for activities like entertainment and lighting. These are not directly connected to the detection of devices that are turned on and off but are mostly based on statistical and time-of-use analyses. Finally, it is noteworthy that accurate consumption breakdown on appliance level requires second or sub-second energy data resolutions, and that is why AMR-data-based NILM solutions break down the consumption on activity level, e.g. instead of recognizing independently each appliance, namely the washing machine, dishwasher and dryer, they group all of them as laundry, white-good appliances, or something similar. Last but not least and in relation to the energy data categorization depending on their granularity preceded, it is important to highlight how different energy data look for different granularities. Below, in Fig. 14.7, energy consumption data from a single house over a day are presented with four different granularities, 1-sec, 10-sec, 1-min and 15-min.

In the first two graphs, several features can be extracted in order consumption events to be identified as an oven (more than 1), washing machine and dishwasher, respectively. However, when moving to 1-min

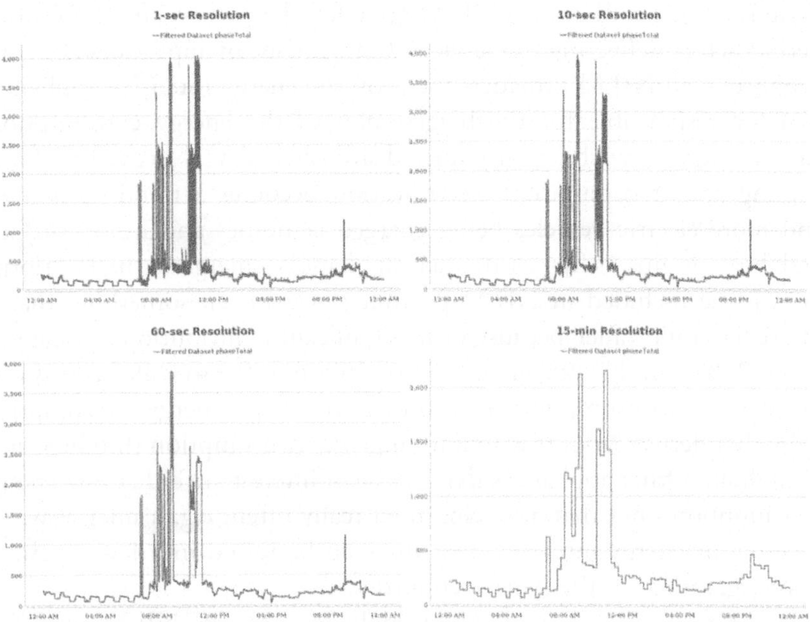

Fig. 14.7 Energy consumption data representation depending on granularity

and especially 15-min resolution data, appliance identification becomes practically impossible since information is actually "lost". So, there is a clear trade-off between data volume and NILM accuracy. On the issue of NILM response and frequency, NILM results are usually reported to a consumer through a mobile app in a convenient and user-friendly way. Depending on the technical characteristics (data means, granularity, etc.) that were mentioned already, the end-user can be informed about the energy consumption breakdown for any installation in real-time, near real-time or after a predetermined short (day) or longer (month) period has passed. Usually, there is a direct connection between the data granularity with the NILM response speed, since more data enable faster and even real-time information generation. However, data resolution is not the only aspect to consider here since simultaneous access to more than one data types, e.g. active and reactive power data, can allow real-time services even for lower granularities. Finally, there is the issue of cloud or edge computing. Living in the so-called cloud era, most of the energy analytics services providers move their complete solutions from local servers in cloud environments as an attempt to reduce costs both for data storage and analysis but also for easier integration with the energy companies backend IT solutions, e.g. billing, CRM, etc. Building service delivery platforms within the pre-existing cloud environment of an energy company is most of the times crucial to be compliant with data protection regulations like GDPR, since the services provider is not the data owner but simply the data controller and there is no additional consent needed for such services to be delivered. On the other hand, moving and running everything on the cloud, especially with high data resolutions and the need to deliver real-time services, can become costly and not scalable making difficult for solution providers to support thousands of customers simultaneously in a cost-efficient way. Because of that, energy insight providers focus on hybrid schemes, where heavier parts of the process can take place on the cloud (e.g. data training) while other more light-weight can run on the edge (e.g. appliance recognition). Although coordination of such schemes is sophisticated and can be complex, it is expected to be the only way to proceed for scalable cost-efficient services for great numbers of end-users.

14.6 Data-Driven Business Models

There is a great variety of services that can be structured on a NILM layer that is accurate enough to gain consumers' trust and covers a great percentage of the total energy consumption of an installation. Such data-driven use cases include:

- Bill breakdown on an appliance or activity level. Such information is critical for a consumer to understand what are the actual appliances or activities that are the most energy-intensive in his/her installation. It is usually the key to understanding how to reduce energy consumption and becoming more energy efficient.
- Insights and recommendations. Energy consumers can be notified with insights and tips that are extracted specifically for them. They can not only have a clear understanding of their energy consumption but also how they compare with other users in their cluster (location, family size, etc.) or with themselves in the past. Such services can be delivered through user-friendly gamification screens in a mobile app, motivating consumers to become more energy efficient.
- Appliance efficiency. Consumption breakdown on appliance level can also reveal how efficient specific appliances within a residence are, after comparing them with international benchmarks or with the other end-users of the same energy company. On the customer side, this proves useful to understand if replacing an appliance with a modern one makes sense, whereas for the energy company such information can enable joined actions with appliance manufacturers in targeting directly the consumers that need a new appliance more than others. Such a service is strongly connected with predictive maintenance since the consumer can see how an appliance consumption increases over time due to malfunctions or ageing and decide to properly and timely maintenance that appliance without having to replace it completely.
- Real-time notifications. Having real-time information about specific—especially energy-intensive—appliances can be directly translated into money savings or quality of life quality, so there are NILM-based services delivered specifically for that. For example, being notified in real-time when your electric vehicle is charged at

80% of its total capacity can lead the end-user to switch it off to prevent over-charging and therefore save battery life. Furthermore, since in some countries the washing machines are installed out of the house, a real-time notification about the end of the washing event can help a consumer to save time. Such real-time triggers, when being part of a smart-home ecosystem, can trigger ON/OFF actions for appliances of interest for the consumers' but also for the grid operators' benefit.

- Elderly care, peace of mind, alerts and home state. Such services can be based on a real-time understanding of an installation's consumption profile and alert people for abnormal behaviours. For example, a consumer can be notified when an appliance (e.g. an oven or an iron) is forgotten on for a duration longer than expected, or someone's next of kin can be sent a push notification when an installation's consumption is absurdly low for more than a predetermined period. Both notifications can even save lives.

14.7 Conclusions

Digital transformation in the energy and utilities sector is not just a trend but the only way for energy companies to differentiate from competition. Focusing on Europe becoming a leader in the area, legislation related to the smart-metering topic already demonstrates tangible results with most of the EU-27 countries presenting significant progress on their deployment.[10] Companies that will manage to exploit data, or the so-called "fuel" of the future, in building data-driven business models will find themselves leading the new era as ESCOs. To that end, cost-efficient, accurate and uninterrupted energy data acquisition—with or without additional hardware—regardless of the (smart) meter interface and its location is a challenge that still requires joint efforts between stakeholders, i.e. energy companies, energy insight solution providers and

[10]"Benchmarking smart metering deployment in the EU-27 with a focus on electricity", European Commission, June 17, 2014, https://eur-lex.europa.eu/legal-content/EN/TXT/PDF/?uri=CELEX:52014DC0356&from=EN.

end consumers. However, there are still significant challenges and obstacles to overcome while moving towards this direction. Access to good quality energy data in terms of types, granularity, and volume, is the key for a reliable energy-disaggregation information layer on which various services and business models can be structured, and the smart-home vision can be realized.

A strong debate is raised regarding energy data acquisition, on who is, or who should be the data owner or controller, and thus more data privacy concerns are raised. Energy companies (DNOs or energy suppliers) have—or will have in the future—access to energy consumption data that can reveal consumers' sensitive information. To protect them, structured "give-consent" or "opt-in" processes should be implemented by policymakers. Strongly related to the above, is the cybersecurity aspect. Additional hardware installed in, or close to residences, introduces further vulnerability either for the consumer or for the grid operators and the—critical for the grid's reliable operation—components and assets. To that end, extensive penetration testing should be a prerequisite whereas R&D efforts funded by the EU can provide an armour against cyber and privacy attacks and data breaches.[11] Moreover, an issue that remains controversial is the question of who should pay for such investments. On the one hand, end-users are usually not willing to cover the additional cost of equipment/services, while on the other hand, a valid business case should be in place for energy companies to proceed with such investments. In parallel, standardization in terms of hardware, software, processes or even terminology is expected to speed up the whole market, especially within groups of countries like EU.[12] Extracting and sharing energy-analytics-based useful information with the end consumers will allow energy companies instead of simply selling a low-margin commodity to upsell products and services and to build stronger customer engagement.

[11]"SU-DS04-2018-2020—Cybersecurity in the Electrical Power and Energy System (EPES)", European Commission, April 2019, https://cordis.europa.eu/programme/id/H2020_SU-DS04-2018-2020.

[12]"Interoperability, Standards and Functionalities applied in the large scale roll out of smart metering", European Smart Grids Task Force, October 2015, https://ec.europa.eu/energy/sites/ener/files/documents/EG1_Final%20Report_SM%20Interop%20Standards%20Function.pdf.

References

Adrian Booth, Eelco de Jong and Peter Peters, "Accelerating digital transformations: A playbook for utilities", McKinsey, January 2018, https://www.mckinsey.com/business-functions/mckinsey-digital/our-insights/accelerating-digital-transformations-a-playbook-for-utilities#.

"Benchmarking smart metering deployment in the EU-27 with a focus on electricity", European Commission, June 17, 2014, https://eur-lex.europa.eu/legal-content/EN/TXT/PDF/?uri=CELEX:52014DC0356&from=EN.

Bert Lutje Berenbroek, "Platform thinking in energy", NET2GRID, accessed February 22, 2020, https://www.net2grid.com/images/n2g_future_of_energy_hlb.pdf.

David Trevithick and Arthur Jouannic, "Guide to choosing a vendor for an energy insight solution: Summary version", Delta-EE, January 2018, https://www.delta-ee.com/EI.

Ian Wright, Sharon Chand and Jian Wei, "Digital transformation in the energy & utilities sector", *Risk & Compliance Magazine*, October–December 2018, https://www2.deloitte.com/content/dam/Deloitte/us/Documents/process-and-operations/us-rc-reprint-digital-transformation.pdf.

"IDC Energy Insights forecasts Western Europe", International Data Corporation (IDC), September 19, 2017, https://www.idc.com/getdoc.jsp?containerId=prEMEA43075617.

"Interoperability, Standards and Functionalities applied in the large-scale roll-out of smart metering", European Smart Grids Task Force, October 2015. https://ec.europa.eu/energy/sites/ener/files/documents/EG1_Final%20Report_SM%20Interop%20Standards%20Function.pdf.

"iONA project", Innogy, accessed February 22, 2020, https://www.iona-energy.com/iONA.

"Looking ahead: Future market and business models", PricewaterhouseCoopers (PWC), accessed February 22, 2020, https://www.pwc.com/gx/en/utilities/publications/assets/pwc-future-utility-business-models.pdf.

Sinan Küfeoğlu, Gaomin Liu, Karim Anaya and Michael G. Pollitt, "Digitalisation and new business models in energy sector", Energy Policy Research Group (EPRG), The University of Cambridge, June 2019, https://www.eprg.group.cam.ac.uk/wp-content/uploads/2019/06/1920-Text.pdf.

"Smart grids and meters", European Commission, accessed February 22, 2020, https://ec.europa.eu/energy/en/topics/markets-and-consumers/smart-grids-and-meters/overview.

"SU-DS04-2018-2020—Cybersecurity in the Electrical Power and Energy System (EPES)", European Commission, April 2019, https://cordis.europa.eu/programme/id/H2020_SU-DS04-2018-2020.
Thomas Pellerin-Carlin, "Putting the consumer at the centre of the European energy system", Notre Europe, Jacques Delors Institute, June 8, 2016, http://www.institutdelors.eu/wp-content/uploads/2018/01/energyconsumer-pellerincarlin-jdi-june16.pdf?pdf=ok.

15

Maintenance: A Key Factor for Energy Efficiency

Vassilios Kappatos

15.1 Introduction

Energy is a critical element of economic growth. Energy resources are limited and become more and more valuable. Energy Efficiency (EE) is a way of managing and limiting the increase in both energy consumption and cost. EE can be achieved through the utilisation of energy-efficient devices or the implementation of strategies that reduce the overall energy consumption. Machinery and electrical devices are the major sources of energy consumption in industries and households. Research works available on machinery and electrical devices advocate that the use of energy-efficient methods can considerably decrease power consumption. The energy and operational efficiency of equipment and systems are affected by potential failures and faults. Energy and operational efficiency

V. Kappatos (✉)
Hellenic Institute of Transport (HIT), Center for Research and Technology Hellas (CERTH), Thessaloniki, Greece
e-mail: vkappatos@certh.gr

M. Mathioulakis (ed.), *Aspects of the Energy Union*,
Energy, Climate and the Environment,
https://doi.org/10.1007/978-3-030-55981-6_15

are carried out through maintenance. Efficiency monitoring is generally performed within a maintenance schedule. Maintenance has gained its importance in every type of equipment and system. Investigations carried out by several universities, institutes and international companies have presented that industries and companies could significantly reduce energy consumption by 5–10% by implementing and focusing on their maintenance strategies.[1] Industries and companies with good and properly scheduled maintenance programmes could provide savings in the range of 5%. The significance of energy efficiency policies within the EU's energy agenda links proper maintenance procedures with the overall success of the (European Union) EU's energy objectives. In February 2015,[2] the European Commission launched a new strategy for a robust Energy Union with a forward-looking climate change policy. The energy Union policy is constructed in five interconnected dimensions, built to reinforce each other, aiming to deliver higher energy security, sustainability and competitiveness. These dimensions refer to Energy security, solidarity and trust, a fully integrated internal energy market, EE that is contributing to the moderation of demand, the decarbonisation of the economy, and research, innovation and competitiveness. EE consists one of the most important factors for European consumers, households and businesses to secure sustainable, competitive and affordable energy. The European Commission's "Energy Union Strategy" is a document meant to pave the way for the creation of an integrated European energy market, where the Member States cooperate to strengthen their energy security, decarbonise their economy and reduce waste in energy consumption. This chapter summarises the European EE and maintenance strategies, indicates the profits of EE and maintenance in industry and society, and explores the interest to maintain equipment for EE.

[1] *PEM Magazine*, "Tips on saving energy using preventive maintenance techniques," https://www.mromagazine.com/2011/05/02/tips-on-saving-energy-using-preventive-maintenance-techniques/ (May 2, 2011).

[2] Energy Union Package, "A framework strategy for a resilient Energy Union with a forward-looking climate change policy" (February 25, 2015).

15.2 Energy Consumption

Energy is one of the most critical elements of human activity. It is a crucial factor for economic growth, presenting a significant correlation between a nation's growth and the availability of energy. Population expansion and economic growth rapidly increase the global demand for energy, especially in emerging market economies. Energy is consumed by different sectors of the economy; industry, households, transport, services (including commercial and public services), agriculture, forestry, etc. Institutions such as the International Energy Agency (IEA),[3] the U.S. Energy Information Administration (EIA),[4] and the European Environment Agency (EEA)[5] record and publish energy data periodically. In the European Union most energy is consumed by the industry sector (31% of final energy consumption), followed by the transport sector (28%), households (25%), services (13%) and agriculture (2%).[6] Worldwide energy consumption increased considerably in 2018,[7] due to the continuous growth of the economy and significant rising demand in the Republic of China. Power generation, strong industrial demand and increasing transport fuel consumption encouraged by a growing vehicle fleet increase the Chinese energy consumption, which shows the highest growth since 2012.

In 2018, the total energy consumption in the United States reached a record high of 2.3 Gtoe, up 3.5% from 2017, mostly due to weather conditions (cold winter and hot summer). On the other hand, in the EU, energy consumption decreased by 1%, because of lower consumption in the power sector. Interpretation of worldwide energy consumption data presents global trends, which help to frame the recent energy problems and encourage movement towards collectively useful solutions. Since

---

[3]International Energy Agency (IEA), https://www.iea.org/.

[4]U.S. Energy Information Administration (EIA), https://www.eia.gov/.

[5]European Environment Agency (EEA), https://www.eea.europa.eu/.

[6]Eurostat, "Shedding light on energy in the EU. A guided tour of energy statistics 2019 edition," https://ec.europa.eu/eurostat/cache/infographs/energy/bloc-3a.html.

[7]Global Energy Statistical Yearbook, https://yearbook.enerdata.net/total-energy/world-consumption-statistics.html.

energy resources are now limited and become more and more expensive, all industrial sectors have been planning and implementing energy management and EE programs with two targets: reduced energy cost and reduced consumption. According to the IEA,[3] improved EE in buildings, industrial processes and transportation, could reduce the world's energy needs in 2050 by one-third, thus helping control global emissions of greenhouse gases.[8]

15.3 Energy Efficiency

EE is a universal concept mentioned in various applications. It is known under different terminologies, namely energy intensity,[9,10] specific energy consumption,[11,12,13] and energy conservation.[14] Indeed, defining the EE is a complex issue. EE refers to the use of energy in an optimum way to accomplish equal service and result, which may have been obtained with a standard efficient system. EE refers to the practice of reducing energy requirements while delivering the required energy output. Europe identifies the transport and building sectors among the most critical ones, highlighting that 75% of the European housing stock is energy inefficient. As the majority of EU gas imports are used for the heating and cooling of buildings, improving their EE would reduce

---

[8] Sophie Hebden, "Invest in clean technology says IEA report" (June 22, 2006).

[9] G. a. Boyd, "Estimating the changes in the distribution of energy efficiency in the U.S. automobile assembly industry," *Energy Economics* 42 (March 2014): 81–87.

[10] L. Peng, X. Zeng, Y. Wang, and G.-B. Hong, "Analysis of energy efficiency and carbon dioxide reduction in the Chinese pulp and paper industry," *Energy Policy* 80 (2015): 65–75.

[11] E. Worrell, R. F. a. Cuelenaere, K. Blok, and W. C. Turkenburg, "Energy consumption by industrial processes in the European Union," *Energy* 19, no. 11 (1994): 1113–1129.

[12] A. Motevali, S. Minaei, M. H. Khoshtaghaza, and H. Amirnejat, "Comparison of energy consumption and specific energy requirements of different methods for drying mushroom slices," *Energy* 36, no. 11 (2011): 6433–6441.

[13] T. Virtanen, M. Tuomaala, and E. Pentti, "Energy efficiency complexities: A technical and managerial investigation," *Management Accounting Research* 24, no. 4 (December 2013): 401–416.

[14] A. Svensson and S. Paramonova, "An analytical model for identifying and addressing energy efficiency improvement opportunities in industrial production systems—Model development and testing experiences from Sweden," *Journal of Cleaner Production* 142 (2017): 2407–2422.

both costs and dependence on external gas suppliers. There are enormous opportunities for efficiency improvements in buildings, transportation and energy generation.[15,16,17] For example, a single-crystalline photovoltaic panel can produce more electrical energy compared to a thin-film photovoltaic panel with the same incident solar power. Saving energy or reduction of energy consumption, are well-known behaviours of EE such as the replacement of incandescent bulbs with fluorescent lamps. Another essential possibility for EE is to remove government-led energy subsidies that promote high energy consumption and inefficient energy use in more than half of the countries in the world.[18] The relationship of EE with economic, social behaviours and environmental issues has been already extensively analysed over the past decades.[19,20] There are many benefits to using EE approaches.[21] Such benefits are established in the theoretical frameworks on EE (codes, standards, label, guideline, the best EE solution handbooks, etc.) and supported by different research centres, international organisations and governments.[20,22,23] EE can be achieved by high technology, improving people awareness etc. The EU has adopted several measures to enhance EE in Europe, while individual EU Member States have set their own indicative national EE targets.

[15] Environmental and Energy Study Institute, "Energy-efficient buildings: Using whole building design to reduce energy consumption in homes and offices," release May 2006.

[16] National Research Council, *Assessment of Fuel Economy Technologies for Light-Duty Vehicles* (Washington, DC: The National Academies Press, 2011).

[17] J. N. Swisher, G. de Martino Jannuzzi, and R. Y. Redlinger, "Tools and methods for integrated resource planning. Improving energy efficiency and protecting the environment," United Nations Environmental Programme, Roskilde, Denmark (1997).

[18] I. Overland, "Subsidies for fossil fuels and climate change: A comparative perspective," *International Journal of Environmental Studies* 67 (2010): 203–217.

[19] F. Apostolos, P. Alexios, P. Georgios, S. Panagiotis, and C. George, "Energy efficiency of manufacturing processes: A critical review," *Procedia CIRP* 7 (January 2013): 628–633.

[20] A. Atmaca and R. Yumrutaş, "Analysis of the parameters affecting energy consumption of a rotary kiln in cement industry," *Applied Thermal Engineering* 66, no. 1–2 (2014): 435–444.

[21] N. Campbell, C. Forbes, and L. Ryan, "Spreading the net: The multiple benefits of energy efficiency improvements," International Energy Program Evaluation Conference, Rome, Italy (2012).

[22] International Organization for Standardization, "ISO 50001:2011 Energy management systems—Requirements with guidance for use," https://www.iso.org/obp/ui/#iso:std:iso:50001:ed-1:v1:en. Accessed: April 2014.

[23] A. Trianni and E. Cagno, "Dealing with barriers to energy efficiency and SMEs: Some empirical evidence," *Energy* 37, no. 1 (January 2012): 494–504.

Europeans could reduce their bills of energy, helping the protection of the environment, using energy efficiently and thereby consuming less. In 2012, under the EE Directive 2012/27/EU, the EU set a 20% energy savings target by 2020 (when compared to the projected use of energy in 2020).[24] In December 2018, the revised EE Directive entered into force (amending Directive EU 2018/2002) updating some specific provisions and introducing new elements. It primarily establishes a headline EU EE target for 2030 of at least 32.5% (compared to projections), with a clause for a possible upward revision by 2023. The 32.5 % target for 2030 translates into final energy consumption of 956 Mtoe and/or primary energy consumption of 1273 Mtoe in the EU-28.[24] The Commission expects that these plans will be pursued in combination with other policy measures under the Europe 2020 flagship initiative for a resource-efficient Europe.

15.4 Maintenance and Types of Maintenance Programs

EE is a strategic priority for the Energy Union, and the EU promotes the principle of "energy efficiency first". EE of any kind of equipment is associated with its proper maintenance. There are several common reasons that equipment can break down. In addition to normal deterioration, other failures may occur, especially when equipment is pushed beyond its design limits or due to operational errors. As a result, equipment downtime, quality problems, energy losses, safety hazards or environmental pollution become the apparent outcomes. All these outcomes have the potential to impact negatively the operating cost, profitability, customer demand satisfaction, and productivity, among other essential performance requirements. Efficiency monitoring is generally performed in a maintenance schedule. Proper maintenance includes functional checks, servicing, repairing or replacing of critical systems, machinery, building infrastructure and supporting services in industrial, business,

[24]European Commission, Energy, https://ec.europa.eu/energy/en/topics/energy-efficiency/targets-directive-and-rules/eu-targets-energy-efficiency.

governmental and residential installations. Maintenance is defined as "the combination of all technical, administrative and managerial actions during the life cycle of an item (e.g. component, machine) intended to retain it in, or restore it to, a state in which it can perform the required function".[25] This combination leads to the support of different strategies depending on whether the item has already failed (corrective strategies) or not (preventive strategies).[26,27] The type of maintenance cannot be equated for each equipment and facility, which depends on the method, cost and critical level. Each equipment or facility has unique and precise needs and therefore demands a specific maintenance strategy to ensure equipment reliability. The following types of maintenance methods are mainly used: (i) corrective maintenance, also known as reactive maintenance or breakdown maintenance, (ii) preventive maintenance, which is regular maintenance performed according to defined schedules, regardless of the condition of the equipment and (iii) predictive maintenance, or condition-based maintenance (CBM) based on the continuous monitoring of equipment and devices, predicting the occurrence of potential failures and faults. *Corrective maintenance* refers to cases when a piece of equipment malfunctions and must be repaired or replaced.[28] *Preventive maintenance* takes place in a well-scheduled manner, avoiding malfunctions and minimising the consequences of equipment failures and faults. The maintenance frequency is defined by operators, end-users and maintenance managers, mainly based on an estimation of asset lifetime and manufacturer's specifications and requirements. A characteristic example of preventive maintenance is heating, ventilation and air conditioning (HVAC), which should be routinely inspected, cleaned and updated as

[25]European Committee for Standardization, "NF EN 13306—Maintenance terminology" (2007).

[26]J. S. Borrero and R. Akhavan-Tabatabaei, "A Markov Decision Process model for optimal policymaking in the maintenance of a single-machine single-product toolset," Winter Simulation Conference (2010): 2594–2605.

[27]E. Alay, K. Duran, and A. Korlu, "A sample work on green manufacturing in the textile industry," *Sustainable Chemistry and Pharmacy* 3 (2016): 39–46.

[28]Infraspeak, "Types of maintenance: What they are and how to apply them," https://blog.infraspeak.com/types-of-maintenance/.

required. *Predictive maintenance's* target is the prediction of a malfunction and faults about to occur in any kind of equipment. When undesirable conditions are detected, and possibly located and characterised, the need for costly corrective maintenance or unnecessary preventive maintenance is eliminated. It is based on the operational status of the equipment, through regular and many times continuous monitoring and testing, using advanced techniques such as infrared, acoustic (partial discharge and airborne ultrasonic), corona detection, vibration analysis, sound level measurements, oil analysis as well as other specific online tests. The most important pros and cons of each maintenance strategy are highlighted below.

15.4.1 Corrective Maintenance

This maintenance approach is ideal for very low priority and low-cost equipment, that is, assets whose absence will not stop the company's operations from continuing to function normally. The problem of implementation of corrective maintenance targets for medium or high priority assets. In the corrective maintenance, no preventive actions are taken; the lifetime of the equipment ends up being shorter than with the other alternative strategies. If the equipment is high-priority and/or high-value equipment, it may lead to unexpected stoppages and possible substantial repair costs.

15.4.2 Preventive Maintenance

This maintenance approach is applicable for high-priority equipment and is perfomed on a regular basis to reduce the likehood of failure. It is performed while equipment is still working so that it does not break down unexpectedly. Preventive maintenance plans, can sometimes be inefficient and result in unnecessary maintenance (including replacement of parts), because they are not based on the actual condition of the equipment, translated to losses in time and money.

15.4.3 Predictive Maintenance

The target of predictive maintenance is to initially detect potential failures and faults while they are still hidden before there is any visible signal, and while the structure or equipment is in service. The maintenance performed is always better-informed, necessary and timely, since the equipment is only subject to maintenance when a malfunction is predicted, reducing the costs and labour time spent on maintenance. The monitoring equipment, staff training, data collection and interpretation, makes the implementation of this strategy very expensive.

15.5 Energy Efficiency with Maintenance

Proper maintenance is essential even for the most efficiently designed systems and equipment with high reaching EE goals. EE improvement can be achieved by the application of EE technologies,[29] operational or production enhancement,[30] and proper housekeeping and maintenance practices.[31] Energy management systems are implemented in almost all industrial systems to control and regulate machinery and electrical devices to improve their EE performance.[32] Machinery and electrical equipment are significant sources of energy consumption in industries and households. It is known that energy-efficient machinery and electrical devices can significantly reduce power consumption. Machinery and equipment are prone to faults, which affect their energy and operational efficiency. A maintenance strategy could not only avoid the failure of the system but also could anticipating the growing up of

[29]M. Pons, A. Bikfalvi, J, Llach, and I. Palcic, "Exploring the impact of energy efficiency technologies on manufacturing firm performance," *Journal of Cleaner Production* 28 (2013): 289–296.

[30]E. L. Plambeck, "Reducing greenhouse gas emissions through operations and supply chain management," *Energy Economics* 34 (2012): 64–74.

[31]A. E. Zohir, "Energy efficiency improvement by housekeeping measures," Thermal Issues in Emerging Technologies, ThETA 3, Cairo, Egypt (December 2010).

[32]A. Hoang, P. Do, and B. Iung, "Integrating energy efficiency-based prognostic approaches into energy management systems of base stations," International Conference on Advanced Technologies for Communications (ATC'14) (2014): 220–225.

Convert HP to KW 1HP (electric) = 0.7457 KW 30 HP x (0.7457 KW/1 HP) = 22.37 KW
In this example, the cooling tower was operated for 22 weeks/year, 5 days/week. Now that the cooling tower fan runs 2 hours less per day, the following reduction in energy consumption is realized:
Total Reduced Operating Hours per Year = (2 Hours/Day) x (5 Days/Week) x (22 weeks/year) = 220 Hours/Year
Total KWh = 22.37 KW x 220 Hours/Year = 4,921.6 KWh/Year
If 1 KWh costs 0.095 $, the annual energy cost is:
4,921.6 KWh x 0.095 $ = 467.66 $/Year

Fig. 15.1 Example of relevant calculations

global energy consumption. Energy consumption reduction can also be achieved by operational or production improvement[30] in a variety of methods including functional or production improvement, economic scheduling of equipment, economical housekeeping, preventive maintenance, predictive maintenance and advanced condition monitoring (fault detection and diagnosis, etc.). The following example[33] describes how preventive maintenance affects EE: If in a cooling tower used for air conditioning the fins and coils are very dirty, heat transfers efficiently through the coils and fins, and cooling tower fans operate extensively to transfer the heat to ambient. Such conditions result in higher energy consumption and reduce the fan's life expectancy. After cleaning the cooling tower's coils and fins, a 30 HP cooling tower fan can operate for two hours less per day, which results in a reduction in energy consumption of 4921.6 KWh/Year, equivalent to 467.55 $/Year (Fig. 15.1).

15.6 Maintenance Strategies for Energy Efficiency

Companies with proper preventive maintenance programs would provide savings of up to 5% in energy consumption. Maintenance techniques targeting EE may result in annual savings of up to 20%, according to

[33]CFMS Consulting Inc., "Preventive maintenance and energy efficiency: Did you know preventative maintenance has a direct impact on energy efficiency?" https://www.cfms.ca/blog/preventive-maintenance-and-energy-efficiency-did-you-know-preventative-maintenance-has-a-direct-impact-on-energy-efficiency/ (April 4, 2013).

a U.S. Department of Energy study.[34] Plants, factories, facilities, and commercial and residential buildings account for nearly 40%[35] of the total amount of American energy consumed. Running faulty equipment is always a critical issue for the energy consumption and operating costs. The solution is to replace or repair malfunctioning equipment. Primarily this has been a matter of scheduling and guesswork. Implementing predictive maintenance strategy instead of a preventive maintenance strategy offers a new way to reduce costs drastically. For example, active HVAC maintenance[36] alone can reduce energy costs by 5% up to 40%, while malfunctioning machines or less-than-optimal equipment function may increase energy spending by 30–60%. The motor-driven equipment, responsible for 64% of total building electricity consumption, is another known culprit.[37] The total damage of ineffective equipment can range from 40 $ to 150 $ billion per year.[38] Regardless the critical importance of the issue, current maintenance decisions are mainly based on conventional indicators such as reliability, availability, direct cost, etc.[39] and only a few studies have been realised to consider the connection between EE and maintenance strategy.[40,41] The research[39] addressed the average EE issue under periodic maintenance as a criterion for the

[34] G. P. Sullivan, R. Pugh A. P. Melendez W. D. Hunt, "Operations & maintenance best practices: A guide to achieving operational efficiency," Pacific Northwest National Laboratory for the Federal Energy Management Program U.S. Department of Energy (August 2010).

[35] Energy Flow Charts, Lawrence Livermore National Laboratory (LLNL), https://flowcharts.llnl.gov/commodities/energy (2017).

[36] Building Efficiency Initiative, "Studies show: HVAC system maintenance saves energy," https://buildingefficiencyinitiative.org/articles/studies-show-hvac-system-maintenance-saves-energy (September 29, 2011).

[37] D. R. Malcolm, "Turn energy waste into profit, Plant Service," https://www.plantservices.com/articles/2006/313/?start=0 (December 12, 2006).

[38] J. C. Augury, "Stop wasting energy: Use predictive maintenance," https://www.embedded-computing.com/industrial/stop-wasting-energy-use-predictive-maintenance (August 18, 2017).

[39] B. Liu, Z. Xu, M. Xie, and W. Kuo, "A value-based preventive maintenance policy for multi-component systems with continuously degrading components," *Reliability Engineering & System Safety* 132 (December 2014): 83–89.

[40] W. Xu and L. Cao, "Energy efficiency analysis of machine tools with periodic maintenance," *International Journal of Production Research* (March 2014): 1–13.

[41] B. Darabnia and M. Demichela, "Maintenance an opportunity for energy saving," *Chemical Engineering Transactions* 33 (2013): 259–264.

maintenance decision-making process at the machine level. The relevant analysis demonstrated the importance of optimising maintenance activities and operating procedures to increase the EE performance of a system.[41] Using the developed efficiency analysis model, it is possible to quantify the impact of maintenance and operating procedures in terms of energy savings or cost-effectiveness and to calculate the cost of conserved energy. It is also possible to estimate the potential reduction of greenhouse gas emissions (CO_2).

15.7 Conclusions

The world is undergoing a radical transformation in the way energy is produced and used, shifting to a cleaner, consumer-centric and more efficient energy system. The European Commission, industry and research organisations are working together to speed up this transformation. Following up on the broader objectives of the Energy Union strategy, the EU 2030 climate and energy policy framework, indicates energy saving as an essential component of EU's energy goals. The European Commission has proposed an action-plan targeting five dimensions, of which one is EE, the other four being the decarbonisation of the EU economy, the security of energy supply, an internal energy market, and research, innovation and competitiveness. EE refers to the mechanisms for saving energy in households and industries. Developing EE as the instrument for providing control of demand, offers the potential for the EU to meet its goals in terms of security of supply, climate change, jobs, growth and competitiveness. It is well known that to ensure EE in equipment and systems, a maintenance approach should be realised and applied. Each equipment or facility has unique and precise needs and therefore demands a specific maintenance strategy to ensure equipment reliability. Understanding the link between maintenance and energy performance is essential to meet the EE sustainability goals.

This chapter aimed to illustrate the important contribution of maintenance facilities in EE and sequentially in the "Energy Union Strategy." This work revealed the benefits of EE and maintenance in both industry and society. The presented results and discussion suggest that proactive maintenance strategies in existing equipment and systems can contribute towards significantly improving energy performance and efficiency. Over the last few years, the industry developed tools to evaluate the relationship between maintenance and energy; however, more studies are needed to accurately quantify the energy savings from varying maintenance strategies as well as the return on investment from EE activities. The European Commission needs to encourage all member states to take full account of the maintenance strategies and management for energy performance, efficiency and savings. Futhermore, funding should be channelled towards projects aimed at overcoming the split incentive barrier in the EE and maintenance sectors. Last but not least, the European Commission needs to guarantee mechanisms and seed funding provided through investment programmes such as the European Fund for Strategic Investment.

References

A. Hoang, P. Do, and B. Iung, "Integrating energy efficiency-based prognostic approaches into energy management systems of base stations," International Conference on Advanced Technologies for Communications (ATC'14) (2014): 220–225.

A. Trianni and E. Cagno, "Dealing with barriers to energy efficiency and SMEs: Some empirical evidence," *Energy* 37, no. 1 (January 2012): 494–504.

A. Atmaca and R. Yumrutaş, "Analysis of the parameters affecting energy consumption of a rotary kiln in cement industry," *Applied Thermal Engineering* 66, no. 1–2 (2014): 435–444.

B. Darabnia and M. Demichela, "Maintenance an opportunity for energy saving," *Chemical Engineering Transactions* 33 (2013): 259–264.

Building Efficiency Initiative, "Studies show: HVAC system maintenance saves energy," https://buildingefficiencyinitiative.org/articles/studies-show-hvac-system-maintenance-saves-energy (September 29, 2011).

CFMS Consulting Inc., "Preventive maintenance and energy efficiency: Did you know preventative maintenance has a direct impact on energy efficiency?" https://www.cfms.ca/blog/preventive-maintenance-and-energy-

efficiency-did-you-know-preventative-maintenance-has-a-direct-impact-on-energy-efficiency/ (April 4, 2013).
D. R. Malcolm, "Turn energy waste into profit, Plant Service," https://www.plantservices.com/articles/2006/313/?start=0 (December 12, 2006).
E. Alay, K. Duran, and A. Korlu, "A sample work on green manufacturing in the textile industry," *Sustainable Chemistry and Pharmacy* 3 (2016): 39–46.
E. Worrell, R. F. a. Cuelenaere, K. Blok, and W. C. Turkenburg, "Energy consumption by industrial processes in the European Union," *Energy* 19, no. 11 (1994): 1113–1129.
Energy Flow Charts, Lawrence Livermore National Laboratory (LLNL), https://flowcharts.llnl.gov/commodities/energy (2017).
Energy Union Package, "A framework strategy for a resilient Energy Union with a forward-looking climate change policy" (February 25, 2015).
Environmental and Energy Study Institute, "Energy-efficient buildings: Using whole building design to reduce energy consumption in homes and offices," released in May 2006.
European Commission, Energy, https://ec.europa.eu/energy/en/topics/energy-efficiency/targets-directive-and-rules/eu-targets-energy-efficiency.
European Committee for Standardization, "NF EN 13306—Maintenance terminology" (2007).
European Environment Agency (EEA), https://www.eea.europa.eu/.
Eurostat, "Shedding light on energy in the EU. A guided tour of energy statistics 2019 edition," https://ec.europa.eu/eurostat/cache/infographs/energy/bloc-3a.html.
F. Apostolos, P. Alexios, P. Georgios, S. Panagiotis, and C. George, "Energy efficiency of manufacturing processes: A critical review," *Procedia CIRP* 7 (January 2013): 628–633.
G. a. Boyd, "Estimating the changes in the distribution of energy efficiency in the U.S. automobile assembly industry," *Energy Economics* 42 (March 2014): 81–87.
G. P. Sullivan, R. Pugh A. P. Melendez W. D. Hunt, "Operations & maintenance best practices: A guide to achieving operational efficiency," Pacific Northwest National Laboratory for the Federal Energy Management Program U.S. Department of Energy (August 2010).
Global Energy Statistical Yearbook, https://yearbook.enerdata.net/total-energy/world-consumption-statistics.html.
I. Overland, "Subsidies for fossil fuels and climate change: A comparative perspective," *International Journal of Environmental Studies* 67 (2010): 203–217.

Infraspeak, "Types of maintenance: What they are and how to apply them," https://blog.infraspeak.com/types-of-maintenance/.

International Energy Agency (IEA), https://www.iea.org/.

International Organization for Standardization, "ISO 50001:2011 Energy management systems—Requirements with guidance for use," https://www.iso.org/obp/ui/#iso:std:iso:50001:ed-1:v1:en. Accessed: April 2014.

J. C. Augury, "Stop wasting energy: Use predictive maintenance," https://www.embedded-computing.com/industrial/stop-wasting-energy-use-predictive-maintenance (August 18, 2017).

J. S. Borrero and R. Akhavan-Tabatabaei, "A Markov Decision Process model for optimal policymaking in the maintenance of a single-machine single-product toolset," Winter Simulation Conference (2010): 2594–2605.

L. Peng, X. Zeng, Y. Wang, and G.-B. Hong, "Analysis of energy efficiency and carbon dioxide reduction in the Chinese pulp and paper industry," *Energy Policy* 80 (2015): 65–75.

B. Liu, Z. Xu, M. Xie, and W. Kuo, "A value-based preventive maintenance policy for multi-component systems with continuously degrading components," *Reliability Engineering & System Safety* 132 (December 2014): 83–89.

A. Motevali, S. Minaei, M. H. Khoshtaghaza, and H. Amirnejat, "Comparison of energy consumption and specific energy requirements of different methods for drying mushroom slices," *Energy* 36, no. 11 (2011): 6433–6441.

N. Campbell, C. Forbes, and L. Ryan, "Spreading the net: The multiple benefits of energy efficiency improvements," International Energy Program Evaluation Conference, Rome, Italy (2012).

National Research Council, *Assessment of Fuel Economy Technologies for Light-Duty Vehicles*. Washington, DC: The National Academies Press, 2011.

PEM Magazine, "Tips on saving energy using preventive maintenance techniques," https://www.mromagazine.com/2011/05/02/tips-on-saving-energy-using-preventive-maintenance-techniques/ (May 2, 2011).

E. L. Plambeck, "Reducing greenhouse gas emissions through operations and supply chain management," *Energy Economics* 34 (2012): 64–74.

M. Pons, A. Bikfalvi, J, Llach, and I. Palcic, "Exploring the impact of energy efficiency technologies on manufacturing firm performance," *Journal of Cleaner Production* 28 (2013): 289–296.

Sophie Hebden, "Invest in clean technology says IEA report" (June 22, 2006).

A. Svensson and S. Paramonova, "An analytical model for identifying and addressing energy efficiency improvement opportunities in industrial production systems—Model development and testing experiences from Sweden," *Journal of Cleaner Production* 142 (2017): 2407–2422.

J. N. Swisher, G. de Martino Jannuzzi, and R. Y. Redlinger, "Tools and methods for integrated resource planning. Improving energy efficiency and protecting the environment," United Nations Environmental Programme, Roskilde, Denmark (1997).

T. Virtanen, M. Tuomaala, and E. Pentti, "Energy efficiency complexities: A technical and managerial investigation," *Management Accounting Research* 24, no. 4 (December 2013): 401–416.

U.S. Energy Information Administration (EIA), https://www.eia.gov/.

W. Xu and L. Cao, "Energy efficiency analysis of machine tools with periodic maintenance," *International Journal of Production Research* (March 2014): 1–13.

A. E. Zohir, "Energy efficiency improvement by housekeeping measures," Thermal Issues in Emerging Technologies, ThETA 3, Cairo, Egypt (December 2010).

16

Conclusions

Michalis Mathioulakis

The Energy Union policy is the driving force behind all the EU's objectives regarding energy and the environment and represents a link—through energy—between the von der Leyen Commission and the Juncker Commission that preceded it. The complex nature of the energy sector, from the production and transportation to the storage, distribution, and supply of energy, creates a demanding and multidimensional conceptual framework that needs an equally multidimensional presentation of its various aspects in order to be utterly understood. In its 14 chapters, this publication aspired to offer such a presentation, combined with a critical look into some of the most essential issues regarding the EU's Energy Union policy and the challenges deriving from its implementation in the European energy markets. Such critical issues and challenges addressed in this book, related to the Geopolitics and Regionality of Energy, Energy Regulation, Climate Change, Renewable Energy,

M. Mathioulakis (✉)
University of Macedonia, Thessaloniki, Greece

Greek Energy Forum, Athens, Greece

M. Mathioulakis (ed.), *Aspects of the Energy Union*,
Energy, Climate and the Environment,
https://doi.org/10.1007/978-3-030-55981-6_16

Gas technologies, and Consumer Empowerment systems. A special focus has also been given throughout most of the individual chapters of the book into the effects the Energy Union policy has in two sensitive regional systems, South-Eastern Europe, and the Eastern Mediterranean. Furthermore, this publication aimed at creating a connection between the EU's fundamental objectives regarding the structure of a common market and the challenges deriving from the European energy security framework on a national and supranational level.

Overall, the themes addressed throughout the book can be summed up in three basic debates. The first is related to the regionality aspect of applied policies, where the particular political and business environment in South Eastern Europe and the Eastern Mediterranean creates a set of unique challenges for the overall success of EU energy policies. The second set of concerns discussed in this book relates to the balance that needs to be achieved between efforts to enhance energy security and efforts to advance the Green Deal objectives. Contrary to what is widely believed, these two policies do not only supplement each other but might also collide. To a large degree this issue is linked with the role natural gas and renewable gases hold in relation to the development of RES technologies in the EU. The third set of issues relates to the technical and regulatory environment and the changes that can promote a fast and reliable path towards the objectives of the Green Deal. The Energy Union's objectives have been examined in relation to the Energy Charter Treaty and the Energy Community initiative. Furthermore a multitude of technical issues has been discussed, spanning from the central role of smart meters in EU's efforts to empower Europeans in their participation in the energy market, to the significance of technical maintenance in enhancing energy efficiency and from the spatial congestion of offshore wind installations to the contribution of new LNG bunkering, hydrogen and biofuel technologies in limiting CO_2 emissions to the 2050 goals.

On the issue of regionality, the debate argued in the book has revealed the difficulties in effectively applying the Energy Union policy in South Eastern Europe and the Eastern Mediterranean due to the challenging political environment in these two sub-systems. Integrating the Balkan countries into the European energy market and infrastructure network remains fundamental to the vision for an cohesive continent-wide energy system where energy will flow freely across borders, based on competition and the best possible use of resources, and with effective regulation of energy markets. While investors are looking into regional advantages for renewable resources in South-Eastern Europe, the Western Balkan countries are late in meeting deadlines and commitments on a set of institutional reforms. The slower Western Balkans move away from coal, the higher the cost of compliance and the energy transition will be in the future. The smaller the steps towards regional energy markets integration, the longer the road towards environmental sustainability, economic growth and membership to the EU. A framework for small-scale installations, located where they are most needed, and where the local resources are available, coupled with increased energy efficiency and with limited, but necessary, investment in the energy transmission and distribution with neighbouring regions, might be the answer. Decentralized energy production also provides significant employment opportunities, as does improving energy efficiency. This switch cannot be triggered by liberalized market forces alone and calls for policy changes and new regulatory frameworks to be put soon in place by the Commission.

As regards to the Eastern Mediterranean, a region with immense potential for the EU's energy security, the geopolitical balance of power with converging and colliding national and energy interests adds a challenging dimension for the effective implementation of the Energy Union strategy. The interests of bordering states in the region intersect with vivid political activity by other international actors that include the United States, Russia, and China, and interact with developments in Libya and the relations between Greece, Cyprus, Turkey, Egypt and Israel. The abundance of gas in Europe and the challenging pricing terrain, combined with US supply and Russian reserves, lead to the assumption, in this book, that currently the EU may not be, in realistic terms, a final market destination for the Eastern Mediterranean hydrocarbons, while a possible solution in the region could be achieved by further integrating the East Mediterranean Gas Forum initiative in the economic-geopolitical strategies of the EU. Overall, the analyses in the book revealed that there are limits for the EU's ambition to raise a unified voice in the international energy stage. This is determined by multiple factors, such as diverging political goals for the MS in the international diplomatic stage, MS special national interests in their foreign relations with third countries and differences in the energy mix due to economic interests or different national indigenous resources. Most importantly, the EU has to provide the right answers promptly. If not, there is a strong possibility that international developments might exceed its ability to resolve overwhelming questions even in its own territory or right outside its borders.

On the issue of the balance between energy security and the advancement of the Green Deal objectives, the debate presented in the book revealed that in order to determine the strategic value of energy infrastructure, we need to focus not on their economic weight and output but rather on their ability to support EU's and member states' security objectives. To that end, specific regulatory instruments that promote in practice the 'external' objectives of the EU's energy security have been identified, namely the provisions for Third Party Access to energy networks, the provisions for Ownership Unbundling of energy infrastructure, and the PCI's initiative. Overall, there is no doubt that since the introduction of the Energy Union Strategy in 2015, the EU's energy security was significantly enhanced. With the new legislation in place, markets can function better, and enhanced regional cooperation can raise gas deliveries during a crisis. The analyses in this book identified that there is considerable improvement of diversification of supplies via LNG and new gas infrastructure. In addition, measures to safeguard the security of gas supply have dramatically improved Europe's resilience to possible gas supply disruptions. At the same time, it seems that negotiations for the Energy Union, especially those for the security of supply, have exhausted the legal limits of the current EU Treaties, as well as, MS tolerance to transfer new powers from the national to the European level. A broader discussion, including all policy sectors (energy, migration, security, defence, etc.) in the highest decision-making level, may be necessary to further reallocate powers between national and European level in energy.

Finding the balance between energy security and the Green Deal objectives inevitably leads to the need in finding a balance between the strong development of RES technologies and the role of natural gas, LNG and renewable gases during the long and challenging transitional period until the 2050 goals are achieved. The predominant task of policymakers, business leaders, and other stakeholders involved in energy is to find ways to accelerate the process for delivering the vast amounts of low-carbon energy that will be needed in the coming decades to support economic growth. The politics of the debate over scrambling to secure hydrocarbon resources versus reducing consumption through efficiency and use of alternatives is emerging as a new focal point in energy policy. Global demand growth, as projected, cannot be sustainably met from the exploitation of conventional resources. As the implications of a global effort to drastically cut emissions can often be positive for energy security issues, a wide range of conflicts will still arise. Critical developments in the energy transition are dramatically speeding up. Many of the global trends that shape the future of energy are, in fact, driven by a multitude of local trends, which will occur at different magnitudes and speeds in specific geographies and sectors. The rapidly increasing share of renewables in electricity generation and its decreasing cost, new storage technologies and the introduction of electric vehicles in the market, are leading towards a new system in which consumers produce, use and sell their electricity; a system that reflects the concept of distributed generation.

The EU has been a pioneer in the deployment of renewable energy. The decarbonization of the electricity sector with renewables has been one of the most transformative changes in Europe's economy. Europe is now set to become carbon neutral by 2050, thereby contributing to the 1.5 °C goal of the Climate Paris Agreement. It will face formidable challenges to do so. Particularly in the decarbonization of other energy uses beyond electricity. However, the deployment of renewable energy still faces many challenges such as increasing the flexibility of the power system, and facilitating, permitting, and plugging the financing gap an electricity market dominated by renewables poses for investment signals. While policy has set the departure from fossil fuels for power generation in motion, the success of renewables is an outcome of a multitude of factors. As seen in this book, Renewables have decreased costs at a remarkable pace. Solar PV panels have decreased costs by 75% since 2010, and onshore wind energy is today the cheapest source of power generation in many parts of the world. This has made a stand-alone business case for investors. Even in times of low oil prices, the cost competitiveness of renewables is seen as a hedge against volatile fuel commodities by many private investors. As renewables scale-up, stakeholders' expectations could become challenging to meet. Unfulfilled expectations and poorly managed impacts could cause project opposition. Despite this, renewables benefit from a positive perception across most of the population in Europe. One of the key issues for integrating renewables in a liberalized energy market is to adapt the rules for electricity trading so that these reflect the intrinsic characteristics of renewable energy generation. Wind speed and solar radiation are variable across different time frames, and their intensity needs to be forecast for energy generation. Therefore, the rules for electricity trading need to address this variability and uncertainty of forecast. This is important as market rules set the way electricity prices are formed, determining the revenues for all power generation stakeholders.

Regarding wind-based power generation in particular, Europe is at the forefront of offshore wind developments with more than 20 GW of cumulative installations by the end of 2019, covering approximately 80% of global installations. The technological trends for offshore wind farms are quite clear when moving to even larger turbines that can harness more energy, further offshore and in deeper waters. Future turbines are expected to have a nameplate capacity between 15 and 20 MW in 2030 with rotor diameters exceeding 250 m. As projects get larger and are developed further offshore, Direct Current technologies will increase their market share. Offshore wind will play an important part in this energy transition, but it must go hand in hand with Europe's new industrial strategy to benefit its society. Under this new industrial strategy focus should be given to supporting skills and innovation, accessing cheap finance for capital-intensive investments and maximizing supply chain efficiencies.

However crucial and urgent the need to achieve the Green Deal objectives, the debate on the relationship between advancing renewables and natural gas technologies presented in this publication, revealed arguments for the need for cooperation rather than collision between the two sectors during the long transition period towards 2050. The ambition to achieve climate neutrality in Europe leads to a plethora of challenges across the gas value chain, from its production and transportation, to the consumption and end-uses. The gas infrastructure business in particular needs to carefully consider future innovative business models and build a strong case for the role of gas in sectoral integration, promoting the use of existing infrastructure in smart combination with renewable electricity to limit costs and enable a just transition. The development of renewable gases across the European Union can contribute to achieving climate neutrality by 2050 and create a new *raison d'être* for European gas infrastructure operators; one that is more sustainable and connected to the EU Green Deal. To that end, bioenergy is also a valuable option for the energy transition area, with positive synergies with other policy priorities, including water and food security, while improving economic growth and stability. Next-generation biofuels can play a key role in transport since they can serve all different modes (road, rail, marine and air). Renewable fuels attract capital due to shrinking investments in the oil sector, overcoming technology barriers and feedstock quest dilemmas.

From a market perspective, the advent of renewable gases poses specific difficult questions regarding the functioning of the internal gas market. Traditionally, the development of the internal gas market has revolved around natural gas. Investments and market developments over the past decades have culminated to the creation of a homogeneous market for natural gas. Favourable feed-in tariffs can contribute to the uptake of biogas/biomethane and hydrogen technologies, however, other solutions are needed as well relating to blending obligations in heating and for industry, akin to the obligations prescribed for the transport sector. To that end, the role of LNG bunkering has also been identified in this book as a critical factor for the shipping industry. The analysis revealed a wide consensus on the fact that LNG is currently the most competitive and proven feasible solution within the alternative fuels' era. LNG has the potential to become the medium for the utilization of renewable gases in the shipping industry, allowing for a smooth transition for shipowners, which will eventually continue to use the same ships, employ the same bunkering assets and rely on the same bunkering methods. Overall it is imperative that the EU legislates on clear definitions for what is considered a renewable gas and its integration to the Green Deal goals.

On the issue of the regulatory and technical environment towards the objectives of the Green Deal, the debate in this book revealed the importance, one the one hand, of the relationship between the Energy Union strategy, the Energy Charter Treaty and the Energy Community initiative, and on the other, of the development of technical tools regarding smart meters and equipment maintenance strategies in the EU. Regarding regulation, there is a pressing need to harmonize laws across borders regarding the required infrastructure and connectivity networks and the promotion and mutual protection of existing regimes for foreign direct investments. The EU needs to develop a genuine foreign energy policy to achieve the effective regional integration of the energy markets in the wider Europe. The EU promotes market access and free competition in the energy-producing countries and encourages new investments in order to secure the adequate and uninterrupted supply of energy resources at affordable prices from these countries to Europe. The Energy Charter Treaty has secured the effective protection of European investors in many countries of the wider Europe. In order to complete the regional energy markets, their operating rules need to be harmonized. To this end, the EU uses various regulatory and action tools to achieve the required degree of harmonization of the operating conditions in these markets. The Energy Community is the most significant one. The development of cross-border energy-transportation infrastructure and interconnection networks is a prerequisite for the integration of the energy markets in the wider Europe. Energy markets cannot be integrated unless they are technically interconnected. Yet, as long as the Energy Community is deprived of enforcement mechanisms and monitoring tools to observe progress, governments in sensitive regions bordering the EU like the Western Balkans have little incentive to avoid procrastination. Disregarding or failing to implement the so-called 'soft measures', only complicates an already complicated technical, institutional, infrastructural and technological transition for areas of EU interest like the Western Balkan energy markets. Given the importance of the convergence of the energy policies in South Eastern European countries with those of the EU for both regions, the investment promotion and protection mechanisms of the Energy Charter Treaty have been analysed in this book for their significance in their potential to encourage renewable energy investments in

South East European countries. Soft law instruments under the ECT can also assist in capacity-building programmes. In addition to the above, the debate in this book revealed the controversy associated with the adoption of lists of projects of common European interest; a list that is governed by the TEN-E regulation. The new TEN-E needs to be aligned with the Green Deal and the EU's long-term climate and energy objectives. The decarbonization value of a project as well as its contribution to sectoral integration, ought to be considered as new criteria. Finally, while the EU Green Deal sets out a cross-cutting approach, the fact remains that legislative proposals will continue to be proposed separately and dealt with by different committees in the European Parliament. Stakeholders also need to take into consideration the trade and industrial competitiveness components of the EU Green Deal. Europe has to compete on a global level and cannot afford to fall behind in the energy technological race.

Regarding the role of technological advancements, the analysis in this publication demonstrated the essential role of smart meters for the empowerment of European consumers. Smart meters roll-out, enables access to residential end-users' energy consumption data. Distribution Network Operators (DNOs), power utilities, and energy retailers are now offered information that allows for more efficient asset and client management. Europe is paving the way on this topic, especially within the Energy Union strategy. This new landscape sparks energy retailers' interest in energy services and data-driven business models. Digital transformation, especially since data significance became more evident, is currently a powerful trend for energy utilities and companies in the EU. More and more digital and 'smart' hardware components replace old-generation equipment spanning from the transmission level to the energy end-users. Regulation and technology lead to smart-meters becoming the necessary tools for the end-users to increase their impact and leverage in the future grid, either by participating in demand response schemes or by simply being able to switch energy providers—even within a day—for their own benefit. In this context, more than 60% of the European utilities have already dedicated digital transformation or innovation teams in place, since they realized that Big (Energy) Data is the pillar of innovation and unlocks variable opportunities for them to become digital, divergent, and decentralized. Advancements in energy data, combined with technologies regarding Energy Efficiency models are areas where EU policies lead the way to the future. On one of the most interesting aspects of energy efficiency has been presented in this publication illustrating the important contribution of maintenance strategies for industrial equipment regarding energy efficiency in the EU. The presented results and discussion suggest that proactive maintenance strategies in existing equipment and systems can contribute towards significantly improving energy performance and efficiency, where proper preventive maintenance programmes would provide savings of up to 5% in energy consumption and maintenance techniques targeting energy efficiency could result in annual savings of up to 20%.

Europe needs to turn the climate and environmental challenges ahead into opportunities that should make the energy transition fair and inclusive for all. The pillars for the success of the Energy Union, and consequently the European Green Deal, will be cooperation across borders, strong human capital, and competent leadership. Aspiring to create the conditions under which European citizens can benefit from secure, sustainable, competitive and affordable energy, the Energy Union strategy objectives constitute an extremely challenging task that involves a structural transformation of the European energy system, including the production, procurement, transportation, distribution, storage, pricing, trading and consumption of energy. It involves structural changes regarding the combination of energy sources used in all human activity, the flow of energy that enters and exits the EU, and the interconnection of the energy policy with EU's Trade, Competition, and External Relations policies. The overall success of the Energy Union as the core policy platform that can steer the EU into its Green Deal goals while maintaining its principal energy security objectives, remains to be seen. The European Union has the resolve, the financial and regulatory tools, and the human capital necessary to lead the successful transformation of its energy sector. The academic and business community needs to keep a critical look into the application of these policies and contribute in an on-going constructive debate on the challenges and opportunities arising from the common effort for a better, cleaner and more sustainable future in Europe.

The manufacturer's authorised representative in the EU is Springer Nature Customer Service Centre GmbH, Europaplatz 3, 69115 Heidelberg, Germany. If you have any concerns regarding our products, please contact ProductSafety@springernature.com

Printed and bound by CPI Group (UK) Ltd, Croydon, CR0 4YY
30/09/2026
02225286-0013